Modeling Embedded Systems and SoCs

Concurrency and Time in
Models of Computation

The Morgan Kaufmann Series in Systems on Silicon
Series Editors: Peter J. Ashenden, Ashenden Designs Pty. Ltd. and Adelaide University, and Wayne Wolf, Princeton University

The rapid growth of silicon technology and the demands of applications are increasingly forcing electronics designers to take a systems-oriented approach to design. This has lead to new challenges in design methodology, design automation, manufacture and test. The main challenges are to enhance designer productivity and to achieve correctness on the first pass. The Morgan Kaufmann Series in Systems on Silicon presents high quality, peer-reviewed books authored by leading experts in the field who are uniquely qualified to address these issues.

The Designer's Guide to VHDL, Second Edition
Peter J. Ashenden

The System Designer's Guide to VHDL-AMS
Peter J. Ashenden, Gregory D. Peterson, and Darrell A. Teegarden

Readings in Hardware/Software Co-Design
Edited by Giovanni De Micheli, Rolf Ernst, and Wayne Wolf

Modeling Embedded Systems and SoCs
Axel Jantsch

Forthcoming Titles

Rosetta User's Guide: Model-Based Systems Design
Perry Alexander, Peter J. Ashenden, and David L. Barton

Rosetta Developer's Guide: Semantics for Systems Design
Perry Alexander, Peter J. Ashenden, and David L. Barton

Component Modeling for FPGA and Board-level Verification
Richard Munden

Modeling Embedded Systems and SoCs

Concurrency and Time in Models of Computation

Axel Jantsch

ROYAL INSTITUTE OF TECHNOLOGY
STOCKHOLM, SWEDEN

MK

MORGAN KAUFMANN PUBLISHERS

AN IMPRINT OF ELSEVIER SCIENCE
SAN FRANCISCO SAN DIEGO NEW YORK BOSTON
LONDON SYDNEY TOKYO

Senior Editor   Denise E. M. Penrose
Publishing Services Manager   George Morrison
Editorial Coordinator   Alyson Day
Series Art Direction, Cover Design, & Photgraphy   Chen Design Associates, SF
Text Design   Rebecca Evans & Associates
Copyeditor   Ken DellaPenta
Proofreader   Sarah Jenkins
Indexer   Nigel d'Auvergne
Printer   The Maple-Vail Book Manufacturing Group

Portions of Chapter 8, Sections 8.1, 8.2, 8.3, 8.4, 8.7, and 8.8, reprinted with permission from "The Usage of Stochastic Processes in Embedded System Specifications" by Axel Jantsch, Ingo Sander, and Wenbiao Wu. *Proceedings of the Ninth International Symposium on Hardware/Software Codesign*, April 2001. Copyright 2001 by ACM Press.

Morgan Kaufmann Publishers
An imprint of Elsevier Science
340 Pine Street, Sixth Floor
San Francisco, California 94104-3205
www.mkp.com

© 2004 by Elsevier Science (USA)
All rights reserved
Printed in the United States of America

07 06 05 04 03   5 4 3 2 1

No part of this publication may be reproduced, stored in a retrieval system, or transmitted in any form or by any means—electronic, mechanical, photocopying, recording, or otherwise—without the prior written permission of the publisher.

Library of Congress Control Number: 2003047408

ISBN: 1-55860-925-3

This book is printed on acid-free paper.

Contents

Foreword xi
Preface xv
Notation xxi

1 Introduction 1
1.1 Motivation   2
1.2 Heterogeneous Models   5
1.3 Separation of Computation and Communication   7
1.4 Systems and Models   8
    1.4.1 System Properties   13
    1.4.2 System Classification Summary   19
1.5 The Rugby Metamodel   20
    1.5.1 Domain, Hierarchy, and Abstraction   21
1.6 Domains   25
    1.6.1 Computation   25
    1.6.2 Data   32
    1.6.3 Time   33
    1.6.4 Communication   34
1.7 Notation   36
1.8 Design Methods and Methodology   36
    1.8.1 Design phases   37
    1.8.2 Design and Synthesis   38
    1.8.3 Analysis   39
1.9 Case Study: A Design Project   39
    1.9.1 Requirements Definitions   41
    1.9.2 System Model   42
    1.9.3 Software: The C Model   43
    1.9.4 Software: The Assembler Model   44
    1.9.5 Hardware: The VHDL Model   44

vi   *Contents*

        1.9.6   Hardware: Synthesized Netlist   44
        1.9.7   Discussion   45
  1.10 Further Reading   45
  1.11 Exercises   46

2 Behavior and Concurrency   47

  2.1 Models for the Description of Behavior   47
  2.2 Finite State Machines   48
        2.2.1   Basic Definition   49
        2.2.2   Nondeterministic Finite State Machines   52
        2.2.3   Finite State Machines with ϵ-Moves   56
        2.2.4   State Aggregation   57
        2.2.5   Regular Sets and Expressions   60
        2.2.6   Finite State Machines with Output   63
        2.2.7   Finite State Machine Extensions   67
  2.3 Petri Nets   69
        2.3.1   Inputs and Outputs   75
        2.3.2   Petri Nets and Finite State Machines   76
        2.3.3   Modeling Templates   80
        2.3.4   Analysis Methods for Petri Nets   85
        2.3.5   The Coverability Tree   91
  2.4 Extended and Restricted Petri Nets   98
  2.5 Further Reading   100
  2.6 Exercises   101

3 The Untimed Model of Computation   107

  3.1 The MoC Framework   108
  3.2 Processes and Signals   111
  3.3 Signal Partitioning   113
  3.4 Process Constructors   115
  3.5 Process Properties   124
        3.5.1   Monotonicity   124
        3.5.2   Continuity   125
        3.5.3   Sequential Processes   128
  3.6 Composition Operators   129
        3.6.1   Parallel Composition   129
        3.6.2   Sequential Composition   130
        3.6.3   Feedback Operator   130
  3.7 Definition of the Untimed MoC   133
  3.8 Characteristic Functions   135
  3.9 Process Signatures   136
  3.10 Process Up-rating   138
        3.10.1 Map-Based Processes   139
        3.10.2 Scan-Based Processes   141

    3.10.3 Mealy-Based Processes   144
    3.10.4 Processes with Multiple Inputs   147
    3.10.5 Up-rating and Process Composition   148
3.11 Process Down-rating   149
3.12 Process Merge   149
    3.12.1 Perfect Match   149
    3.12.2 Rational Match   154
3.13 Rugby Coordinates   155
3.14 The Untimed Computational Model and Petri Nets   155
3.15 Synchronous Dataflow   158
    3.15.1 Single-Processor Schedule   161
    3.15.2 Multiprocessor Schedule   166
3.16 Variants of the Untimed MoC   174
3.17 Further Reading   176
3.18 Exercises   177

4 The Synchronous Model of Computation   181

4.1 Perfect Synchrony   182
4.2 Process Constructors   185
4.3 Feedback Loops   188
4.4 Perfectly Synchronous MoC   196
4.5 Process Merge   197
4.6 Clocked Synchronous Models   199
4.7 Extended Characteristic Function   201
4.8 Example: Traffic Light Controller   206
4.9 Rugby Coordinates   209
4.10 Validation   209
    4.10.1 A U-Turn Section Controller   209
    4.10.2 Monitors   211
    4.10.3 Validation Strategies   215
4.11 Further Reading   216
4.12 Exercises   217

5 The Timed Model of Computation   223

5.1 Introduction   224
5.2 Process Constructors   227
    5.2.1 Timed MoC Variants   230
    5.2.2 Representation and Distribution of Global Time   231
5.3 Discrete Event Models Based on δ-Delay   236
    5.3.1 The Two-Level Time Structure   236
    5.3.2 The Event-Driven Simulation Cycle   239

viii   Contents

    5.4 Rugby Coordinates   239
    5.5 Applications   240
    5.6 Further Reading   241
    5.7 Exercises   241

6   MoC Interfaces   243

    6.1 Interfaces between Domains of the Same MoC   244
    6.2 Interfaces between Different Computational Models   246
          6.2.1 Strip-Based Interface Constructors   246
          6.2.2 Insert-Based Interface Constructors   248
    6.3 Integrated Model of Computation   249
    6.4 Asynchronous Interfaces   253
    6.5 Process Migration   257
    6.6 Applications   265
    6.7 Further Reading   267
    6.8 Exercises   268

7   Tightly Coupled Process Networks   269

    7.1 Nonblocking Read   270
    7.2 Blocking Read and Blocking Write   274
    7.3 Oversynchronization   276
    7.4 Rugby Coordinates   280
    7.5 Further Reading   280
    7.6 Exercises   281

8   Nondeterminism and Probability   283

    8.1 The Descriptive Purpose   284
          8.1.1 Determinate Models   284
          8.1.2 Nondeterminate Models   285
    8.2 The Constraining Purpose   292
    8.3 The σ Process   293
    8.4 Synthesis and Formal Verification   294
    8.5 Process Constructors   295
          8.5.1 select-Based Constructors   295
          8.5.2 Consolidation-Based Constructors   296
    8.6 Usage of Stochastic Skeletons   297
          8.6.1 The Descriptive Purpose   298
          8.6.2 The Constraining Purpose   299
    8.7 Further Reading   299
    8.8 Exercises   300

9 Applications — 303

- 9.1 Performance Analysis 305
  - 9.1.1 Untimed Analysis 305
  - 9.1.2 Timed Analysis 312
- 9.2 Functional Specification 312
  - 9.2.1 Example: Number Sorter 314
  - 9.2.2 Example: Packet Multiplexer 316
  - 9.2.3 Role of a Functional Specification 317
- 9.3 Design and Synthesis 320
  - 9.3.1 Design Problems and Decisions 321
  - 9.3.2 Consequences for Modeling 323
  - 9.3.3 Review of MoCs 325
- 9.4 Further Reading 326
- 9.5 Exercises 327

10 Concluding Remarks — 333

Bibliography — 335

Index — 343

Foreword

When it first appeared in the mid-1990s, System-on-Chip (SoC) was arguably just a marketing term. At that point, the semiconductor fabrication process technology had achieved the scales of 350 and 250 nm, allowing the integration of only relatively simple digital systems. But, by the turn of the millennium, with 180, 150 and 130 nm processes available, many design teams were building true SoC devices.

These devices are systems in every sense of the word. They incorporate programmable processors (often with at least one RISC and one DSP), embedded memory, function accelerators implemented in digital logic, complex on-chip communications networks (traditional master–slave buses as well as network-on-a-chip), and analogue interfaces to the external world.

The integration of programmable processors into every SoC, and the growing importance of software in the products into which they are embedded, emphasizes, to a much greater degree than ever before, the role of embedded software in SoC. This includes the whole software development tool chain (compilers, debuggers, assemblers, etc.) as well as the development, porting and reuse of software, including real-time operating systems, device drivers, middleware or end applications.

But the design theories, methods and tools for designing, integrating and verifying these complex systems have not kept pace with the advanced semiconductor fabrication processes that allow us to build them. The well-known International Technology Roadmap for Semiconductors (ITRS) and analyses point to an ever-widening gap between what we are capable of building, and what we are capable of designing. Many techniques have been proposed and are being used to close that gap, including design tool automation and integration from RTL to Graphic Data System II (GDSII), integrated testbenches, IP reuse, platform-based design, and, last but not least, system-level design techniques.

Commercial tool providers, large design companies, and academic researchers are all doing a good job of developing the theory and practice of many of these techniques. The one that they are failing on most conspicuously is system-level design. Yet the underlying systems attributes of large complex IC designs makes this area the most important one to develop. Raising design team productivity, maximizing design reuse, and reducing the risk and effort involved in the design of SoCs absolutely demand that the center of design move from its current focus on RTL-based hardware design to true system level design.

Unfortunately, system-level design has no universal set of principles underlying it that correspond to the widely accepted design methodologies for digital logic, moving from RTL to GDSII via logic synthesis, and for standard cell placement and routing. Rather, system-level design consists of a series of design domain concepts and tools that lack universal applicability. For example, mastering dataflow design using one set of tools provides little grounding for understanding control-dominated system modeling using state machines.

One key reason for the lack of integrated system modeling approaches lies in the fact that heterogeneous systems consist of multiple design domains that are best modeled using different models of computation. This heterogeneous modeling requirement is quite new to digital IC designers. In fact, their traditional design methodology of "synthesis, place and route" completely ignores it, reducing everything to the single discrete event model of computation and concurrency supported by commercial simulators.

In order to design systems and Systems-on-a-Chip, a designer must know how to correctly transform specifications into verifiable implementations, to build executable system models, and to understand the impact of different computational models on design and implementation choices. In addition, the industry as a whole needs to adopt agreed upon system level modeling theories and abstractions in order to build better tools for modeling and design in the SoC era.

Modeling Embedded Systems and SoCs: Concurrency and Time in Models of Computation, specifically addresses these issues. A comprehensive and rigorous examination of models of concurrency and computation, it introduces formalisms for precise definitions and analysis for building executable system models today. Axel Jantsch's experience in teaching this material at the Royal Institute of Technology, Stockholm has helped him to create an excellent text for graduate-level courses. The extensive bibliography and discussion of applications broaden the book's usefulness beyond the classroom however. Industrial practitioners especially will appreciate that it establishes concepts important for choosing the best models to use for particular systems. In fact, anyone who needs a solid grounding in the theory underlying various system modeling approaches will find the book essential.

After an introduction that explains the importance of understanding models of computation and concurrency in system design and introduces basic system modeling concepts, Chapter 2 presents the important models of Finite State Machines and Petri Nets. Chapter 3, the heart of the book, introduces the author's formal modeling framework and notation in an introduction to untimed models of computation (MOC), with extensions to include the embedding of synchronous dataflow (one of the most useful models) in this analytical framework. Chapter 4 adds major concepts of synchronous models, building a base for Chapter 5's introduction to timed models of computation. Chapter 6 examines the issue of interfacing different models, whether using the same or different MOC. Chapter 7 elaborates tightly coupled process networks. Chapter 8 addresses the important notions of determinism and nondeterminism and the use of stochastic processes as a more tractable substitution for nondeterministic models. The last major chapter in the book, Chapter 9, covers the use of the various models in performance analysis and functional specification, and also discusses the links between modeling, design and synthesis. The Rugby analytical framework introduced by Jantsch in earlier papers provides the underlying framework with which he provides a context for the various models and their use models.

I recommend this book highly to those who are new to this subject or who need to better understand its formal underpinnings. Understanding this material provides the foundation for advanced system level modeling and implementation. System-level tool developers will find these concepts an excellent basis for determining which modeling approaches will be supported, and why. Jantsch's book is a major advance in the material available to students, researchers and practitioners alike in this important and vital area.

Grant Martin
Fellow, Cadence Laboratories
Berkeley, California
February 2003

Preface

This book is the result of a desire to systematically understand and present various models of computation developed during the 1980s and 1990s. These models have been elaborated, analyzed, and applied in the context of digital hardware and embedded software design. All of them have in common a strong emphasis on issues of time.

During the 1980s several new languages appeared that were based on the *perfect synchrony assumption*—that neither computation nor communication takes any noticeable time and that the timing of the system is solely determined by the arrival of input events. These so-called synchronous languages (Esterel, Signal, Argos, Lustre, StateCharts, etc.) were developed for either embedded software or digital hardware design. Since then a few of them, in particular Esterel, have enjoyed the continuous support of a growing user community.

The *synchronous design style* for hardware design, developed in the 1960s, has been extraordinarily successful and has dominated the development of digital hardware so completely that several generations of hardware designers have been educated under the implicit assumption that this is the only feasible way to design hardware.

Several *dataflow models* and languages were developed in the 1970s. With the definition of restricted models, such as synchronous dataflow, cyclo-static dataflow, and boolean dataflow, between 1985 and 1995 and their application to the design of signal processing systems, sophisticated tools with a large and growing user community evolved. Today dataflow models are firmly established in the designer communities of a number of application domains.

Discrete event models also have a long history. They have been used for modeling and simulation in a large variety of application areas. Discrete event models are a very general modeling technique: any kind of system can be modeled and simulated with any degree of accuracy. They are used to model objects as diverse as tiny semiconductor transistors and the communication patterns of the global Internet. Any entity of interest to scientists and engineers has been subject to simulation as a discrete event system. In hardware design a variant, the *delta-delay-based discrete event model*, has become popular with the advent of languages such as VHDL and Verilog.

The representation of time, at different abstraction and accuracy levels, plays a prominent role in all these models. The treatment of time might be considered to be among the most characteristic distinguishing features of these models. In fact a main conjecture of this book is that the same concepts and formalism can be used to model discrete event, synchronous, and dataflow models by simply varying the abstraction level for the representation of time. Thus, time, synchronization, and communication are central themes in this book. Further, the relationship between the communication part and the communication part of a process is important; the clean separation of the two parts is a guiding principle in developing the formalism used throughout this book.

As a result, this book presents one basic formalism to represent the seemingly very different computational models for discrete event, synchronous, and dataflow models. An advantage of this approach is that these models can be studied and compared within the same framework, thus making the essential differences and similarities apparent without allowing syntactical and superficial differences to obscure the deeper relationships.

How to Read This book: A Roadmap

In addition to the desire of a better understanding of the subject matter, the book is also the response to the demands of teaching a course. Draft versions of this book have been used since January 2001 in the course "System Modeling" for students of the System-on-Chip master's program at the Royal Institute of Technology in Stockholm. It has been found pedagogically useful to introduce the students to the basics of system modeling, finite state machines, and Petri nets before the unifying formalism of the book is explained and applied to the main models of computation and concurrency. Moreover, some chapters not only reflect the inherent logic of the subject but also are partially there in their current form and presentation due to requirements of that particular curriculum, of which "System Modeling" is a part. Consequently, depending on the background and interest of the reader, the book can be read in different ways. Figure P-1 outlines four alternative paths through the book.

- Reader A follows the book in both the selection of topics and the course of presentation.

- Reader B is mostly interested in the new material not covered in other books. She is either already familiar with or else not very interested in the basic concepts of modeling, finite state machines, and Petri nets. As a matter of fact, a good understanding of the topics of Chapters 1 and 2 is not a prerequisite for the rest of the book. Petri nets are used again at the end of Chapter 3, in Sections 3.14 and 3.15, but you can safely skip these sections without jeopardizing your understanding of the rest of the book. Also, the Rugby model, introduced in chapter 1 is used as a notational roadmap during the rest of the books. Most chapters have a section at the end, named "Rugby Coordinates", which locates the discussed model of computation within the Rugby model. Although these sections are intended to be an orientation aid for the reader, they are not required to understand the other sections. Thus, these sections can be skipped together with the discussion of the Rugby model in Section 1.5.

FIGURE P-1

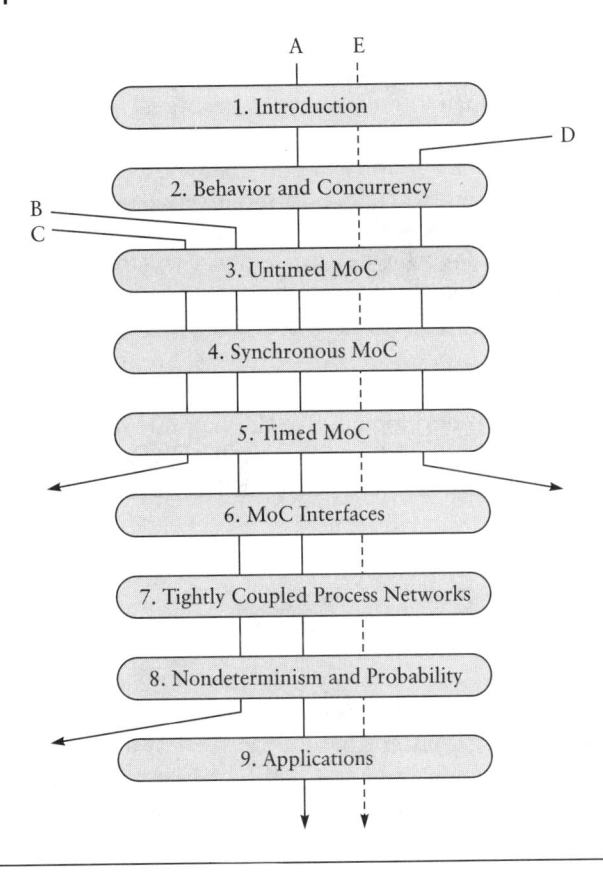

Different paths to read this book.

- Reader C is, like Reader B, only interested in the new material but prefers to focus fully on the discussion of the models of computation (MoC) in the unifying framework. Perhaps he is short of time or is just not interested in more advanced topics like MoC integration or nondeterminism. Chapters 3, 4, and 5 are in a sense the core of the book, and focusing first on them to obtain an understanding of the main ideas and lines of thought is an easily justifiable approach.
- Reader D is, like Reader C, only interested in the unifying presentation of the main models of computation but is not very fluent in finite state machines and Petri nets and thus reads Chapter 2 as a preparation.

Other alternative reading paths and path combinations are feasible. Chapters 6, 7, 8, and 9 are almost independent of one another and can be read in any order corresponding to the reader's preferences, after Chapters 3 through 5 have been perused.

It is also possible to benefit from the book without penetrating all the formal details and peculiarities of the presented formalism.

- Reader E is interested in an intuitive understanding of the book's main concepts but not so much in their mathematical presentation. He may follow any of the reading paths outlined above but may even in the core Chapters 3 through 5 skip, or browse superficially, the following sections without losing the thread or jeopardizing the chance to intuitively grasp the main ideas of succeeding chapters: Sections 3.5 (theoretical properties of processes), 3.6 (process composition and feedback loops), 3.10.2–3.10.5 (up-rating of processes other than the simplest possible), 3.11 (process down-rating), 3.12.2 (process merging for complex cases), 3.15.2 (multiprocessor schedules), 4.3 (feedback loops in synchronous process networks), and 4.7 (extended characteristic functions in the clocked synchronous model). These sections are interesting and required for the formal development of the framework and its properties, but are not required for an intuitive understanding. Again, Chapters 6 through 9, which partially also involve a more formal treatment, may be selected or skipped individually.

Who Should Read This Book

As mentioned earlier, this book has been used as a textbook for "System Modeling" in the System-on-Chip master's program at the Royal Institute of Technology in Stockholm. Thus, a full graduate course can be based solely on this book. It can also be used as complementary material for a course that attempts to give a comprehensive view of different important models of computation. For instance a graduate course in embedded system design could use Chapters 3 through 5 to introduce the untimed, synchronous, and timed models of computation.

Since the book gives a systematic and consistent account of a number of important computational models, it is suitable for *graduate students, researchers, and teachers* in the areas of design of electronic and embedded systems who are interested in the tasks of modeling, specification, and validation. With the development of a unifying framework, in which different computational models can be studied and compared, and with the presentation of new viewpoints on issues like oversynchronization, nondeterminism, and stochastic processes, we hope to provide a contribution that advances the understanding in this field.

To the *practicing engineer* the formal treatment may be an obstacle to the quick extraction of the main message. However, with the earlier suggestion of reading path E, we hope to ease the penetration of the subject even for those who are not enthusiastic about mathematical formalisms. Moreover, many small examples and interesting application cases are interwoven in the text, some of which may even be interesting for their own sake. For instance, at the end of the discussion of the untimed computational model we present scheduling techniques for synchronous dataflow models, and after the presentation of the synchronous models we acquaint the reader with validation techniques based on monitors. In Chapter 7 we give practical suggestions on how to realize determinate behavior in the presence of a nondeterministic underlying model, and in Chapter 8 we give an alternative to nondeterministic models based on stochastic processes.

Hence, we hope that students, teachers, researchers, and engineers will consider the theoretical foundation worthy of a careful study to gain new insights and a better understanding, and that they will find the illustrating examples and applications of immediate practical use.

Acknowledgments

Writing this book took four years. During this time many people have helped and contributed in different ways. It is in fact difficult to correctly trace back all ideas, contributions and influences and I am afraid I am not able to do justice to everybody because memories tend to change over time and distort the real cause of events. Anyway, I will try hard to properly and gratefully list all inputs since they together significantly shaped the final result and made it a much better book than I could have written otherwise.

At the Royal Institute of Technology in Stockholm I find the former Electronic Systems Design Lab (ESDlab) and now the Laboratory of Electronics and Computer Science (LECS) to be a very pleasant and inspiring research environment. I had the privilege to discuss with and benefit from many people there. Historically, my first attempt to find common, deep principles beneath a vast and confusing variety of languages, methods, tools and methodologies resulted in the formulation of the Rugby model. Numerous discussions with Shashi Kumar, now at Jönköping University, Sweden, and Ahmed Hemani, now at Acreo AB, Stockholm, Sweden, have been very exciting and considerably deepened my understanding in this subject. The work of the graduate students Andreas Johansson (now at Spirea AB, Stockholm), Ingo Sander and Per Bjuréus (now at Saab Aerospace, Stockholm) has been a rich source of inspiration and ideas and forced me more than once to reconsider basic assumptions and opinions. In particular, Ingo Sander's work on ForSyDe has countless connections to many of the topics and concepts presented in this book. Thanks to Zhonghai Lu for his implementation case studies, to Wenbiao Wu, to Ashish Singh who helped me better understand some of the formal properties of the framework and to Tarvo Raudvere, who in countless hours developed examples and exercises for the associated course. Some of his exercises also made it into this book. The discussions with Luc Onana, Seif Haridi and Dilian Gurov helped me better understand some of the mysteries of fixed point theory. With many others at LECS I had many fruitful and exciting discussions with no direct connection to the subject of this book but which nonetheless have influenced it. Among those are Johnny Öberg, Peeter Ellervee, now at Tallinn Technical University, Mattias O'Nils, now at Mid Sweden University, Sundsvall, Sweden, Abhijit K. Deb, and Björn Lisper, now at Mälardalen University, Västerås, Sweden. Finally, I am very grateful to Hannu Tenhunen for creating a very inspiring environment and for his never-ending stream of ideas and suggestions on a variety of technical and non-technical issues.

Many thanks to the technical reviewers Edward A. Lee, University of California at Berkeley, California, Grant Martin, Cadence Laboratories, Berkeley, California, and Perry Alexander, University of Kansas, who read a draft of the manuscript and provided many valuable suggestions for improvement. Unfortunately I could not implement all their ideas and proposals. I particular I would like to thank also Grant Martin for reading so carefully the final manuscript and for spotting many small and some big technical mistakes, which would not have been detected otherwise.

A warm thank you goes to Alyson Day and Denise Penrose at Morgan Kaufmann Publishers for their strong support and encouragement throughout the publishing process, and to Richard Cook at Keyword Publishing Services, Essex, UK, for ironing out all the grammar, style and format errors that I managed to introduce. Needless to say, any technical and language errors that remain in the text despite their heroic effort are of my origin.

Although it has been observed many times, it is still true and urgent to state that the endeavor of writing such a book is only possible through the strong support and many sacrifices of one's family. I apologize for spending so many hours with this book rather than with them and I am most grateful to Ewa, Simon and Philip for support and encouragement during the last four years.

Notation

$\mathbb{N}$	Set of natural numbers: $1, 2, \ldots$
$\mathbb{N}_0$	Set of natural numbers including 0: $0, 1, 2, \ldots$
$\mathbb{Z}$	Set of integers: $\ldots, -2, -1, 0, 1, 2, \ldots$
$\mathbb{R}$	Set of real numbers
iff	if and only if
E	Set of all events
$\dot{E}$	Set of all untimed events
$\bar{E}$	set of all synchronous events
$\hat{E}$	Set of all timed events
S	Set of all signals
S^n	Set of all signals with length $n \in \mathbb{N}_0$
$\dot{S}$	Set of all untimed signals
$\bar{S}$	Set of all synchronous signals
$\hat{S}$	Set of all timed signals
$s \oplus s'$	concatanation of signal s with s'
$\bot$	"Bottom" event, i.e., nothing is known about its value
$\sqcup$	"Absent" event, i.e., no event occurs at the concerned time
$\bar{E}_\bot$	Set of events including $\bot$
$\bar{S}_\bot$	Set of signals that can carry $\bot$ events
$\bar{S}_\bot^\infty$	Set of infinite signals that can carry $\bot$ events
$\langle\rangle$	Empty signal or sequence

Notation	Description
$\langle \sqcup \rangle^\infty$	Infinite siganl with only absent events
$\langle \bot \rangle^\infty$	Infinite signal with only $\bot$ events, i.e., it is undefined everywhere
$s \sqsubseteq s'$	means that signal s is a prefix of signal s'
$s[i]$	denotes the ith event in signal s
$x \preceq y$	x is "weaker" than y, i.e., x contains less information than y or as much information as y
$x \prec y$	x is strictly "weaker" than y, i.e., x contains less information than y
$x = \text{LUB}(X)$	x is the least upper bound of the partially ordered set X
FB$_\text{P}$	the feedback operator based on the prefix order $\sqsubseteq$ of signals
FB$_\text{S}$	the feedback operator based on the Scott order $\preceq$ of signals
$\neg$	Logic "not" operation
$\wedge$	Logic "and" operation
$\vee$	Logic "or" operation
$\forall x \in X : A$	For all elements x from a set X the formula A is true
$\{x : A\}$	Set of x for which formula A is fulfilled
$\wp(X)$	Power set of set X, i.e., the set of all subsets
$X \setminus Y$	Subtraction operation over sets, i.e., the result is the set of elements from X that are not in Y
I-V	Current-Voltage: a device is characterized by equations describing current and voltage relations

chapter one
Introduction

After a general discussion of systems, modeling, and heterogeneity, we introduce in this chapter some basic concepts of modeling such as state, state space, time, *and* event. *We establish that an initial state, the state equations, and the output equations constitute a complete model of a system and allow a prediction of future output provided we can observe the system's inputs. However, we admit that any model is subject to limitations due to limited scope and accuracy of input observations; the chosen domains for input, state, and output variables; the chosen time model; and so on. For instance in a discrete time model we cannot take into account effects that require a continuous time model and differential equations to represent. Or if our model of a logic circuit does not include temperature as a parameter because we assume that the logic behavior is temperature independent, the model will only be valid and useful in a temperature range where this assumption is indeed correct. Consequently, a model is only useful or correct with respect to a purpose and a task. We point out that a model can be minimal with respect to a task if it contains exactly the information relevant and useful for the task.*

Based on characteristics of state and output equations, several system properties are discussed, such as linearity, time invariance, *and* determinism. *We also distinguish between* event-driven *and* time-driven *simulation mechanisms.*

In the second part of the chapter we use the Rugby *model to organize modeling elements in terms of* domains *and* abstraction levels. *The Rugby model identifies four domains:* computation, communication, data, *and* time. *All domains can be modeled at several abstraction levels, ranging from detail-rich physical levels to highly abstract system levels. For instance the time domain, which plays a prominent role throughout this book, can be modeled as continuous time, as discrete time, as clocked time, or as a causality relation. We thoroughly elaborate on the concept of abstraction and its differences with hierarchy. The Rugby model is used in later chapters to categorize models and models of computation with respect to the abstraction level used in each domain. We designate this as the* Rugby coordinate *of a particular model. We illustrate this approach at the end of this chapter with a network terminal application. All models of the network terminal design are placed in the Rugby model by assigning Rugby coordinates. Moreover, the design and synthesis steps are also characterized by Rugby coordinates by giving the starting and end points of the design activity.*

1.1 Motivation

Many different activities are required to carry a complex electronic system from initial idea to physical implementation. *Performance modeling* helps to understand and establish the major functional and nonfunctional characteristics of the product. *Functional modeling* results in a specification of the functional behavior of the product. *Design and synthesis* refine the product specification into a sequence of design descriptions that contain progressively more design decisions and implementation details. *Validation and verification* hopefully ensure that the final implementation behaves as specified. Many other activities, such as requirements definition, test vector generation, design for test, static timing analysis, technology mapping, placement and routing, solve a specific problem and thus contribute to a working product.

All these activities operate on models and not on the real physical object. One obvious reason for using a model is that the real product is not available before the development task is completed. Today's pressing demands for short product development cycles usually prohibit the manufacturing of a complete prototype as part of the development. Even if one or a few prototypes can be built, they cannot replace the hundreds or thousands of different models that are routinely developed for an average electronic product.

We need so many different models because of the fundamental trade-off involved in model selection and model design. To appreciate this, let's adopt a working definition that will be useful to keep in mind when reading this book.

Model   *A model is a simplification of another entity, which can be a physical thing or another model. The model contains exactly those characteristics and properties of the modeled entity that are relevant for a given task. A model is* minimal *with respect to a task if it does not contain any other characteristics than those relevant for the task.*

First, note that a model relates to another entity. This other entity may be another model; hence we can have a sequence of models, each simpler than the previous one. The entity could also be a real or conceived physical object. If we make a model of a product under development, the product does not yet exist. This is a rather fundamental usage of models by humans. Whenever a human consciously makes something, she has a model of this object in her mind, before the creation process can start. Without such a model in mind, no *planned* activity would be possible.

Second, the model is a simplification of this other entity. It is simpler because certain properties and features are systematically missing in comparison to the original entity.

Third, the model is related to a task and an objective. Whether a model is useful and interesting depends on what we intend to do with it. Any simplification of an entity leads to a model of that entity, but which properties are missing and which are present decides what the model can be used for and what it is useless for. If a model of an integrated circuit does not contain delay information, we cannot use it to derive the clock frequency of the final product. If a network of components does not contain geometry information, we cannot estimate the physical size of the manufactured circuit. Only if the network of components contains behavior information about the components will we be able to simulate it and analyze the behavior of the system. In almost all cases the model is therefore designed carefully for a particular task. Ideally, a model should be *minimal*—it should contain all relevant properties but omit all irrelevant properties. Irrelevant properties usually make the task more difficult—and can even make it impossible. For a microprocessor with a few hundred million transistors, a netlist of discrete gates with the precise delay information for all gates and all interconnects may be a useful model to verify that the device will work at a particular clock frequency. But it is unsuitable for verifying that the operating system will correctly boot because the simulation of a few seconds could easily take weeks or months. To validate that the hardware works properly with operating system software, a different model of the hardware is needed. In this context the physical delays of all gates are irrelevant and could be removed from the model. A useful model could be an instruction set simulator that only exhibits the correct behavior for each instruction as it is observable from the outside. But care has to be taken to understand exactly which characteristics are relevant and which are not. For instance, is it necessary for the instruction set simulator to mimic the instruction decoding pipeline of the processor? In most cases it is probably not, but in some cases it may be important, for example, for the correct handling of interrupts. But because the inclusion of the instruction pipeline may slow down the simulation speed of the instruction set simulator by an order of magnitude, the best solution might be to use a crude model without a pipeline for the bulk of the simulations, and to use a more accurate model with a pipeline for validation of the interrupt-handling process. This example illustrates the trade-offs in the development and selection of models. On the one hand, they should contain all relevant characteristics, but on the other hand, they should be as simple as possible. Since we have so many different tasks during the development of a product, it is no surprise that we need to develop a great number of different specialized models.

Another subtle aspect of modeling is specific for design activities. When we model a device that does not yet exist, but will exist as result of the design activity, we have not only freedom to decide on the characteristics of the model, but we also have some freedom to decide the precise properties of the resulting product. This is different from the situation when we model an existing object that is defined in all its details down to the quantum physics level. This difference affects how a model is interpreted and used. A specification model *constrains* the final product. It defines a range or a set of possible behaviors, and all implementations that fall into this range are acceptable. Consequently, models that allow the efficient formulation of functional and nonfunctional constraints are desirable.

A particularly interesting situation is when a model[1] or a modeling language either automatically guarantees the compliance with specific constraints, or allows the efficient determination of whether constraints are met. For instance, the finite state machine model allows only systems with a finite state space to be modeled. It is simply not possible to write down a finite state machine with an unbounded or infinite state space. Thus, every correct implementation of a finite state machine will have a finite state space that can be calculated statically. This is an example of a model that guarantees a specific property for all its concrete models.

Another example is synchronous dataflow, which provides an efficient procedure to compute the buffer sizes between communicating processes. It guarantees that, for any concrete synchronous dataflow model, we can efficiently determine if the buffers between the processes are finite and, if so, what their size is. Hence, even though the model allows the description of a system with infinite buffer requirements, we can statically find out if this is the case.

Such formal features are desirable because they guarantee certain properties or allow us to deploy certain efficient design or analysis procedures. However, in general we can observe that they reduce the expressiveness of models. Again, there is a trade-off involved, and the decision of which model to select depends on the problem we have to solve.

It is obvious by now that there is no best or ideal model or modeling language due to the diversity of tasks and problems we have to address, each of which has different requirements for the models. And this is also the main motivation for this book: because designers of electronic systems are confronted with a variety of different tasks, languages, and modeling concepts, it is of utmost importance to understand the major and essential features of the different modeling concepts. Only if the superficial and accidental features of a modeling language or model are not confused with the important and essential concepts, is it possible to assess the limits of a language and to determine which language to select for a particular design task.

It is not feasible to address all interesting aspects of system modeling in a single book. We will therefore concentrate on *models of computation* (MoC)[2], that is, on the aspects that are relevant for the communication, synchronization, and relative timing of concurrent processes. Even in this narrower field, we cannot give a complete account of all concurrency models, and we do not even attempt to do this. Instead we prefer to develop the field in a systematic way by introducing a common notation that allows us to compare and relate some important models of concurrency. Then we discuss major

---

[1] Note that the term "model" has two distinct meanings. First, it denotes a particular instance or object, for example, a description in VHDL. Second, it denotes a set of modeling concepts such as the "finite state machine model" or the "dataflow model." We use the word "model" in both meanings, but if there is any danger of confusion, we use the terms "concrete model" and "metamodel" to distinguish the two meanings.

[2] We use the expressions "model of concurrency" and "model of computation" as synonyms, both abbreviated as "MoC." Since we focus mostly on aspects of interaction and synchronization of concurrent processes, the term "model of concurrency" is more accurate for our purposes, while "model of computation" is broader, including purely sequential models like algorithms and state machines. In the literature the term "model of computation" is well established and more widespread than "model of concurrency."

design activities and relate them to the different concurrency models. This should give you a basic understanding of which modeling concepts are important for a particular application. Again, the list of applications is incomplete. This is partially explained by our focus on functionality and time. Many design activities require the consideration of non-functional aspects such as geometry, power consumption, signal integrity, and so on.

1.2 Heterogeneous Models

Today the idea that a unified language or internal representation cannot meet all requirements of a system modeling and specification tool has been firmly accepted. By "system" we mean embedded electronic systems, such as those built into mobile phones, cars, planes, toys, robots, and so on. They may be distributed or integrated onto a single integrated circuit. The requirements for a language to model such systems are just too diverse to be supported neatly by one single language. The diversity is due to several factors:

1. The applications consist of parts with different characteristics. For instance, dataflow-dominated parts require different modeling techniques and styles than control-dominated parts.

2. Different objectives of modeling activities require the inclusion or exclusion of certain information. For instance, at an early design stage the overall performance and buffer requirements in relation to the data samples to be processed has to be analyzed. For this purpose detailed timing information in terms of nanoseconds or clock cycles may unnecessarily complicate the analysis. But for a detailed analysis to establish the maximum clock frequency and to find the critical path, detailed and accurate timing information is necessary. In general, different activities such as specification, design, synthesis, verification, and performance analysis put different requirements on the modeling techniques and languages.

3. Different communities, who historically have dealt with different parts and aspects of systems, have developed different languages, tools, and methodologies. It must be acknowledged that the investment in education, expertise, tools, and libraries in these different communities is so significant that there is just no other choice than to continue working with them. For instance, Matlab-related tools and libraries are so well suited for the analysis and development of algorithms that users will not drop them even though these tools do not support the development of control-dominated protocols or the implementation of the algorithms in hardware.

Amid the difficulties in finding and establishing a single, unifying notation, researchers have started to link and integrate different well-established languages. The most influential project is Ptolemy at the University of California, Berkeley (Buck et al. 1992; Davis et al. 1999). Ptolemy is a simulation environment that allows the linking of different languages and simulators together. It defines a number of computational

FIGURE 1-1

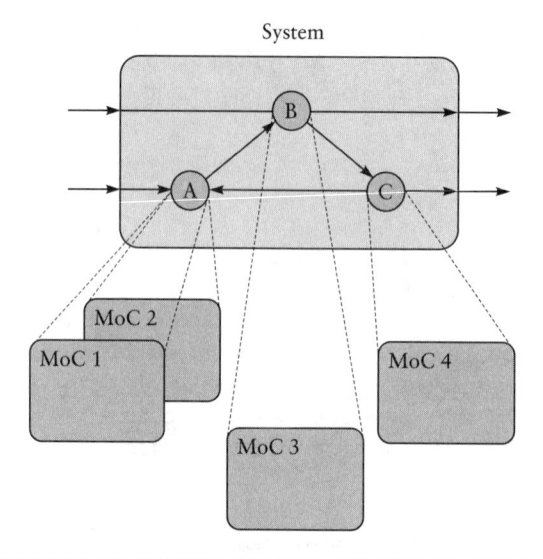

Different parts of a system are modeled in different computational models and integrated in an MoC framework.

models and the interfaces between them. It does not impose restrictions on what is allowed inside an MoC domain.[3] Only when the different domains exchange data or synchronize do they have to comply with the interfaces and protocols defined by Ptolemy. Many other, less general frameworks successfully connect two or more languages with their simulation environments.

In Figure 1-1, a system model consists of three parts, A, B, and C, which are all described in different computational models. The framework integrates these and allows, for instance, joint simulation runs by defining communication and synchronization interfaces. These interfaces could be defined for every pair of MoCs such that every point-to-point connection is defined separately. Adding a new MoC to the framework would require defining and implementing interfaces from the new MoC to every other existing MoC. Alternatively, the framework could provide a communication and synchronization medium with interfaces to which all MoCs have to connect. The framework mediates all communication between different MoC processes. A sender process just dispatches a message into the medium of the framework, which in turn delivers the message, after appropriate protocol conversions, to the receiving process. In this scenario, adding a new MoC to the framework requires defining and implementing only the interface between the new MoC and the framework. This latter approach is taken by Ptolemy.

Figure 1-1 also shows that a process may be modeled in more than one MoC for different purposes. Process A is modeled both in MoC 1 and MoC 2. The two models

---

[3] Be aware, that we use the term "domain" with three different meanings. First, a domain may be a set of processes that are modeled according to a particular MoC. Second, it can denote a data type such as the domain of integers or real numbers. Finally, in the Rugby model it denotes the four domain lines: computation, communication, data, and time.

can be used for different purposes: for example, one for high-level functional validation and the other for detailed timing analysis. This idea can be systematically developed into the concept of *domain polymorphism*, which allows the development and use of different MoC models for the same process and thus defines clearly how the same process interacts in different MoCs.

Its many advantages notwithstanding, the heterogeneous coupling of different languages and computational models has a few fundamental problems. The strict separation of different system parts, embodied by the usage of different languages and their respective tools, is a high barrier for synthesis, design, analysis, and verification tools. Once a division between different MoC domains has been decided upon and frozen into concrete design languages, it is difficult to move behavior from one domain to another. Thus, the overall design space is greatly reduced by this decision. Also, this barrier impedes performance analysis and formal verification tools, which already have difficulties coping with the growing size of designs even when they are represented in a uniform and suitable notation.

It is unclear today if these problems of heterogeneous models will eventually be overcome, or if they will be gradually replaced by more homogeneous models. But solutions will, in any case, rest on a thorough understanding of the underlying computational models and their interaction.

1.3 Separation of Computation and Communication

A major guiding principle recurring throughout this book is the separation of different concerns, which is also explained thoroughly by Keutzer et al. (2000).

In Sections 1.5 through 1.9, the Rugby model identifies four separate domains (computation, communication, data, and time), motivated by the possibility of analyzing these domains independently from one another. For instance, many popular design and synthesis steps can be formulated as transformations in only one of these domains and do not affect the other domains at all. A case in point is the scheduling task in the behavioral synthesis of hardware, which takes place solely in the time domain.

Throughout the rest of the book, we concentrate on the separation of communication from computation. We start out in Chapter 2 by introducing both in a pure and relatively simple form. On the one hand, a finite state machine models computation in a very general way but without any concurrency and communication involved. On the other hand, a Petri net can be considered as a process network with processes that exhibit the simplest possible behavior: they only communicate by consuming and emitting tokens. Thus finite state machines and Petri nets can be used to study computation and communication separately. In the chapters that follow, a formal framework is developed that separates the communication and synchronization aspects of processes from their internal behavior. Essentially, a process is divided into a *process core*, capturing its computation, and a *process shell*, capturing its communication with other processes.

On a conceptual level the advantage of this separation is an improved understanding of the different issues. When dealing with communication, synchronization, and concurrency problems, you do not need to bother with how the computation of processes is represented. It may be modeled as a finite state machine, as an algorithm, as a mathematical function, or as a set of logic constraints. Different processes may also be modeled in different formalisms.

On a practical level it means that components can be developed independently, integrated easier, and reused to a much higher extent than otherwise possible. A prime example is the Internet Protocol stack, which determines how computers communicate with each other. An implementation of this stack can be designed and implemented independently of the computers it connects. It can be implemented in hardware, software, or a mixture of both and can be used to connect PCs, workstations, and mainframes running Unix, Windows, or any other kind of operating system. On the level of applications, abstract communication protocols can be devised (e.g., based on message passing or shared memory), again independently from how the behavior of the application is expressed. They can, however, be used to facilitate communication between almost arbitrary applications. Furthermore, they can be implemented on a variety of lower-level communication protocols, such as an on-chip bus protocol or on the Internet Protocol stack.

Hence, in summary, the separation of communication from computation has far-reaching, positive effects on both our ability to penetrate and understand the essential issues and on our capability to design and develop large and complex systems.

1.4 Systems and Models

What is meant by a *system* always depends on the context of discourse. What is a system in one context may be a tiny component in another context and the environment in a third. For instance, an integrated circuit (IC) is a very complex system to IC designers, but it is only one out of many components of the electronic equipment that goes into an airplane. The electronic system is only a small part of the airplane, which in turn is only one element in the air traffic organization of a country.

Even in a given specific context, we are usually content with an intuitive understanding of what we mean by the word "system," and we rarely try to give a formal definition. Nonetheless it is worthwhile to consider definitions found in dictionaries.

A *system* is

- "an aggregation or assemblage of things so combined by nature or man as to form an integral of complex whole" (*Encyclopedia America*)

- "a regularly interacting or independent group of items forming a unified whole" (*Webster's Dictionary*)

- "a combination of components that act together to perform a function not possible with any of the individual parts" (*IEEE Standard Dictionary of Electrical and Electronic Terms*)

All three definitions emphasize that a system consists of simpler components. These components cooperate or interact and together exhibit a property or behavior that is beyond any of the individual parts. And precisely this is our main objective in this book: to analyze the interaction of components to form a system behavior.

In Chapter 2 we will first discuss finite state machines, which are a prominent technique to model an entity as one whole without considering or assuming an internal structure. Since this is not sufficient for our purposes, we go on to discuss Petri nets. Petri nets are among the first and most important techniques to model and analyze concurrency and interaction of entities. Starting with these two complementary views and techniques, we devote the rest of the book to integrating them.

But first we introduce some basic concepts of system theory. A system that we want to model has some behavior. If we ignore for a moment the internal structure of the system, we can observe only the inputs and outputs of the system. To model the system behavior, we therefore have to relate its outputs to its inputs, which we do by means of a mathematical function.

The inputs and outputs can assume different values over time, so we represent them as functions over time. The set of *input variables*

$$\{u_1(t), \ldots, u_p(t)\} \qquad t_0 \leq t \leq t_f$$

represents the values of p inputs in the time period $[t_0, t_f]$. Similarly, we define the set of *output variables*

$$\{y_1(t), \ldots, y_m(t)\} \qquad t_0 \leq t \leq t_f$$

for m outputs. Often we use column vectors $\vec{u}(t)$ and $\vec{y}(t)$ to represent input and output variables, respectively. Hence we write

$$\vec{u}(t) = \begin{bmatrix} u_1(t) \\ \vdots \\ u_p(t) \end{bmatrix}$$

$$\vec{y}(t) = \begin{bmatrix} y_1(t) \\ \vdots \\ y_m(t) \end{bmatrix}$$

for p input and m output variables. For instance, if the system is an electrical circuit, the variables u and y may represent voltage or current levels. If we are dealing with a piece of software, they may stand for character strings or more complex data types.

We assume we have the ability to control and set the input variables, while the system controls the output variables. We think of the output variables as being the *response* of the system to our input *stimuli*.

The core of our model of the system is a set of functions that relate the inputs and outputs. We use one function for each output.

$$y_1(t) = f_1(u_1(t), \ldots, u_p(t))$$

$$\vdots$$

$$y_m(t) = f_m(u_1(t), \ldots, u_p(t))$$

FIGURE 1-2

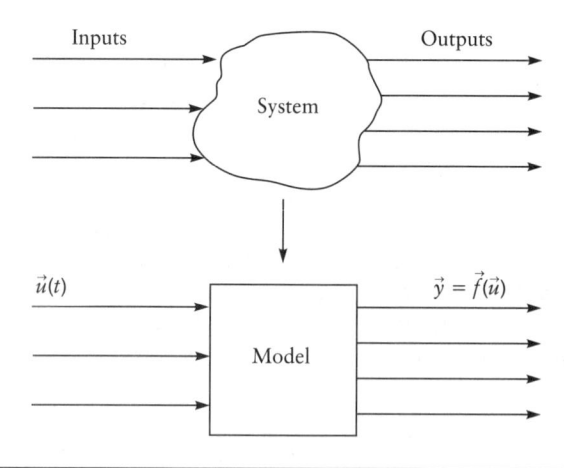

A mathematical model of a real system.

In vector form we write

$$\vec{y}(t) = \vec{f}(\vec{u}(t)) = \begin{bmatrix} f_1(u_1(t), \ldots, u_p(t)) \\ \vdots \\ f_m(u_1(t), \ldots, u_p(t)) \end{bmatrix} \quad (1.1)$$

where $\vec{f}$ is the column vector of the functions $f_1, \ldots, f_m$.

Equation (1.1) is a model of the real-world object, the system illustrated in Figure 1-2. It represents only a few aspects of the system but ignores most properties. First, the input and output variables should not be confused with the inputs and outputs of the system, even though we often use the words "input" and "output" when we actually mean "input variable" and "output variable." For instance, if the system input is a wire and we represent the voltage with our input variable, we ignore all the other properties that determine the real wire (e.g., the current, the temperature, the material, the shape, etc). Another model may choose to represent some other aspect because it is used for another purpose, but any model will be very selective in what properties it considers. Second, the function $\vec{f}$ cannot capture the system behavior completely and accurately, not even for the selected properties only. For example, if $\vec{f}$ relates an input and output voltage, it will do so only for certain conditions, such as a particular temperature range, a certain radiation level, a certain chemical environment, and so on. Even for conditions for which the model is valid, it reflects reality with limited accuracy. Thus, $\vec{f}$ does not take all influencing parameters into account.

Note also the asymmetry of a system and its model. For every real-world system we can build a model—in fact, an infinite number of different models. But there are models that do not correspond to any real system. For instance, let $u(t) = a_t$ represent strings of characters a_t appearing at the input at time t and $y(t) = \text{repeat}(\text{'x'}, -\text{length}(a_i))$

FIGURE 1-3

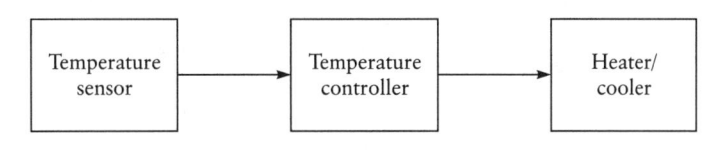

```
┌────────────┐      ┌────────────┐      ┌────────────┐
│ Temperature│ ───▶ │ Temperature│ ───▶ │  Heater/   │
│   sensor   │      │ controller │      │   cooler   │
└────────────┘      └────────────┘      └────────────┘
```

A temperature controller.

be the output at time t, with length(a) giving the length of string a and repeat(c, n) returning the string consisting of n characters c. Obviously, strings with negative length do not exist, and thus there is no real system corresponding to this model. Note, however, that there is an interpretation step involved, which allows us to judge that there is no real system corresponding to the model. If we interpret both u and y as strings and functions length( ) and repeat( ) as functions over strings, a real system is infeasible. But if we interpret input and output variables as vectors and length( ) and repeat( ) as appropriate functions over vectors, a real system corresponding to our model is very well conceivable. Whether a system exists for every *consistent* mathematical model given a proper interpretation is a question beyond the scope of this book.

Example 1.1 Consider a temperature controller with one input from a temperature sensor and one output to a heating/cooling device as shown in Figure 1-3. The controller receives information about the current temperature, compares it to a reference temperature, and sends a control signal to a heating/cooling device to adjust the temperature to the reference temperature. Depending on our objectives we can conceive of different models of this controller. Figure 1-4 shows three possibilities denoted as TC1, TC2, and TC3.

The input to TC1 is a temperature value expressed in degrees Celsius in the range $[0°, 50°]$ and its output is a control signal between -1 and $+1$. A control value in the range $[-1, 0)$ requests cooling while a value in $(0, +1]$ requests heating; a value of 0 turns off both the cooler and heater.

So for TC1 we have $u(t) = T_t$, representing the temperature at time t, and $y(t) = (R_T - u(t))/50$, where R_T is the reference temperature (Figure 1-5). $y(t)$ is 0 when $u(t)$ equals the reference temperature, otherwise it will approach -1 or $+1$ for large deviations from R_T.

The second model, TC2, has input and output voltages representing the temperature on the input and the control signal on the output. The voltage range on the input is $[0\,V, 5\,V]$ and represents a temperature range of $[0°, 50°]$; hence, $u(t) = V_t$. On the output a voltage range of $[-5, +5]$ represents the control values between -1 and $+1$, and we have $y(t) = (R_V - u(t))$, where R_V is the voltage level representing the reference temperature.

FIGURE 1-4

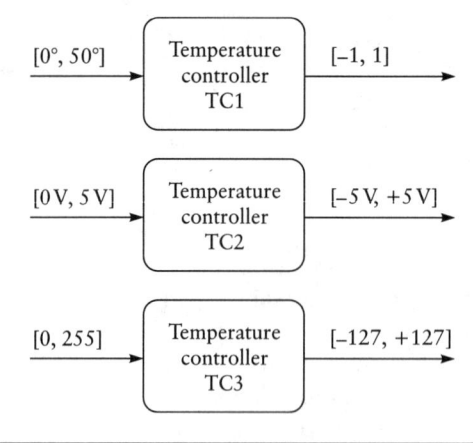

[0°, 50°] → Temperature controller TC1 → [−1, 1]

[0 V, 5 V] → Temperature controller TC2 → [−5 V, +5 V]

[0, 255] → Temperature controller TC3 → [−127, +127]

Three different models of the temperature controller.

FIGURE 1-5

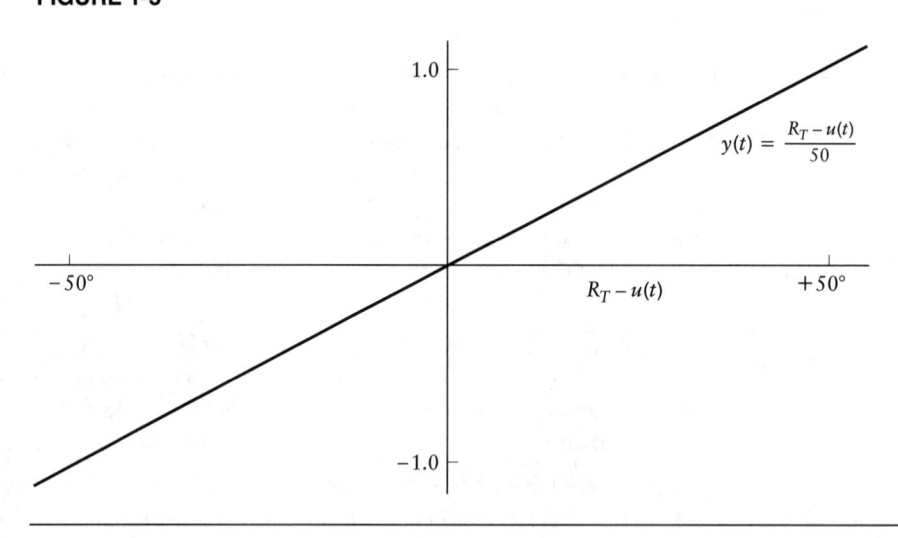

$$y(t) = \frac{R_T - u(t)}{50}$$

$R_T - u(t)$ versus the control output for TC1.

The third model, TC3, excludes the analog-to-digital converter at the input and the digital-to-analog converter at the output. It is a purely digital model, and both the input and output are represented by 8-bit words. The input is thus a number between 0 and 255, and the output is between −127 and +127. We have $u(t) = N_t$ and $y(t) = (R_N - u(t))/4$, where R_N is the number representing the reference temperature. We also assume that the time is discrete; that is, t is drawn from a discrete set such as the integers.

1.4.1 System Properties

Based on some properties of the next-state and output functions we can identify special cases of systems that have desirable features for analysis and modeling.

State-less and State-full Systems

The three models TC1, TC2, and TC3 are quite different and can be used for different purposes. Some of the differences we will take up later. But they all have the property in common of being a state-less model.

Definition 1.1 A *state-less system* is one where the output $y(t)$ is independent of past values of the input $u(t'), t' < t$ for all t.

In all three models the output $y(t)$ is a function of only the input at time t, $y(t) = f(u(t))$, with no other hidden parameters. The reference temperature is assumed to be a constant, and we have not considered if and how it can be changed. Our controller does not take into account how the temperature is changing, and this may in fact be a shortcoming. For instance, if the temperature is increasing rapidly and approaching the reference temperature, our controller will still turn on the heater, with the effect that the temperature will increase beyond the reference point, which in turn will switch on the cooling system. Consequently, the temperature will oscillate around the reference point, and the heater/cooler will be permanently active. One possible remedy is to take the change of the input into account as well.

Definition 1.2 A *state-full system* is one where the output $y(t)$ depends on the current input value $u(t)$ and on at least another input value $u(t')$, with $t' < t$, $y(t) = f(u(t), u(t'))$.

Since $u(t')$ is not available at the input at time t, this requires some internal memory where past input values are stored.

Example 1.2 We derive a new model, TC4, by changing TC3 such that a quickly rising temperature should have a decreasing effect on the control output. On the other hand, a quickly falling temperature should increase the control output, thus decreasing the cooling or increasing the heating effect. If we consider the difference between $u(t)$ and $u(t-1)$, we note that

$$u(t) - u(t-1) < 0 \quad \text{falling temperature}$$
$$u(t) - u(t-1) = 0 \quad \text{constant temperature}$$
$$u(t) - u(t-1) > 0 \quad \text{rising temperature}$$

By simply subtracting this from our original function, we get

$$y(t) = \text{mm}\left(\frac{R_T - u(t)}{4} - (u(t) - u(t-1))\right)$$

where $\text{mm}(x) = \min(\max(x, -127), 127)$ contains the output in the proper range.

There are different ways to model state-full systems. For systems such as TC1 in Example 1.1, where the input and output variables and the time are continuous, we often see differential equations. For systems such as TC4 with discrete domains for inputs, outputs, and time, we often have difference equations or an explicit internal state. The latter is the most common case we will encounter.

Time-Varying and Time-Invariant Systems

In a time-invariant system the output function does not explicitly depend on the value of the time variable. It may depend on the current input value $u(t)$ as well as on all past input values $u(t'), t' < t$, but the actual value of t is not relevant. Thus, if the entire input history is shifted in time by a constant value, the output will be shifted by the same amount but otherwise be identical. Hence, we have the following definition.

Definition 1.3 A model M with $\vec{y}(t) = \vec{f}(\vec{u}(t))$ is *time-invariant* if, supplied with input variables $\vec{u}'(t) = \vec{u}(t - \tau)$, it defines output variables $\vec{y}'(t) = \vec{f}(\vec{u}'(t)) = \vec{y}(t - \tau)$ for any τ.

Thus, if the entire input is shifted in time, the output is shifted by the same amount, as illustrated in Figure 1-6.

Definition 1.4 In contrast, a *time-varying* model is one for which this property does not hold, that is, where the output function explicitly depends on t: $\vec{y}(t) = \vec{f}(\vec{u}(t), t)$.

We will deal exclusively with time-invariant systems.

System State

For state-full systems the output $\vec{y}(t'), t' \geq t_0$, cannot be predicted based only on the input $\vec{u}(t')$ because the output depends also on past inputs before t_0. This means that the system maintains some additional information internally

Definition 1.5 The *state* of a system at time t_0 is the information required such that the output $\vec{y}(t)$ for all $t \geq t_0$ is uniquely determined by this information and the inputs $\vec{u}(t), t \geq t_0$.

The state of a system at time t_0 is called the *initial state*.

FIGURE 1-6

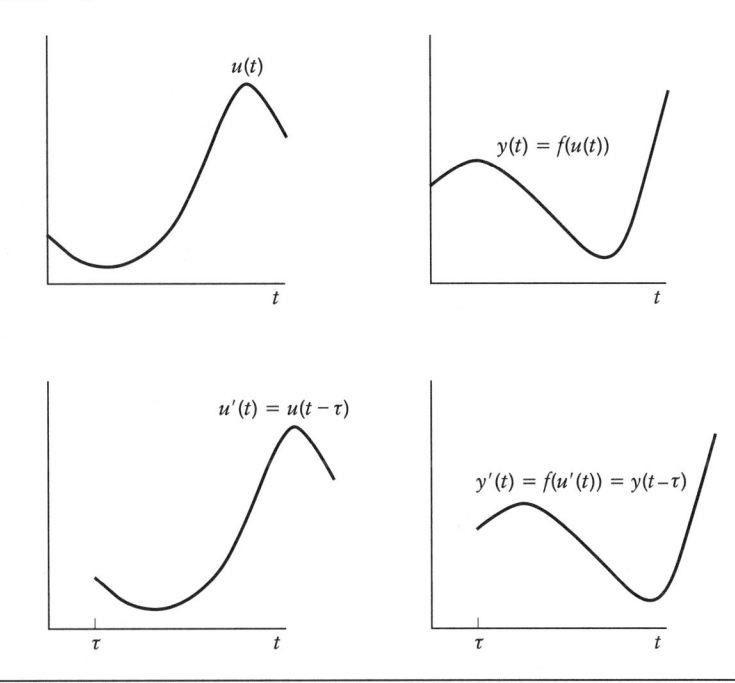

A time-invariant model.

The equations required to specify the state for all $t \geq t_0$, given the initial state and the input $\vec{u}(t), t \geq t_0$, are called *state equations*.

The *state space* of the system is the set of all possible values of the state.

Just as for inputs and outputs, the state may, in general, be a vector. We usually denote the state vector by $\vec{x}(t)$, the initial state by $\vec{x}_0$, and the state space by X.

The form of the state equations depends on the type of system under consideration. For a *continuous state* and *continuous time* system, the state space X and the set of time values are continuous sets (e.g., the real numbers). In that case the state equations are a set of differential equations:

$$\dot{\vec{x}}(t) = \vec{g}(\vec{x}(t), \vec{u}(t), t)$$

and the system is completely defined by

$$\begin{aligned}\dot{\vec{x}}(t) &= \vec{g}(\vec{x}(t), \vec{u}(t), t) \\ \vec{x}(0) &= \vec{x}_0 \\ \vec{y}(t) &= \vec{f}(\vec{x}(t), \vec{u}(t), t)\end{aligned} \qquad (1.2)$$

16 chapter one Introduction

For a discrete time system (i.e, the time values are integers), the state equations are difference equations, and the system is defined by

$$\vec{x}(t+1) = \vec{g}(\vec{x}(t), \vec{u}(t), t)$$
$$\vec{x}(0) = \vec{x}_0 \qquad (1.3)$$
$$\vec{y}(t) = \vec{f}(\vec{x}(t), \vec{u}(t), t)$$

For state-less systems the state vector and the state equations are not required, and for time-invariant time-discrete systems the equations (1.3) specialize to

$$\vec{x}(t+1) = \vec{g}(\vec{x}(t), \vec{u}(t))$$
$$\vec{x}(0) = \vec{x}_0 \qquad (1.4)$$
$$\vec{y}(t) = \vec{f}(\vec{x}(t), \vec{u}(t))$$

which is the kind of system we will mostly deal with.

Linear and Nonlinear Systems

One more interesting property of system models is based on the kind of functions $\vec{f}$ and $\vec{g}$ used.

Definition 1.6 A function $f : A \to A$ is linear if and only if

$$f(a_1 x_1 + a_2 x_2) = a_1 f(x_1) + a_2 f(x_2) \text{ for all } a_1, a_2, x_1, x_2 \in A$$

A function $\vec{f} : A^n \to A^n$ is linear if and only if

$$\vec{f}(a_1 \vec{x}_1 + a_2 \vec{x}_2) = a_1 \vec{f}(\vec{x}_1) + a_2 \vec{f}(\vec{x}_2) \text{ for all } a_1, a_2 \in A, x_1, x_2 \in A^n$$

where A^n is the set of vectors of length n with elements of A.

A system modeled by equations (1.2), (1.3), or (1.4) is linear if and only if both functions $\vec{f}$ and $\vec{g}$ are linear.

The class of linear systems is in fact rather small. For instance, a simple function such as $f(x) = x^2$ is not linear. However, as we will see, there are still many interesting systems falling into this category.

For linear systems we can formulate equations (1.3) in matrix form:

$$\vec{x}(t+1) = \mathcal{A}(t)\,\vec{x}(t) + \mathcal{B}(t)\,\vec{u}(t)$$
$$\vec{y}(t) = \mathcal{C}(t)\,\vec{x}(t) + \mathcal{D}(t)\,\vec{u}(t) \qquad (1.5)$$

where

$\mathcal{A}(t) = n \times n$ matrix
$\mathcal{B}(t) = n \times p$ matrix
$\mathcal{C}(t) = m \times n$ matrix
$\mathcal{D}(t) = m \times p$ matrix

n = number of state variables
m = number of output variables
p = number of input variables

For linear, time-invariant systems we get the equations

$$\vec{x}(t+1) = \mathcal{A}\,\vec{x}(t) + \mathcal{B}\,\vec{u}(t)$$
$$\vec{y}(t) = \mathcal{C}\,\vec{x}(t) + \mathcal{D}\,\vec{u}(t) \tag{1.6}$$

with constant matrices $\mathcal{A}$, $\mathcal{B}$, $\mathcal{C}$, and $\mathcal{D}$.

Deterministic, Stochastic, and Nondeterministic Systems

A discussion of modeling properties cannot be complete without covering issues of deterministic and nondeterministic behavior. So far we have only used fully deterministic models; that is, given a precisely defined input, we get a unique, precisely defined output. There are two complications that force us sometimes to deviate from this nice situation.

First, the input may not be precisely defined. For instance, the arrival time of inputs to an electrical circuit can vary due to variations in delays on wires. Both the content and the arrival time of messages in a telecommunication system are hardly precisely defined when we design the system. The execution time of tasks in a multiprocessing operating system depends on the computer hardware and on the input data of the different processes, among many other factors.

Second, the system that we intend to model may not always react with the same outputs when confronted with the same inputs. For instance, the communication times in distributed computer systems are notoriously hard to predict and may vary widely due to physical effects such as varying delays on wires and in communication units, and due to interference with other systems that share the same resources. Taking all of these factors into account is formidable due to the huge number of details involved. But ignoring them means that our system behaves nondeterministically with respect to the factors taken into account.

In many of these cases a probability distribution captures the nature of the input and the system behavior much better than a precisely defined function. If the inputs are modeled as random variables with a probability distribution, the system model must relate random input variables to random variables describing the outputs. If the system's behavior is modeled as a stochastic process, even precise inputs will result in random output variables.

In some cases it is either not possible or not desirable to assume some probability distribution. For instance, the distributed electronic system in a car is a safety-critical system. Even though the delay of messages between different parts of the system can be modeled as a stochastic process, the overall system must perform correctly and deterministically for all possible delays. In such a case it may be most reasonable to assume nothing is known about the delay of messages and to show that the system still works properly. We say the delays of messages are nondeterministic, and as a result the entire system may also behave nondeterministically.

Definition 1.7 A system model is *deterministic* if $\vec{f}$ and $\vec{g}$ in equations (1.2) through (1.6) are functions in the sense that they evaluate a given argument always and unambiguously to the same result.

A system model is *stochastic* if at least one of its output variables is a random variable.

A system model is *nondeterministic* if a given input may result in different outputs.

Stochastic and nondeterministic processes are powerful concepts, but they significantly complicate the analysis. We will devote Chapter 8 to these issues.

Events

While a state is associated with a time period (e.g., the system can be in the same state for a very long time), an event "occurs" at a particular time instant. An event is uniquely bound to its occurrence time and cannot occur more than once. When we sloppily say that an event is occurring several times, we actually mean the occurrence of different events of the *same type* at different time instants.

Virtually anything that implies a change can be an event. The arrival of a message, the change of a signal value, a counter exceeding a given threshold value, a tank becoming full, a time period that elapsed are all possible events.

An event is inherently discrete—in fact, binary: it either occurs or it does not occur. A continuous variable cannot represent an event. For instance, the velocity of a body cannot be an event. If we define an event that occurs when the body reaches the speed of sound, it is something quite different from the velocity of the body.

Even though an event is binary—it occurs or it does not—we can associate a value with an event. We often consider the "value of an event," but what we actually mean are two separate things. We have a variable or a message that has a value. Then we associate an event with a particular time instant in the lifetime of the variable. For instance, a process receives a data packet. The appearance of the packet at the input of the process may be denoted by an event. The packet lives on after the event has occurred; it may perhaps be re-emitted by the process, thus generating another event. To describe this situation we sometimes say that the process receives an event and the value of that event is the data packet.

Time-Driven and Event-Driven Systems

The principal triggering mechanism motivates another classification. In a *time-driven system*, the advance of time causes the system, or parts of it, to become active. When the time variable is increased, all events occurring at the current time are applied to the appropriate inputs, and the system and its components are activated, consuming these events. If no event occurs at a particular time instant, a "null" event is assumed, which causes no state change.

In an *event-driven system*, the next event to occur is selected and activates the system. The time variable is updated according to the time the next event occurs.

This property can be thought of as a property of the simulation mechanism, or as a modeling style, rather than an inherent property of the system or the system model.

In principle both mechanisms can be applied to a given model and should exhibit the same behavior, although they may feature very different simulation performance. If events occur very regularly, a time-driven mechanism may be very efficient; if they occur rarely with irregular and long time intervals in between, an event-driven simulation mechanism is more efficient. That means that some systems are better and more naturally modeled by one mechanism than by the other. In that sense we can consider this property to be a property of the system under consideration and of the system model.

1.4.2 System Classification Summary

We have discussed a set of important system properties that categorize the different kinds of systems we may encounter (see Figure 1-7). We summarize these properties briefly.

- *State-less and state-full systems* A state-less system does not maintain an internal state. The output at a given time is a direct map from the input at that time without additional influencing factors. A state-full system maintains an internal state, and the output depends on both the state and the current input.

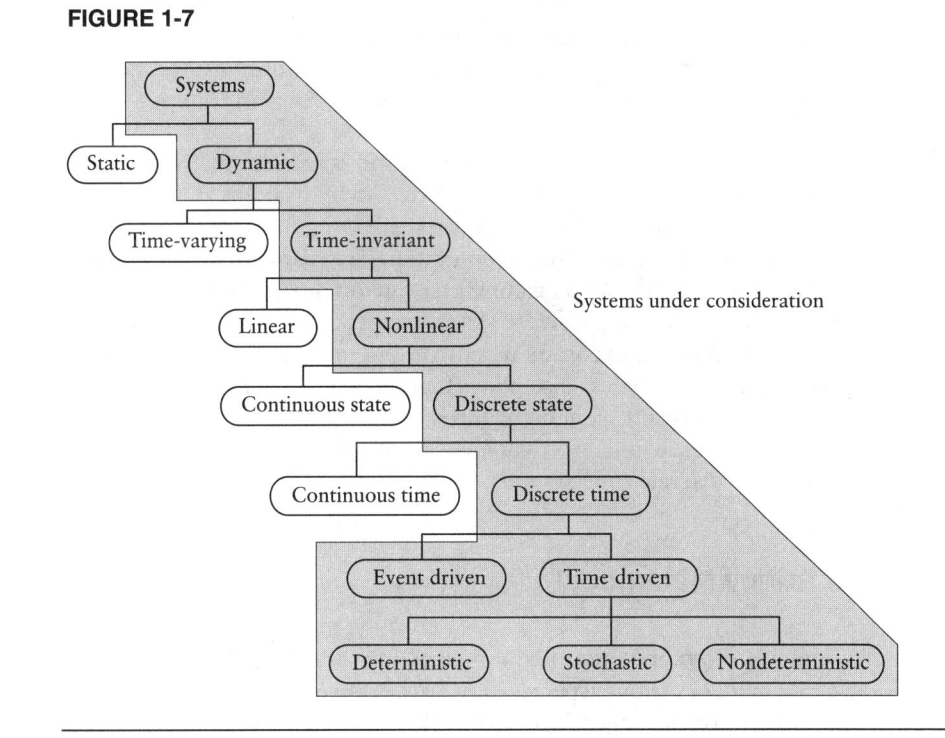

FIGURE 1-7

Classification of system models.

- *Time-varying and time-invariant systems:* A time-varying system explicitly uses the value of the current time to determine the next state or the output. The same input history applied at different times may lead to different results. For time-invariant systems, the same input history, even if applied at different times, will always result in the same output.

- *Linear and nonlinear systems:* Linear systems are invariant with respect to addition and multiplication, which greatly facilitates the analysis. They represent a small but important class of systems.

- *Continuous-state and discrete-state systems:* For continuous-state systems the state values are elements of a continuous set, such as the real numbers. In discrete-state systems the state values are from a discrete set, such as the integers or natural numbers. Typically, the inputs and outputs of continuous-state systems are also continuous variables, while the inputs and outputs of discrete-state systems are discrete variables.

- *Continuous-time and discrete-time systems:* For continuous-time systems the time variable is a continuous variable, while for a discrete-time system it is a discrete variable.

- *Event-driven and time-driven systems:* In event-driven systems the occurrence of events is the main triggering mechanism of activity, while in time-driven systems the advance of time is controlling the activities.

- *Deterministic, stochastic, and nondeterministic systems:* Deterministic systems produce one unambiguous output for a given input. Stochastic systems produce a probability distribution for the output. Nondeterministic systems may produce different outputs with unknown probability for a given input.

Different disciplines deal with different kinds of systems. For instance, an important part of control theory studies state-full, time-invariant, linear, continuous-state, continuous-time systems. In electrical engineering, analog designers usually work with nonlinear, continuous-time, continuous-state systems, and people developing analog/digital and digital/analog converters are on the borderline between continuous-state, continuous-time and discrete-state, discrete-time systems. Software engineers and computer scientists concentrate mostly on discrete-state, discrete-time systems. They are also among the very few disciplines that on certain occasions deal with time-varying systems, as the Y2K problem highlighted.

In the rest of the book, we will focus mostly on state-full, time-invariant, nonlinear, discrete-state, discrete-time systems.

1.5 The Rugby Metamodel

The models of computation and concurrency, which will be discussed in the next chapters, differ in various aspects. Some of the differences are superficial and irrelevant, while others are profound and have major consequences. For instance, one important difference between Petri nets and finite state machines, which will be discussed first in Chapter 2, is decomposability. Petri net descriptions can be easily partitioned into smaller, interacting Petri nets, and they can just as easily be assembled into bigger models. The size of a Petri net P composed of two smaller nets P_1, P_2 is equal to the sum of the sizes of its subnets: size $(P) = \text{size}(P_1) + \text{size}(P_2)$. In contrast, the partitioning of a finite state machine is nontrivial, and the composition of two state machines leads to an explosion of the state space because the size of a finite state machine F composed of two smaller state machines F_1 and F_2 is a multiplicative function: $\text{size}(F) = \text{size}(F_1) \times \text{size}(F_2)$, when done in a straightforward way without optimization. Furthermore, the proof of equivalence between a finite state machine and a set of interacting smaller machines is a computationally expensive problem. The reason for this difference is that states are represented explicitly in state machines, while they are represented only implicitly in Petri nets. This difference is profound and has severe implications for a variety of applications.

Apparently, the way a model represents states (i.e., explicitly or implicitly), is an important parameter in characterizing a modeling concept. But it is definitely not the only parameter. Is there a complete list of parameters that fully characterizes the profound aspects of a modeling concept and allows us to predict all the important properties solely from the values of these parameters? Unfortunately, today we are not aware of such a parameter list, and we do not have a general scheme that allows us to derive the concrete properties of a computational model from a few, very general principles. However, we can identify a number of analytical concepts that have been used to discuss, teach, and compare different modeling styles, techniques, and languages. A prominent example is *abstraction*, which is used to order models and modeling concepts from low abstraction levels to high abstraction levels. Although a precise definition of "abstraction" is lacking, and different authors use the term in different ways with different assumptions, it is widely used because it conveys important differences between models of computation or between concrete models in an intuitive way. An excellent example is a transistor, which can be modeled as a switch or as a set of differential equations. The switch model is at a higher abstraction level because it contains much less detail and magnifies only a few characteristics of the transistor, such as that the transistor has two stable states. As would be expected from a good abstraction, these characteristics turn out to be very useful in a number of important applications. For instance, they can be used to simulate the behavior of very large transistor networks.

In the following sections we introduce a metamodel, the Rugby[4] model, which provides a way to analyze and classify different models of computation. Even though it is not complete, in the sense that we can derive all interesting properties of a model of computation from the coordinates it has in the Rugby model, it will prove to be very useful in understanding some of the fundamental differences between different models.

1.5.1 Domain, Hierarchy, and Abstraction

The Rugby model uses two important concepts, namely, domain and abstraction, which need some clarification because different authors have used these terms in sometimes

---

[4]The model derives its name from the similarity of its visual representation (see Figure 1-11) to the shape of a rugby ball, with the domain lines forming the seams.

FIGURE 1-8

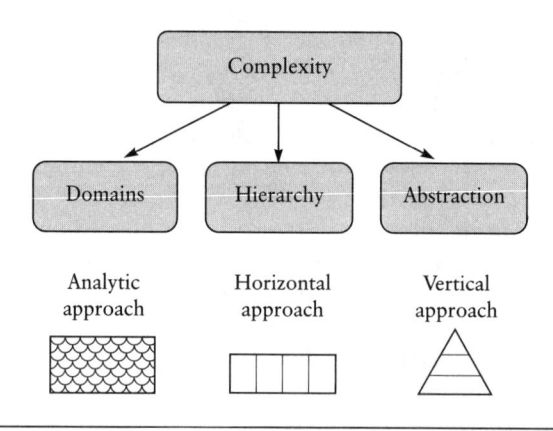

Ways to handle complexity.

very different ways. In particular, we have to be careful in distinguishing abstraction from hierarchy. They are two different means to handle complexity as illustrated in Figure 1-8. Hierarchy partitions a systems into smaller parts. Abstraction replaces one model with another model that contains significantly less detail and information. Both reduce the amount of information and detail that must be considered for a particular purpose.

Consider the 32-bit adder in Figure 1-9. Even though we can use the icon labeled "Add 32" in our design, it is in fact modeled as a network of full adders. Thus the "Add 32" icon and the network of full adder components represent the same entity but at different hierarchical levels. But the meaning of the "Add 32" box is only defined in terms of the full adder network. A simulator would in fact use the full adder network to compute the output of the "Add 32" box. So it is really only one model, but sometimes we prefer to hide some of its details, for instance the internal structure of the "Add 32" box.

In contrast, consider the model in Figure 1-10. It also represents an addition of numbers, but it is quite distinct from the "Add 32" box in Figure 1-9. As a model it does not use a netlist of simpler components, but it uses algebraic formulas instead . It can be considered to be a model at a higher abstraction level because it does not bother with how the numbers are represented (e.g., as bit vectors), and it does not specify how the result is computed and what partial results are computed on the way. It employs directly algebraic operations rooted in a mathematical theory. However, note that the variable z in Figure 1-10 is *implicitly* constrained to a range between 0 and $2^{32} - 1$, while x and y are not. In summary, the add function is a model at a higher abstraction level than the "Add 32" box, which is a model at a higher hierarchy level than the network of full adders.

A third way to tackle complexity is the analytic division of models into domains. Unlike hierarchy and abstraction, it does not lead to physically separate models or parts, but it is an analytic means to study different aspects of a model individually. For instance, we can reason about the data types used in a model or the representation of time, without consideration of the computation performed. This may be very useful for

FIGURE 1-9

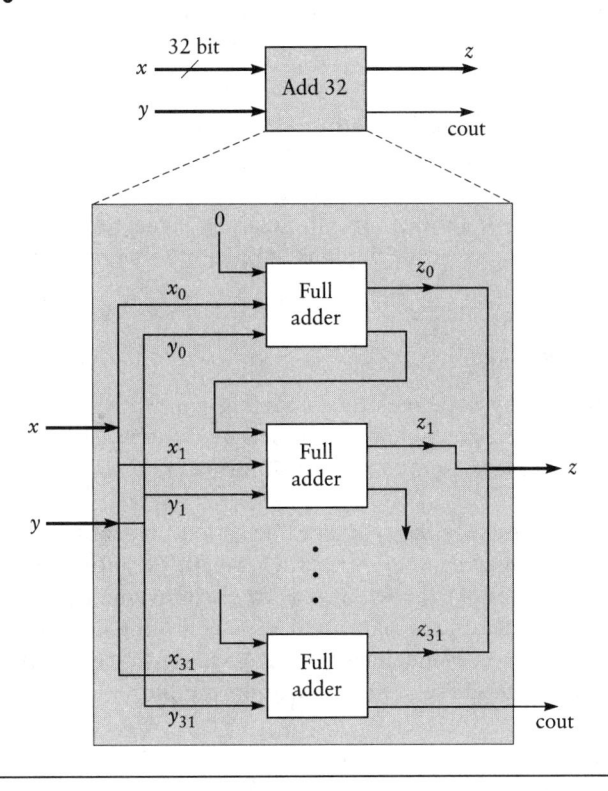

A 32-bit adder is modeled as a network of full adder components.

FIGURE 1-10

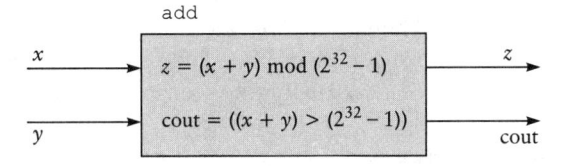

add

$z = (x + y) \bmod (2^{32} - 1)$

$cout = ((x + y) > (2^{32} - 1))$

A 32-bit adder is modeled in terms of algebraic formulas.

capturing the essence of a particular problem. The selection of discrete or continuous time has far-reaching consequences and has to be analyzed with respect to the objectives of the entire modeling effort and considering the fundamental nature of the problem at hand. However, we cannot simulate these aspects separately. Thus every concrete model will contain several domains and all metamodels and languages must be able to represent time, data, and computation in some way.

Even though a mathematical definition of these terms is impossible, we will try to be as precise as possible and give working definitions.

Hierarchy

A *hierarchy* is a (possibly recursive) partitioning of a design model such that the details of each part are hidden in a lower hierarchical level.

Hierarchy defines the amount of information presented and visible at a particular hierarchical level of a model. At all hierarchy levels the same modeling concepts are used. The motivation for hierarchy is to hide information when it is not needed and to display details when they are useful.

Abstraction

An *abstraction level* defines the modeling concepts and their semantics for representing a system. The type of information available at different levels is different. A higher level ignores some irrelevant information of a lower level or encodes it using different concepts.

Abstraction defines the type of information present in a model. Unlike hierarchy, abstraction is not concerned with the amount of information visible, but with the semantic principles of a model. In general, the movement from high to low abstraction levels includes a decision-making process. By making design decisions and increasing information about implementation details, we replace more abstract models with less abstract models, until the system is manufacturable.

Domain

A *domain* is an aspect of a model that can logically be analyzed independently from other aspects.

A domain focuses on one design aspect. Real models always contain several aspects or domains, but different models may emphasize one domain more than another. Models that focus on one particular domain use modeling notations and constructs to model the design aspect of concern explicitly. Other design aspects may be part of the models implicitly. Whereas hierarchy and abstraction simplify the design, domain partitioning helps the developers of tools and methodologies to cope with the complexity. The domains considered in this chapter are computation, communication, data, and time.

Although hierarchical partitioning is mostly a manual endeavor, the definition of abstraction levels and the transformations between them is behind most of the advances in design automation. While hierarchy is a general and important concept, it is not explicit in the Rugby model. However, it is assumed that hierarchy is possible at all abstraction levels of all domains.

Rugby focuses solely on design modeling rather than design process modeling, which will help us in our attempt to understand the different models of computation from a pure modeling perspective in Chapters 3 through 5. When we try to relate the computational models to specific application areas, we cannot expect direct assistance from Rugby. However, we will see that design phases and their related applications are rather closely associated with specific modeling abstractions.

FIGURE 1-11

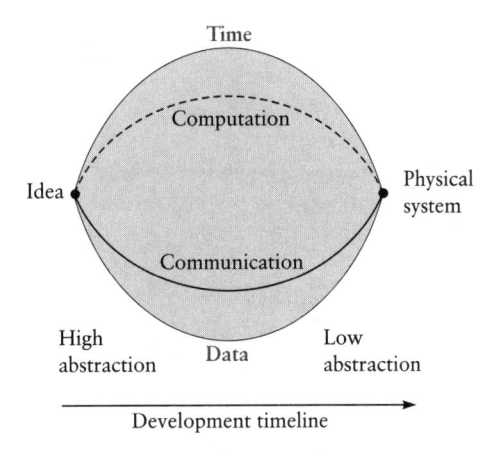

The Rugby metamodel.

1.6 Domains

Rugby considers four domains, computation, communication, time and data (see Figure 1-11), which we discuss in turn. Figure 1-12 magnifies part of the domain lines of the Rugby model and shows the abstraction levels from abstract requirements definitions to a concrete mixed HW/SW implementation. It illustrates that domain lines can split when design activities specialize. However, each split must have a corresponding join during system integration, which is not shown in the figure.

1.6.1 Computation

The computation domain focuses on the way the results are computed independently of the exact data types and the timing of the computation. Since communication is also a separate domain, computation in the Rugby model does not deal with issues of parallelism and concurrency. Hence, it is used in a narrower sense than the terms "behavior" or "functionality," which sometimes are used in a similar context. Computation is concerned with the relationship of input and output values, that is, the behavior as it is observable from the outside.

Transistor and Logic Gate Level

At the transistor level, models are based on differential equations representing I-V characteristics. At the logic block level, models are based on boolean functions. In fact, the logic gate is a prime example of a successful abstraction, so we will analyze it in a bit more detail.

FIGURE 1-12

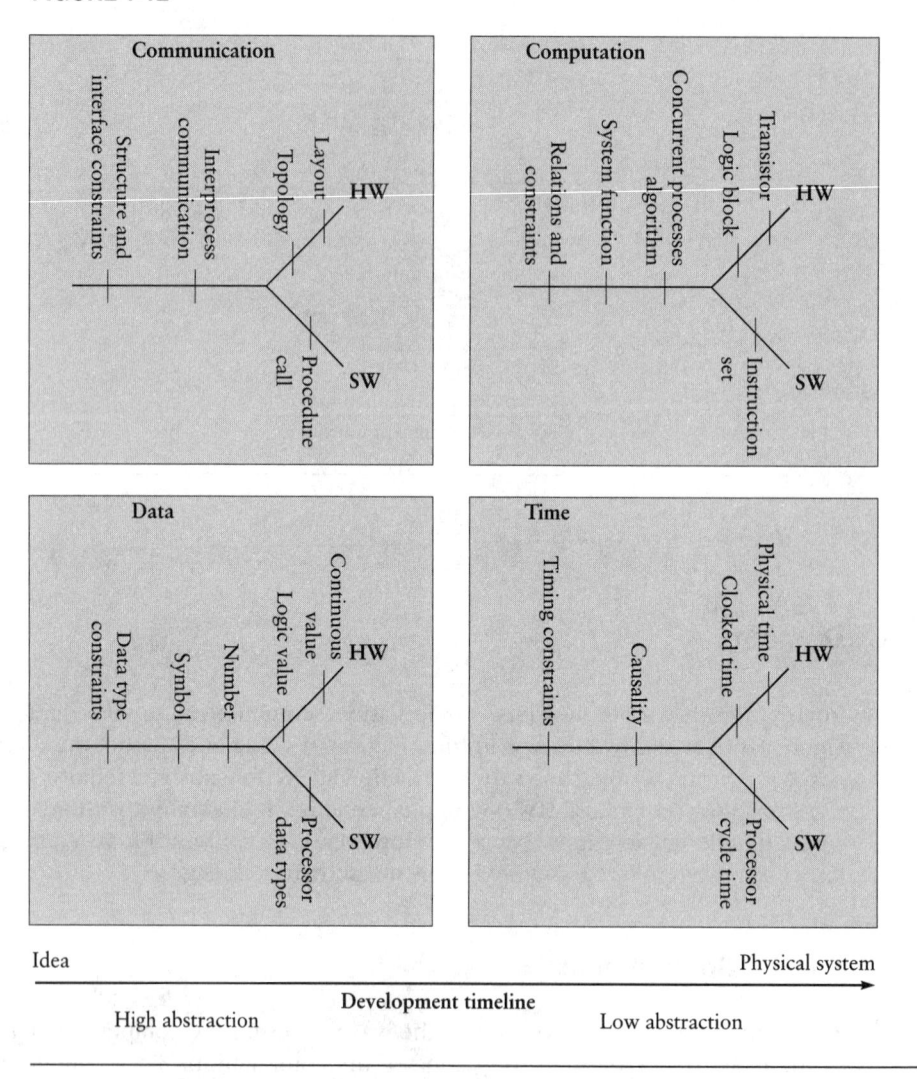

The four domains in the Rugby metamodel.

Consider the model of an MOS transistor in Figure 1-13. It calculates the current between drain and source as a function of the voltage between drain and source and the voltage between gate and source. Even without further investigation, it is obvious that the number of details could be overwhelming. Spice-like circuit simulators work with models of this kind. Consequently they can simulate circuits with a few or up to a few hundred transistors.

In contrast, the model in Figure 1-14 is very simple. It only distinguishes between two values on the inputs, the drain and the gate, and between three values on the output, the source. If the gate is closed, gate = 0, the output is undefined. If the gate

FIGURE 1-13

$V_{DS} > V_{GS} - V_T$ (conducting state):

$$I_D = \frac{k'_n W}{2L} (V_{GS} - V_T)^2 (1 + \lambda V_{DS})$$

$V_{DS} < V_{GS} - V_T$ (subthreshold state):

$$I_D = \frac{k'_n W}{L} ((V_{GS} - V_T) V_{DS} - \frac{V_{DS}^2}{2})$$

where

$V_T = V_{T0} + \gamma (\sqrt{|-2\phi_F + V_{SB}|} - \sqrt{|-2\phi_F|})$
V_{DS} = drain-source voltage
V_{GS} = gate-source voltage
V_T = threshold voltage
V_{SB} = substrate-bias voltage
V_{T0} = threshold voltage for $V_{SB} = 0$
I_D = drain-source current
k'_n = process transconductance parameter
W = channel width of transistor
L = channel length of transistor
λ = channel length modulation
ϕ_F = Fermi potential

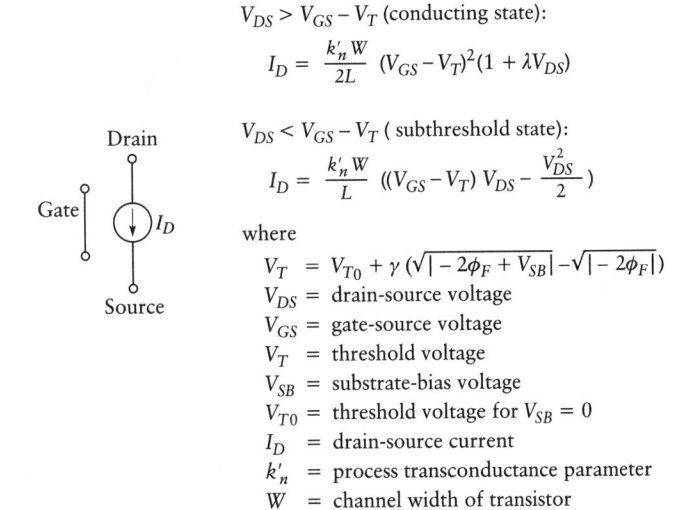

An MOS transistor model (Rabaey, 1996, p. 47).

FIGURE 1-14

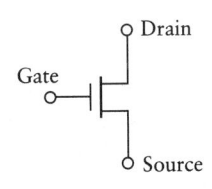

Gate	Drain	Source
0	0	undefined
0	1	undefined
1	0	0
1	1	1

A transistor modeled as a switch.

is open, gate = 1, the output equals the input. Because this model is a simple, suitable abstraction of a transistor, it represents not only MOS transistors, like the model in Figure 1-13, but any kind of transistor-like device, be it a bipolar transistor or a relay. Thus it can be used to model any kind of transistor, and it can be used to analyze and simulate large networks of transistors. The downside is that it is less accurate and it cannot be used to answer questions on power consumption, switching time, and temperature dependence.

However, the switch model of the transistor is not very convenient as a basis for complex circuits because many of the possible transistor networks are electrically

FIGURE 1-15

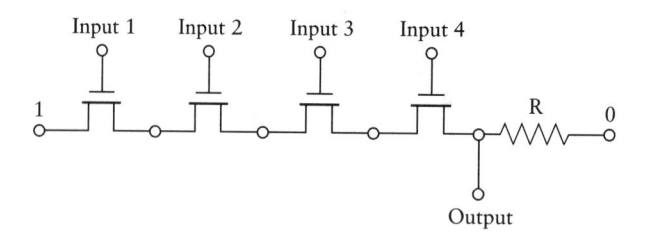

A network of transistors implements a four-input AND gate.

infeasible. For the switch-based transistor model to work, the gate-source voltage must be clearly either above or below the threshold level V_T. If this is not the case, the switch-based model becomes unreliable and a poor model for a network of transistors. Unfortunately, connecting transistors in arbitrary ways will frequently violate this assumption, and the circuit behavior will deviate from the prediction of the switch-based model. Consider the transistor network in Figure 1-15. According to our switch-based model, this network would implement a four-input AND gate, but we can observe two problems. First, it is impossible with our switch to assert a definite output value when the input is 0. In fact, it is impossible to model an inverter. To rectify this, we have introduced the resistor R in Figure 1-15, which connects the output to 0. Hence, if any of the inputs is 0, the output will be pulled to 0. By using detailed knowledge about the electrical properties of transistors, we can dimension transistors and resistors so that the AND gate in Figure 1-15 would work properly. However, this is a deviation from our switch-based model of Figure 1-14. The second problem is due to a voltage drop at every transistor. No matter how we dimension the resistor, the scheme cannot work for an arbitrary number of inputs. If we stack together too many transistors in a row, the gate-source voltage will get too close to the threshold voltage for some transistors. Thus, the switch-based model does not scale very well, and we run into difficulties if we assemble more than a few transistors—difficulties that can only be understood with the more detailed model of Figure 1-13. In summary, the switch-based transistor model is not a stable foundation for modeling large transistor networks.

As a remedy we can restrict the transistor network to a small number of patterns that can be combined in arbitrary networks without violating the assumptions of the switch-based transistor model. These patterns have been termed *gates*. The simplest of these gates, the inverter, is shown in Figure 1-16. Other gates realize the simple logic operations AND, OR, NAND, NOR, and XOR for a few inputs. The resulting abstraction level, based on gates rather than a transistor model, is a much more reliable basis for large networks. It has the following properties, which in general characterize useful abstraction levels:

1. The primitive elements are defined by simple models, for example, small truth tables in this case.

2. The primitive elements can be implemented in a wide range of technologies, such as all technology generations of CMOS, NMOS, bipolar, and so on. This is

FIGURE 1-16

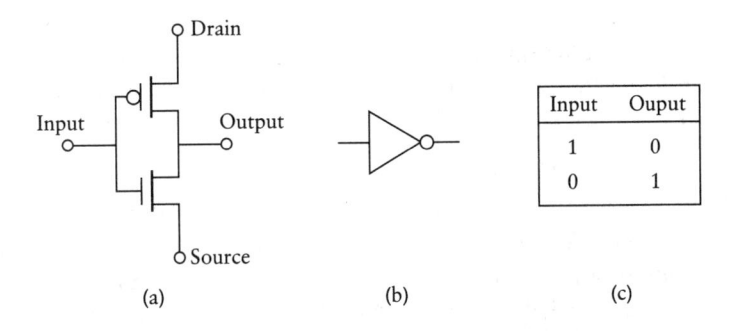

(a) (b) (c)

(a) An inverter model based on the switch model of the transistor with the assumption that Source = 0 and Drain = 1. (b) The symbol for the inverter as an abstract logic gate. (c) The truth table defining the functionality of the inverter.

the basis for effectively separating the lower abstraction level from the higher level; the same network of logic gates can be implemented in any of the target technologies.

3. The model holds even for arbitrarily large networks of primitive elements. The designer can assemble very complex networks and still rely on the assumption that the behavior of the system is described by a simple concatenation of the models of the primitive elements, the truth tables.

The Instruction Set Level

The instruction set level is the lowest abstraction for software. Although some computational concepts like sequencing, branching, and subroutines are similar to the algorithmic level, it is considered less abstract because the control elements are typically more primitive. In addition, and very importantly, it provides an abstraction layer around the details of the underlying processor architecture. The Intel x86 instruction set is an excellent example. The instruction set remained for almost two decades, while the underlying processor architecture and implementation changed radically. Hence, the freezing of the instruction set as an abstraction layer allowed for a continuous evolution of the processor and at the same time provided compatibility with programs written for earlier versions of the processor. We can again observe the three key properties of a successful abstraction level: (1) simple primitive elements, (2) which can be implemented in a wide range of different ways, and (3) which can be used to build arbitrarily complex systems.

The algorithmic level

At the algorithmic level, models are based on control primitives such as "sequence," "parallel," "if," and "loop," and operators that manipulate data objects. Parallelism at this level is expressed in terms of concurrent processes, where each process is typically described by an algorithm. While the instruction set level makes a model independent of a particular processor version, the algorithmic level makes a model independent of any particular implementation style. An algorithm can be implemented not only on any processor but also on an FPGA (field programmable gate array) or ASIC (application-specific integrated circuit) device. Due to its complete implementation independence, the algorithm and the state machine models have traditionally been used for investigations of fundamental questions of computation and its complexity.

System Functions and Relations

At the system function level, the system is described from a purely external view without consideration of the partitioning into parallel activities of the system. The difference from the algorithm level is that an algorithm gives a detailed sequence of steps, while a function gives only the functional dependences but no detailed recipe for how to arrive at a result. According to this view, a function in a C or VHDL program is an algorithm and not a "pure function."

The difference between the "system functions" and the "relations and constraints" level is analogous to the difference between a function and a relation in mathematics. Let's illustrate this rather abstract discussion with a concrete example.

Figure 1-17 illustrates the three abstraction levels: algorithm, system function, and relation. At the algorithmic level, the control and sequencing are detailed and explicit. The algorithm defines the sequence of steps. If it can express parallelism, it is explicit.

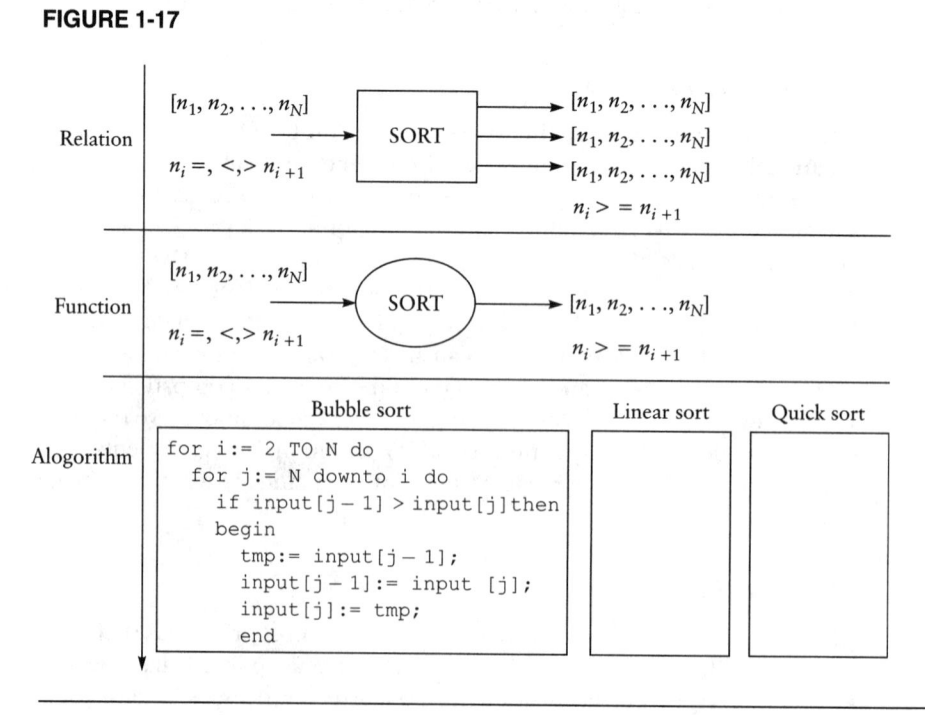

FIGURE 1-17

Relation	$[n_1, n_2, \ldots, n_N]$ → SORT → $[n_1, n_2, \ldots, n_N]$, $[n_1, n_2, \ldots, n_N]$, $[n_1, n_2, \ldots, n_N]$		
	$n_i =, <, > n_{i+1}$		$n_i >= n_{i+1}$
Function	$[n_1, n_2, \ldots, n_N]$ → SORT → $[n_1, n_2, \ldots, n_N]$		
	$n_i =, <, > n_{i+1}$		$n_i >= n_{i+1}$
	Bubble sort	Linear sort	Quick sort

Alogorithm	``for i:= 2 TO N do``
	`` for j:= N downto i do``
	``  if input[j-1] > input[j] then``
	``  begin``
	``   tmp:= input[j-1];``
	``   input[j-1]:= input[j];``
	``   input[j]:= tmp;``
	``  end``		

Algorithms, functions, and relations.

FIGURE 1-18

$$sort :: Integer \rightarrow [Integer]$$

$$sort \quad [\,] = [\,]$$

$$sort \quad (x : xs) = (sort(selectLT\ x\ xs)) ++ [x]$$
$$++ (sort(selectGE\ x\ xs))$$

A functional definition of sorting.

Algorithms are based on control primitives such as "sequence," "parallel," "if," and "loop," and operators that manipulate data objects.

At the system function level, control, sequencing, and parallelism are to a large extent implicit. Only the data dependences define the sequence; thus the sequence of statements is only partially ordered. However, how the result is computed is exact and determinate. For a given input, only one result can be computed. Consider the functional definition of sorting in Figure 1-18.

[ ] denotes the empty array; (x : xs) splits the array into the first element x and the rest of the array xs. (selectLT x xs) returns an array that contains only those elements of xs that are less than or equal to x. (selectGE x xs) returns an array with only those elements of xs that are greater than or equal to x. ++ is an array concatenation operator.

This recursive definition precisely defines the result. For a given input, there is only one possible output. However, this functional definition does not provide any control, sequencing, or communication information, which is typically present at the next lower level of abstraction, the algorithmic level. The functional model can be implemented in a purely sequential or in a massively parallel manner. The algorithmic level would also define the sequence of operations and the synchronization and communication details in the case where a parallel implementation is adopted.

Figure 1-19 shows two algorithms that are directly derived from the function in Figure 1-18. `sort_s` is a fully sequential solution. It takes the first element of the input array, x, selects the elements that are less than x, sorts them, selects elements that are greater than x, sorts them, and finally puts the pieces together into the result array.

In contrast, `sort_p` uses the operator `par`, which executes the parts separated by "|" in parallel. Thus it sorts the different subparts of the array in parallel and assembles the result at the end.

Relations are even more abstract than functions in that they allow for several different results. Relations only relate input values with output values, and for a given input many different outputs may be acceptable. The relation only defines properties that a result must comply with for a given input.

Consider the relational definition of a sort in Figure 1-20, which maps an integer array to another integer. This definition only states that there must be the same elements in A and in B and that the array B is sorted with the smallest elements coming first. However, if two elements are equal, the relation does not specify in which order they appear in the result. This may be significant for sorting records that are different even if they have the same key.

FIGURE 1-19

```
Array sort_s (iarray) {                     Array sort_p (iarray) {
  Array oarray = EArray;                      Array oarry, oarray2 = EArray;
      // initialized to the empty array       int x = iarray[0];
  int x = iarray[0];                          par{oarray = sort_p(selectLT (x,iarray))
  oarray = sort_s(selectLT (x,iarray));          |oarray2 = sort_p(selectGT (x,iarray))
  append(orray,x);                            }
  append(oarry,sort_(selectGT(x,iarray)));    append(orray,x); append(oarray, oarray2);
  return (oarray);                            return (oarray);
}                                           }
              (a)                                            (b)
```

Both algorithms, (a) a sequential sorting algorithm `sort_s` *and (b) a parallel algorithm* `sort_p`, *are derived from the purely functional description in Figure 1-18.* Array *is a data type representing an integer array. The function* append *appends an element or an entire array at the end of the array given as the first argument.* par *is an operator that executes the parts in the following block in parallel.*

FIGURE 1-20

sort (IntArray A) –> (IntArray B)

Precondition: true

Post condition: $\forall a \in A : a \in B$

$\wedge \forall b \in B : b \in A$

$\wedge \forall i, j \in \text{Integer}, b_i, b_j \in B : i < j \Rightarrow b_i \leq b_j$

A relational definition of sorting.

1.6.2 Data

Data is an important aspect of all models, and going from one representation of data to another representation often marks a distinctive step in the design flow. For instance, signal processing algorithms are often developed with real numbers based on the assumption of infinite accuracy. At some point, however, a finite, fixed-point representation of the data has to be selected in order to implement the algorithm in hardware or on a DSP with limited resources.

Idealized data types, such as real or integer numbers, or tokens, are useful for investigating the principal properties of an algorithm or a functionality not blurred by the maneuvers necessary to deal with concrete, implementation-oriented data types.

However, in some applications (for instance, database-dominated applications), the data, the data types, and the relation between them is the focal point that requires careful investigation and modeling, undisturbed by how exactly the data is processed. Therefore, elaborate techniques for modeling data alone have been developed, such as entity-relationship diagrams.

The data domain in the Rugby frame captures various abstraction levels at which data can be modeled.

The *continuous value* level is based on real numbers and is used to quantify physical units like voltage, frequency, and so on. The *logic value* level is based on mathematical logic and is used to represent boolean and logic expressions. It corresponds to the logic block level in the computation domain.

In software the lowest abstraction level is based on data types of the target processor, which are typically bits, bytes, and words of varying length.

The *number* level uses ideal mathematical numbers without any concern for implementation issues. As mentioned earlier, ideal mathematical numbers, be they reals, integers, or rational numbers, are preferably used in early design phases to concentrate on the basic functionality and algorithms. Later on, when the principal questions about the algorithm have been solved to the satisfaction of the designers, they can concentrate on finding a cost-efficient data representation and deal with accuracy questions and the corner cases of overflow and underflow.

Symbols are used to further abstract from the detailed properties of data. For instance, for the design of a telecommunication network it is interesting to examine the flow of packets without any concern for the detailed contents of the packets. To this end, a model may only distinguish between three kinds of packets: high-priority packets, low-priority packets, and control packets for configuration and maintenance of the network itself. Hence, three different symbols suffice to model the data packets, which in reality may contain hundreds or thousands of bits.

1.6.3 Time

Time is a crucial design characteristic that deserves independent analysis. Many electronic systems are reactive real-time systems or have hard real-time constraints. But even in systems without real-time requirements, time is of the utmost concern if the system contains concurrent activities. For purely sequential models, such as a sequential algorithm, the model of time can be simple and implicit: Each step takes a finite amount of time, and each step can only start after the previous step has finished. This is the model of time that assembler programs or conventional C programs assume.

Consider a very simple system with four concurrently active gates as shown in Figure 1-21. Without an assumption about the timing behavior of these four gates, we cannot know the overall behavior of the system. A simple question like "Given all zeros at the input, what are the output values?" cannot be answered without considering time. Under the assumption that the gates react immediately without observable delay, the answer is $x = 0, y = 1$. However, if the gates react with a delay, we may see a sequence of different output patterns until the effects from the inputs have fully propagated to the outputs.

There are again different levels of accuracy in representing time. Furthermore, it turns out that the representation of time is not bound to particular kinds of computation, but is rather independent. We will even see that the way time is handled is perhaps the

FIGURE 1-21

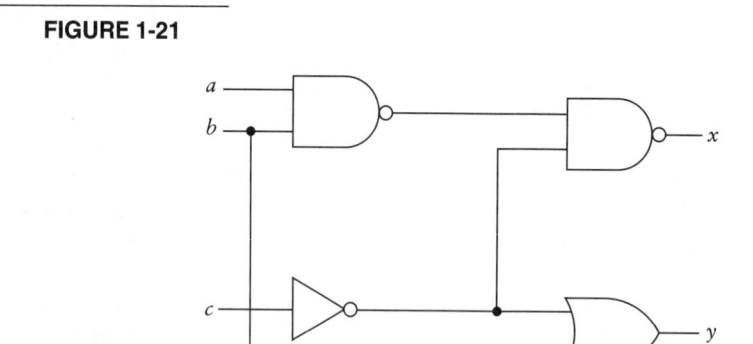

A simple netlist with four concurrent activities.

most fundamental difference between the various models of computation that we discuss in the following chapters. For this reason, we treat the issue of the representation of time rather briefly now, because we will get back to it in much greater detail in the following chapters.

The *physical time* level uses physical time units and is based on physical principles. Propagation delay, as it appears in simulation languages for digital systems, is a simplification of the physical time and could be viewed as a separate abstraction level.

At the *clocked time* level, all activities are related to a clocking scheme and are based on concepts of digital time. While there is no concept analogous to physical time in software, the processor cycle time corresponds to the clocked time in hardware.

At the *causality level*, the total ordering of events is replaced by a partial ordering defined by generation and consumption of data and by explicit control dependencies. At the highest level, time is expressed through *performance and time constraints* like data rate or frames per second.

1.6.4 Communication

A separate communication domain is desirable given the prevalence of parallelism and concurrency in virtually all models above a certain level of complexity. Complex systems are naturally modeled as communicating concurrent processes. Refining these abstract communications to intracomponent and intercomponent communication primitives is a major part of the design effort. Components may be arbitrarily complex, such as ASICs, processor cores, memories, and so on.

The communication domain is concerned with the connections and interactions between design elements. For hardware, the layout level is based on the principles of geometry and uses physical units to describe geometric parameters. The topological level is only concerned with the presence or absence of connections between design elements. A netlist is a common representation of topology. For instance, Figure 1-21 describes which gate is connected to which other gate. The model does not indicate how big or how long a connection is, if it is one wire or several, nor what kind of data

FIGURE 1-22

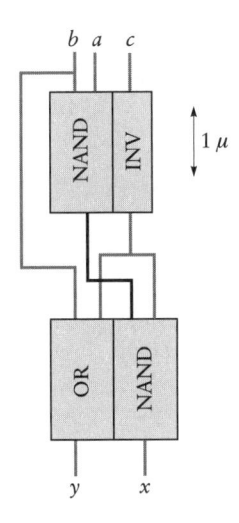

A layout representation of the netlist in Figure 1-21.

or signal is traveling over the connection. The model simply designates the connections between elements. In contrast, the model in Figure 1-22 provides detailed geometric information about the size and shape of the elements and the interconnecting structures.

In software there is no concept equivalent to the topology and layout levels of hardware. Since the main structuring element in software programs is the procedure or function, the communication mechanism between them constitutes the lowest abstraction level, which is denoted by *procedure call* in Figure 1-12. It is applicable both to assembler programs and to higher-level programming languages such as C or Pascal.

The *interprocess communication* level is concerned with mechanisms and protocols of communication between concurrently active entities. The implementation of communication channels and the physical structure of the design are irrelevant at this level. The communication network may be arbitrarily complex but still be transparent to the communication mechanism of the communicating processes. For instance, processes A and B in Figure 1-23 may communicate via a send-message/receive-message mechanism. Thus, in the models of A and B we only have statements like `send(data,id_B)` and `receive(data,id_A)`, where `id_A` and `id_B` are process identifiers. The communication network, which provides the communication service, is transparent. Consequently, the send-message/receive-message primitives can be implemented in different ways with different performance and cost characteristics without any effect on the models of A and B.

At the highest level, only the interface and communication constraints are expressed. These constraints may concern a range of properties such as electrical properties, number of wires, frequency, protocol type and protocol family, and so on. Like all constraints, they allow for a number of different solutions and are typically used in early design phases to express the requirements and constraints on the system from the environment.

FIGURE 1-23

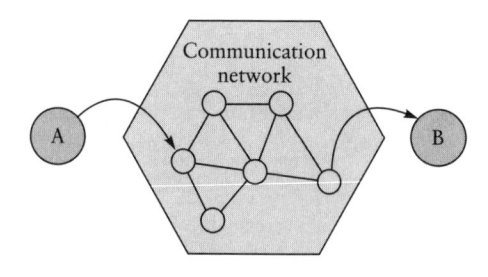

Two processes A and B communicate via a communication network, which may be transparent to the two processes.

1.7 Notation

For the sake of clarity and conciseness in the rest of the book, we introduce a few conventions. Table 1-1 lists all the abstraction levels of Rugby together with two designations, a long form and a short form. The long form will be used to unambiguously denote a Rugby abstraction level, and the short form will mostly be used in Rugby coordinates to define models of computation. A *Rugby coordinate* is a 4-tuple, one element for each domain. It is ordered as follows: ⟨Computation domain, Communication domain, Data domain, Time domain⟩. For instance ⟨Tran, Lay, CV, PhyT⟩ denotes a computational model that only covers the lowest hardware abstractions in all four domains. ⟨RC, IC, DTC, TC⟩ is a model that can only express constraints and will only be useful in requirements engineering. We allow the wild card "*" to represent all abstraction levels in a domain, and ranges and sequences to represent several abstraction levels. For instance [Alg-Tran] would denote the levels Algorithm, LogicBlock, and Transistor, and the sequence [Alg,LB,Inst] would denote Algorithm, LogicBlock, and InstructionSet in the computation domain. The underscore "_" is used to indicate that a particular domain is not relevant.

For example, the Rugby coordinate ⟨[Alg-LB], [IPC-Top], [Sym-Log], PhysicalTime⟩ defines the timed model of computation, as we will see in Chapter 5.

Design activities transform models, and they often cross abstraction levels on the way. We characterize design activities by a pair of Rugby coordinates, (fromCoordinate → toCoordinate). For instance, place and route tools would be denoted by (⟨_,Top,_,_⟩ → ⟨_,Lay,_,_⟩).

1.8 Design Methods and Methodology

The abstraction levels are related to design tasks and the design process. One could even formally define individual design methods by specifying the abstraction levels in the four domains of the input and output of the method. A design methodology can

1.8 Design Methods and Methodology

TABLE 1-1: *Notation used for abstraction levels in Rugby.*

Computation domain		
Relations and constraints	RelationConstraints	RC
System functions	SystemFunctions	SF
Concurrent processes, algorithms	Algorithm	Alg
Logic block	LogicBlock	LB
Transistor	Transistor	Tran
Instruction set	InstructionSet	Inst
Communication domain		
Structural and interface constraints	InterfaceConstraints	IC
Interprocess communication	InterProcessComm	IPC
Topology	Topology	Top
Layout	Layout	Lay
Procedure call	ProcCall	PC
Data domain		
Data type constraints	DataTypeConstraints	DTC
Symbol	Symbol	Sym
Number	Number	Num
Logic value	LogicValue	LV
Continuous value	ContValue	CV
Processor data types	ProcessorDataTypes	PDT
Time domain		
Timing constraints	TimingConstraints	TC
Causality	Causality	Caus
Clocked time	ClockedTime	CT
Physical time	PhysicalTime	PhyT
Processor cycle time	ProcessorCycleTime	PCT

be defined by giving the order refinements from higher to lower abstraction levels. In the following we do not give strict definitions but we relate abstraction levels to design methods and phases as they are in industrial use to date.

1.8.1 Design phases

Even though there is no unique and generally accepted way to organize design projects, we can observe that all larger projects are separated into phases. We somewhat arbitrarily assume four design phases: requirements definition, specification, design, and implementation. Figure 1-24 shows the common usage of abstraction levels in the different design phases. It is common practice that a design project progresses from abstract models in the beginning to more refined and concrete models as more and more design decisions are taken. This picture is a blunt idealization because in practice any design phase will use a mixture of abstraction levels, mixing abstraction levels

FIGURE 1-24

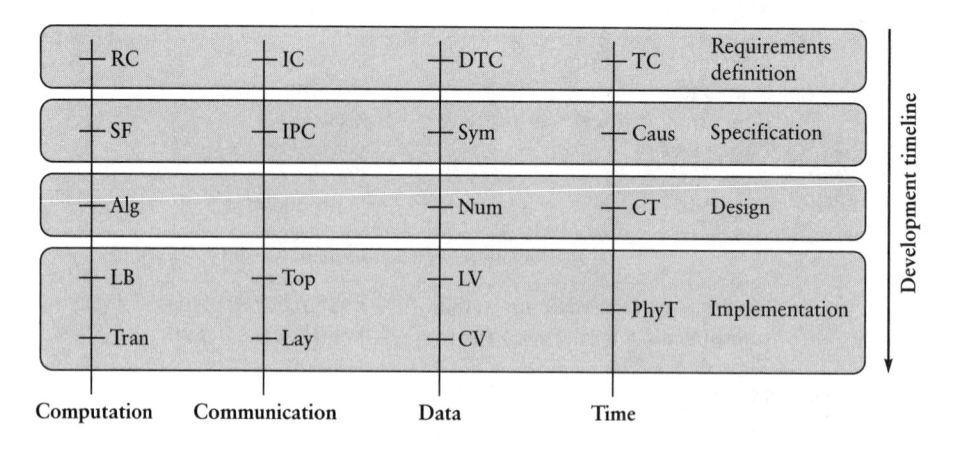

	Computation	Communication	Data	Time	
Requirements definition	RC	IC	DTC	TC	
Specification	SF	IPC	Sym	Caus	
Design	Alg		Num	CT	
Implementation	LB / Tran	Top / Lay	LV / CV	PhyT	

The occurrence of abstraction levels in different design phases.

from different domains and going back and forth between levels. Even within concrete design descriptions, we usually find a blend of different abstraction levels. Nonetheless, Figure 1-24 is a useful way to structure the design process because all methodologies without exception, and more or less consciously, traverse the domain lines from higher to lower abstraction levels. They only differ significantly in the definition of the individual steps. Figure 1-24 further helps to explain methodologies because it gives a suitable frame for analyzing what essentially happens in different phases—whatever the names of the phases may be.

We refrain from a detailed elaboration of this topic here, but we want to illustrate briefly the relationship between abstraction levels in models with the design flow by giving a few examples of design activities and a case study.

1.8.2 Design and Synthesis

Figure 1-25 places some design activities on the domain lines. These activities typically make design decisions and thus refine the design into a model at a lower abstraction level. Given a topological description, place and route tools decide on the geometry. In terms of the Rugby notation, this would be expressed as a transformation $(\langle\_,\text{Top},\_,\_\rangle \rightarrow \langle\_,\text{Lay},\_,\_\rangle)$. Given a description that exposes the causal dependences between operations, scheduling presses the operations into a strict time grid $(\langle\_,\_,\_,\text{Caus}\rangle \rightarrow \langle\_,\_,\_,\text{CT}\rangle)$.

Most activities operate in more than one domain. System partitioning not only decides on the partitioning of the functionality into concurrent activities, but it often also involves a refinement of the interfaces. Technology mapping implicates refinement in all four domains.

Figure 1-25 gives a very incomplete and superficial picture of design and synthesis activities, but it should indicate that knowing which domains and abstraction

FIGURE 1-25

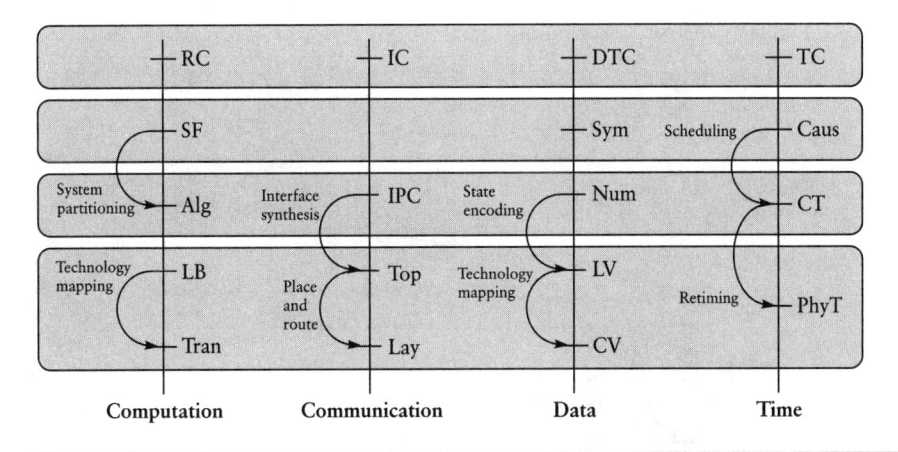

Several design activities described in terms of the abstraction levels they use.

levels are used by a tool or method conveys a fundamental insight into its potential and limitations.

1.8.3 Analysis

Analysis activities do not refine models. They check for consistency or produce estimates. As with design activities, specific analysis techniques involve specific abstraction levels in specific domains. In early design phases, they check given requirements for their mutual consistency or infer more detailed constraints from given requirements. They may also check the feasibility of a particular proposed solution. In that case, the proposed solution has to be modeled at a lower abstraction level, and consequently the shape of the feasibility analysis activity has to be extended to lower abstraction levels in one or several domains.

A performance analysis of an algorithm could involve the Rugby coordinates ⟨Alg, _, [Sym,Num], PhyT⟩, as shown in Figure 1-26. Alternatively, if the analysis is content with a clock-level accuracy, its coordinates would be ⟨Alg,_,[Sym,Num],CT⟩ instead. The performance analysis of the same algorithm could be more accurate if it uses more detailed models. For instance, if it uses information about the concrete representation of the data and the corresponding operators, its results may be more reliable no matter whether they are expressed in terms of clock cycles or physical time. The respective Rugby coordinates would be ⟨Alg,_,LV,CT⟩ and ⟨Alg,_,LV,PhyT⟩.

1.9 Case Study: A Design Project

As an illustrative example we use the Rugby model to describe various steps used in the *Network Terminal* (NT) design project. The NT provides the interface between

FIGURE 1-26

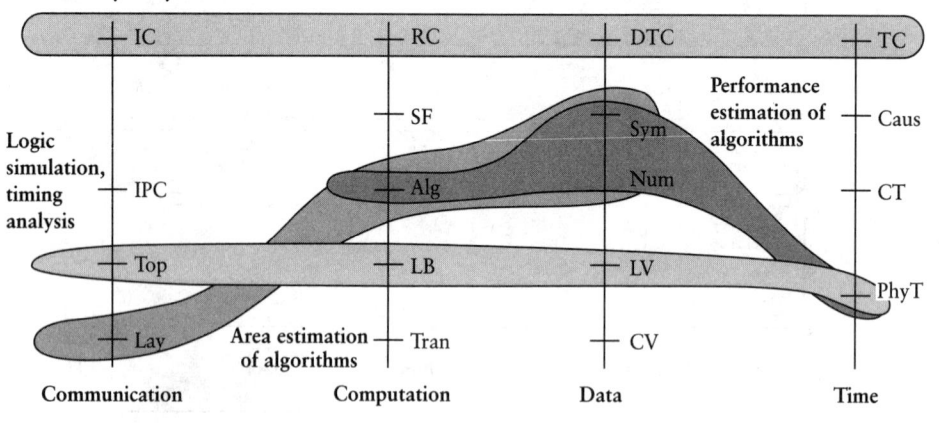

Several analysis activities described in terms of the abstraction levels they use.

FIGURE 1-27

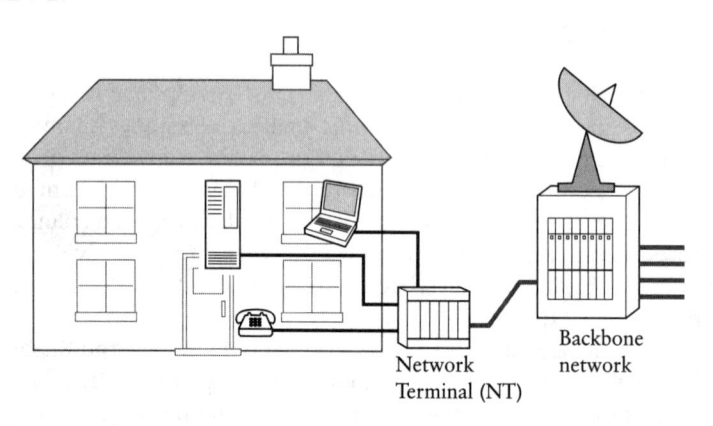

The Network Terminal connects various devices in the customer's premises to the backbone network.

the public access telecom network and the private *customer premises network* (CPN), which may be a private household or an office. Physically it is a device that is installed at the customer's premises, and it allows a user to access the distribution network with its various services (Figure 1-27). The NT connects to the access network through a distribution network interface block and provides interfaces to several different interfaces in the customer's facilities, for example, an Ethernet interface, interfaces for plain old telephones and cable TV.

FIGURE 1-28

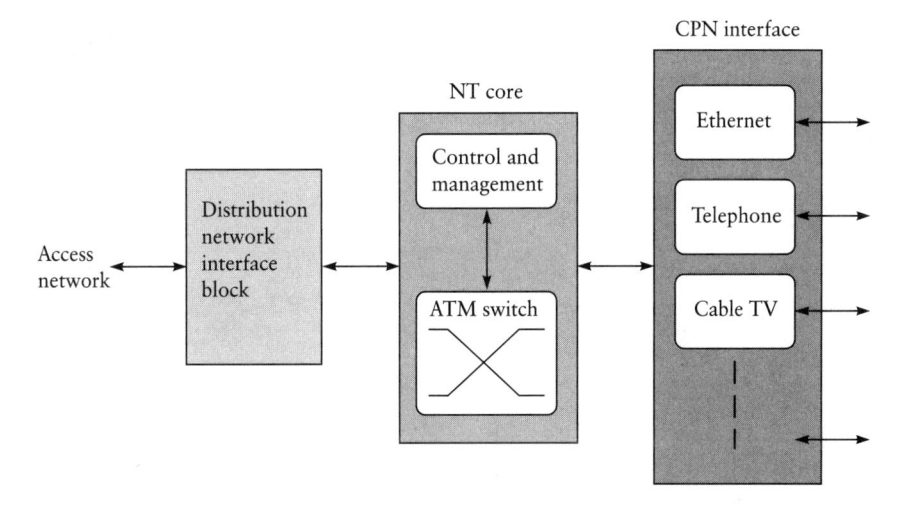

The Network Terminal consists of an interface to the backbone network, a switch, and several specialized interfaces.

The traffic model is based on the *Asynchronous Transfer Mode* (ATM) protocol suite (Prycker 1995). In ATM, *virtual channels* can share the same physical transport medium like a fiber. The different CPN interfaces are connected to different virtual channels in the access network. The ATM switch, as seen in Figure 1-28, routes data packets from the different virtual channels in the access network to the respective CPN interface and vice versa. In ATM, data packets are called *cells* and have a fixed length of 53 bytes.

Several models have been used or developed during the design of the NT, each of which is characterized in the following sections using the Rugby model. It was a very typical design project in terms of the kinds of models developed and tools used, which makes it a good example for us.

Table 1-2 summarizes several NT models, from the requirements definitions to implementation models.

1.9.1 Requirements Definitions

The requirements definitions are the input to the development activities. They consist of several more or less relevant documents, for instance, ITU[5] documents describing protocol standards, demands from the marketing department, field and test engineers, and others. A significant part is informal based on experience from similar products. These requirements include functional constraints, interface constraints, data constraints, and performance constraints.

---

[5]International Telecommunication Union.

TABLE 1-2: *Models in the Network Terminal design project.*

Models		Abstraction levels			
		Computation	Communication	Data	Time
Requirements		RC	IC	DTC	TC
System		Alg	IPC	Sym	Caus
HW	VHDL	Alg	Top	LV	CT
	Netlist	LB	Top	LV	PhyT
SW	C	Alg	PC	Sym	CT
	Assembler	Inst	PC	PDT	PCT

Examples of functional constraints are that ATM cells received at an input have to be transmitted at one of the outputs according to some information in the header of the ATM cell, or that OAM (operation and maintenance) cells have to processed in a particular way depending on the type of the cell and the state of the connection. Many aspects are not defined at this level, including the presence of some of the functions that are optionally suggested in ITU documents.

The number and types of the interfaces to the environment are part of the requirement, again leaving many details open.

The layout of ATM cells is defined quite accurately by ITU documents, including the meaning of bit patterns in the header, but other data formats, (e.g., for setting up and releasing connections and other signalling information) are described only vaguely.

Performance requirements are given in terms of cell rates, processing latency, and cell loss ratio.

1.9.2 System Model

From the initial requirements, a first executable model of the system is developed. Its abstraction level is squarely determined by the choice of the language, which in our case is SDL (Specification and Design Language) (Ellsberger et al. 1997). SDL is based on concurrent processes that communicate with each other via an asynchronous message-passing mechanism.

The behavior of the processes is described based on an extended finite state machine. Thus the level of abstraction in the computation and the communication domain of the model is directly determined by the language.

In the data domain, symbols are used (e.g., `atm_cell`, `oam_cell`, `fault_management_cell`) rather than bit patterns, even in those cases where the requirements are more detailed. Thus, for the sake of model clarity and design and simulation efficiency, not all the information from the requirements is used at this level.

The timing abstraction in the system model is causality because it is very natural in SDL just to use dependences of events and actions, and no detailed timing is necessary to describe the basic functionality. Again, not all the information from the requirements definitions, such as performance constraints, is used in the system model.

The most important design decision during development of the system-level model is the partitioning into concurrent processes, which determines the system architecture. It also lays out the framework for the HW/SW partitioning, which is performed at the process-level granularity, leading to the next lower-level models in C and VHDL.

1.9.3 Software: The C Model

The system is implemented partially in hardware and partially in software. The most demanding performance requirements are imposed on the ATM switch. Hence, it is implemented in hardware, while most of the control to set up connections and monitor the traffic ends up in software.

C programs constitute the first software models, which are compiled into assembler and machine code. The finite state machine description of SDL processes is directly translated into algorithms expressed in C. This step does not involve any major decisions because it does not cross an abstraction level. Consequently it can be fully atomated. However, moving from an SDL process network to C programs involves a decision in the communication domain. Shall the concurrent activities be retained, or shall the processes be transformed into a single sequential program? If the concurrent activities are maintained in the C programs, an operating system has to be included in the implementation to schedule the processes and manage the interprocess communication. In our case study, the concurrency is eliminated and the SDL process network is transformed into a single sequential C program. This is a major effort requiring that the SDL processes are statically scheduled and the communication is implemented as part of the C program.

The data abstractions are based on symbols derived from the SDL model and on numbers and strings, which are readily available data types in C. You could argue that C has a poor abstraction level with respect to numbers because the data type of the underlying processor shines through, imposing range restrictions that suggest that the abstraction is instead at the processor data type level. However, in many cases the available data types in the language and in appropriate libraries provide a good approximation to the ideal numbers.

The abstraction in the time domain is a clocked time. Even though the cycle time of the imaginary clock is not defined and may vary significantly for different cycles, the strictly sequential nature of a C program realizes de facto a clocked time structure where each C statement is unambiguously assigned to one clock cycle, thus imposing a total order on the execution of all statements. Since the SDL processes are already modeled in a sequential way, the major design effort is due to the concurrent processes, which must be arranged sequentially, that is, a static schedule must be defined and implemented.

In summary, the essence of the step from the SDL system model to the C program can be concisely described in the framework of the Rugby metamodel and facilitates an understanding of the process. The main design decisions and the major effort concern exactly the refinement steps across abstraction levels, which happen in the communication domain (InterProcessComm → ProcCall) and the time domain (Causality → ClockedTime). However, the analysis also illustrates that the refinement activities themselves are not necessarily separated along domain lines.

1.9.4 Software: The Assembler Model

The compilation of the C code into assembler is typically a fully automated process, which makes it very predictable. The computational elements are the instructions of the target processor, which also determine the time abstractions and the data types. No major design decision is made during this step.

1.9.5 Hardware: The VHDL Model

The developed VHDL model is synthesizable with Synopsys's design compiler. Thus, it is RTL code with a clocked time and algorithms describing the computations inside the processes. The algorithms are written in a style that is interpreted by the design compiler as finite state machines. Consequently, the description of the processes is very similar to the SDL models in terms of computation, control flow, and dataflow. The principal difference comes from the different level of abstraction in the time domain: causality in the SDL model and clocked time in the VHDL model.

The second important difference concerns the interprocesses communication, which is significantly refined in the VHDL model. In the VHDL model the topology is determined, describing precisely which port of one process connects to a port in another process. The asynchronous message-passing mechanism in SDL is refined into handshake or finite FIFO-based protocols. Note that the InterProcessComm abstraction level can also be realized in VHDL. However, the VHDL model of the NT switch did not use arbitrary global signals for communication but only explicitly defined ports that determine the connectivity between processes very precisely, sometimes down to the individual wire. This observation illustrates two aspects. First, the distinction between two abstraction levels is sometimes blurred. Second, the Rugby coordinates provide information about a model that is not apparent from the choice of the modeling language.

The data types have been refined in many places into bit vectors, either by using the information from the requirements definition or by making design decisions about the representation of data. In some places symbols are still used, leaving the decision to the subsequent synthesis step.

In summary, the main difference between the SDL and the VHDL model is in the communication and the time domains. Consequently, most of the effort for developing the VHDL model was spent refining communication and scheduling operations and communication events.

1.9.6 Hardware: Synthesized Netlist

The result of synthesizing the VHDL model with Synopsys's design compiler is a technology-mapped netlist. The computation elements are logic blocks, and the communication abstraction is a topology between the logic blocks as defined by the netlist. Note that for the VHDL model the communication abstraction is also a topology, which denotes the connection between processes. In the synthesized netlist this topological model has been extended into the processes leading to a topological hierarchy. All symbols have found a concrete bit vector representation, and the timing has been refined into physical time units based on the delay of the elementary blocks.

In the next step of placement and routing, the topology would be refined into a geometric model with physical units for the elementary cells and the wires, and the physical timing model would be extended to the interconnects.

1.9.7 Discussion

The models in this case study can be characterized naturally with the domains and abstraction levels of the Rugby model. Table 1-2 gives a concise and comprehensive view of the different models. To convey the same amount of information without the Rugby framework, we would need lengthy explanations, which would introduce many opportunities for misunderstanding. Thus, the Rugby model simplifies communication and understanding.

In addition to the information about the individual models, Table 1-2 provides insight into the methods used for refinement of one model into the next. For instance, the main differences between the SDL and VHDL models are in the communication and time domains, where the main effort in this refinement step has been spent. This is apparent at a glance from the Rugby coordinates. Some effort was made to represent symbols in bit vectors. In the computation domain the mapping from SDL state machines to VHDL state machines is straightforward without any need for design decisions. Note that this information is not inherent in the choice of the languages involved. If we had used Synopsys's behavioral compiler instead of the design compiler to synthesize the VHDL code, the VHDL model would have been at a different abstraction level in the time domain because scheduling is a major task of the behavioral compiler, that is, transforming a design from a causality level to a clocked time level in the time domain.

Another example is the refinement of the communication domain in the software part, where Table 1-2 reveals that concurrent processes are merged into a sequential program.

1.10 Further Reading

Cassandras (1993) gives an excellent account of system analysis and simulation based on discrete event models. It covers the basics of system and control theory as well as advanced stochastic techniques based on Markov processes and queuing theory.

Another book that covers a similar range of topics but from a different angle is Severance (2001).

For a detailed discussion of the Rugby model and other similar metamodels, refer to Jantsch et al. (1999, 2000); Gajski and Kuhn (1983); and Ecker et al. (1996).

1.11 Exercises

1.1 A main light controller in a lecture room has four inputs:

1. A dimmer that can be set to any value between "no light" and "full light."

2. A motion sensor, that notifies the controller when no motion has been detected in the room in a specific time period. If no motion has been detected, the light should be switched off. If motion is detected, the light should be switched on again.

3. When the overhead projector is on, the light should be reduced by 50%.

4. When the front light is on, which is also controlled by a dimmer, the main light should be reduced in linear proportion to the front light.

   a. Make a continuous time, continuous state model of the main light controller.

   b. Make a discrete time, discrete state model of the main light controller, that is, a model of the controller between the analog/digital and the digitial/analog converter.

   c. Classify the models in terms of Figure 1-7.

1.2 An audio-processing system receives a regular input signal and emits a signal after a delay D. The output signal is an exact copy of the input signal, except that signal values above a maximum level L are reduced to L at the output.

   a. Make a continuous time, continuous state model of the audio-processing system.

   b. Make a discrete time, discrete state model of the audio-processing system, that is, a model between the analog/digital and the digitial/analog converter.

   c. Classify the models in terms of Figure 1-7.

1.3 VHDL descriptions are typically categorized into "behavioral level" and "register transfer level" descriptions. Describe the difference between the two in terms of the Rugby notation.

1.4 How can the synthesis step from register transfer level VHDL to a netlist of gates

   a. excluding and

   b. including technology mapping be described in the Rugby terminology? (Technology mapping takes as input a netlist of gates, and replaces each gate with a layout block of this gate from a library. In addition, the layout of the wires connecting the gates is generated.)

   c. How can placement and routing be described in the Rugby terminology?

1.5 In software, how can the steps of compilation, from C to assembler language, and assemblage, from assembler language to machine code, be described in the Rugby terminology?

chapter two
Behavior and Concurrency

First we introduce finite state machines. We closely follow the first chapters of the book by Ullman (1979). In later chapters we assume that individual processes are modeled as finite state machines. Although other modeling notations could be used as well for this purpose, finite state machines constitute a fundamentally important technique that should be well mastered before dealing with process networks. Moreover, the chapter introduces the deep relationship between state machines and accepted or generated symbol sequences—useful for our purposes considering that processes can be viewed as accepting and generating event sequences continuously.

In the second part of the chapter we introduce Petri nets. Here we follow the book by Cassandras (1993, Chapter 2), enriched by material from Peterson (1981). Petri nets are useful for studying concurrency in its purest form.

In later chapters we combine the modeling of individual processes and the modeling of concurrency when we deal with process networks. At the end of Chapter 3, we will also see an example of how to abstract process networks into Petri nets for particular applications to solve specific problems (Sections 3.14 and 3.15).

2.1 Models for the Description of Behavior

We distinguish between computation and communication according to the notation of the Rugby model. That means that there are no explicit concurrent activities, and there is only very simple communication between the parts of a computation; for instance, parameters and arguments are passed between functions and procedures. How this data transfer takes place is not specified. Different parts of the computation may be simulated or executed in parallel to each other as long as data dependencies allow this. But again this is not specified and not part of the computational model.

There are several important languages and notations for representing purely sequential behavior, for example, sequential algorithms, imperative languages such as C and Pascal, and state machines. We first introduce finite state machines (FSM) as an important representative of this class of formalisms. They have been a very influential vehicle for theoretical studies of sequential behavior. They have also been used as a basis for a number of modeling and design languages, and many tools have been developed with various textual and graphical user interfaces. Thus, a large body of literature and knowledge exists and here we can give only a very basic introduction. In later chapters we use FSMs as the principal way to model processes.

In the second part of this chapter we introduce concurrency by means of Petri nets. Similar to FSMs, Petri nets are a well-established formalism often used for theoretical studies and as a basis for modeling languages and tools. All activities are concurrent, but the individual processing nodes are extremely simple. Furthermore, all data is reduced to the simple and abstract concept of a token. Thus, Petri nets are an ideal device to study issues of concurrency in a pure form without being confused by the complex behavior of processes or complex data transformations.

Based on this introduction to basic models of computation and concurrency, in later chapters we will merge both to study general process networks.

2.2 Finite State Machines

As we noted in the previous chapter, the inputs to a system alone often do not contain sufficient information to predict the system's behavior. In order to model such systems, we need to postulate some internal state. If the amount of information in the state is always finite, we have a finite state system. Many interesting systems can be viewed as finite state systems, such as the human brain and any kind of computer.

We informally introduce some basic notation and concepts of finite state machines with an example (Hopcroft and Ullman 1979).

A man with a goat, a wolf, and a cabbage wants to cross a river. He has a boat that is too small to carry all of them. In fact the boat can carry only two of the four. The problem is that both the wolf and the goat are hungry, and if not attended, the wolf would eat the goat and the goat would eat the cabbage. So the question is how to transport everything to the other side of the river without losing anything.

We model the problem by distinguishing between the state of the system and the transition from one state to another. What is the relevant information that characterizes the particular state of the system? Apparently it is on which side of the river the man, the wolf, the goat, and the cabbage are. We represent this by a string consisting of the letters M, W, G, and C and a hyphen (-). The letters stand for "man," "wolf," "goat," and "cabbage," respectively, and the hyphen designates the river. For instance the string "MW-GC" designates the state where the man and the wolf are on one side of the river and the goat and the cabbage are on the other side, with the consequence that the cabbage becomes the goat's lunch. We want to avoid these kind of states. An action of the man is represented by one of the letters w, g, or c, indicating that the man brings either the wolf, the goat, or the cabbage to the other side of the river. Actions trigger state transitions, and therefore they are considered to be inputs to the state machine. We can now draw a diagram as shown in Figure 2-1 to represent a sequence of states and transitions.

FIGURE 2-1

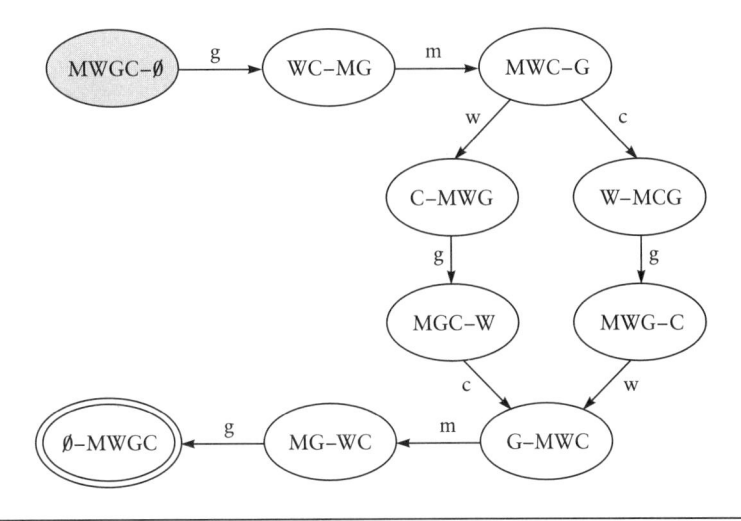

A state transition diagram for the man, wolf, goat, and cabbage problem.

The nodes of the diagram denote the states, and the arcs denote the transitions. The shaded ellipse is the starting state, and the double-lined ellipse is the final state. There are many solutions to the problem, but the finite state machine shown in Figure 2-1 represents the two shortest solutions. The actions of the man—that is, the inputs to the state machine—trigger state transitions. Hence, we can say that the state machine in the initial state "MWGC-Ø", when receiving the sequence of inputs $\langle g, m, w, g, c, m, g \rangle$, will end up in the final state "Ø-MWGC".

2.2.1 Basic Definition

Definition 2.1 A *finite state machine* is a 5-tuple

$$(\Sigma, X, g, x_0, F)$$

where

- Σ is a finite *alphabet*
- X is a finite *set of states*
- g is a *state transition function*, $g : X \times \Sigma \to X$
- x_0 is the *initial state*, $x_0 \in X$
- F is the *set of final states*, $F \subseteq X$

The word "finite" in finite state machine refers to the fact that the set of states X is a finite set.

The example in Figure 2-1 is formally written as a state machine $M = (\Sigma, X, g, x_0, F)$, where

$$\Sigma = \{m, w, g, c\}$$
$$X = \{(x, y) \mid x, y \subseteq \{M, W, G, C\} \text{ and } y = \{M, W, G, C\} \setminus x\}$$
$$x_0 = \{(\{M, W, G, C\}, \emptyset)\}$$
$$F = \{(\emptyset, \{M, W, G, C\})\}$$

$$g((\{M, W, G, C\}, \emptyset), g) = (\{W, C\}, \{M, G\})$$
$$g((\{W, C\}, \{M, G\}), m) = (\{W, C, M\}, \{G\})$$
$$g((\{M, W, C\}, \{G\}), w) = (\{C\}, \{M, W, G\})$$
$$g((\{M, W, C\}, \{G\}), c) = (\{W\}, \{M, G, C\})$$
$$g((\{C\}, \{M, W, G\}), g) = (\{M, G, C\}, \{W\})$$
$$g((\{W\}, \{M, G, C\}), g) = (\{M, W, G\}, \{C\})$$
$$g((\{M, G, C\}, \{W\}), c) = (\{G\}, \{M, W, C\})$$
$$g((\{M, W, C\}, \{G\}), w) = (\{G\}, \{M, W, C\})$$
$$g((\{G\}, \{M, W, C\}), m) = (\{M, G\}, \{W, C\})$$
$$g((\{M, G\}, \{W, C\}), g) = (\emptyset, \{M, W, G, C\})$$

We represent a state as a pair of sets, and we have used the set subtraction operator $\setminus$ in the definition of X. This is equivalent to the representation as strings, but we avoid introducing the basic operations on strings.

There are a few things to note. First, g is a function, which means that in a given state and with a given input there is only one possible next state; thus, the finite state machine is deterministic. However, g is not defined on all state-input pairs, and therefore how the machine behaves in certain states given certain inputs is not defined. For example, $g((\{M, W, G, C\}, \emptyset), m)$ is not defined. Second, F is a set, which means in general we can have more than one final state. Even though g in our example is not defined on the final state for any input, this need not be the case. In general, the state machine is allowed to leave a final state given that there is some input and an appropriate definition of g. Actually, we have not yet said what our intention is with the final states. To do this, we first generalize the transition function g such that it can accept a sequence of inputs.

If Σ is a final set of symbols, we define Σ^* to be the set of strings formed of elements of Σ. Σ^* contains the empty string as well as all infinite strings. For instance, let $\Sigma = \{a, b, c\}$; then "a", " ", "cab", and "$a \cdots$" are all elements of Σ^*. By convention we denote the empty string by the symbol ϵ. We use the operator $+$ to concatenate strings; thus, $r + s = rs, \forall r, s \in \Sigma^*$. $\text{length}(r)$ denotes the length of a string r, and we have $\text{length}(\epsilon) = 0$.

Let $g : X \times \Sigma \to X$ be a state transition function. Then $g^* : X \times \Sigma^* \to X$ is defined as follows, $\forall x \in X, a \in \Sigma, r \in \Sigma^*$:

$$g^*(x, \epsilon) = x$$

2.2 Finite State Machines

$$g^*(x, a) = g(x, a)$$
$$g^*(x, \text{``}a\text{''} + r) = g^*(g(x, a), r)$$

g^* operates on a sequence of inputs and yields the same result as if g were applied on each input separately. For instance, in the above example we would get $g^*((\{M, W, G, C\}, \emptyset), \text{``}gm\text{''}) = (\{M, W, C\}, \{G\})$ and $g^*((\{M, W, G, C\}, \emptyset), \text{``}gmwgcmg\text{''}) = (\emptyset, \{M, W, G, C\})$. For convenience we will often not explicitly distinguish between g and g^* and just write g for both functions if it is unambiguous.

We say a finite state machine $M = (\Sigma, X, g, x_0, F)$ *accepts* an input string $r \in \Sigma^*$ iff $g^*(x_0, r) \in F$.

We also say the *language* accepted by M, designated $L(M)$, is the set of strings accepted by M; that is, $L(M) = \{r \in \Sigma^* \mid g^*(x_0, r) \in F\}$.

Thus, in our example the machine M accepts strings "gmwgcmg" and "gmcgwmg" but no other strings; the language accepted by M is $L(M) = \{\text{``}gmwgcmg\text{''}, \text{``}gmcgwmg\text{''}\}$.

Example 2.1 Consider the finite state machine $M = (\Sigma, X, g, x_0, F)$ in Figure 2-2, where

$$\Sigma = \{0, 1\}$$
$$X = \{x_0, x_1, x_2, x_3\}$$
$$F = \{x_0, x_3\}$$

$$g(x_0, 0) = x_2, \quad g(x_0, 1) = x_1$$
$$g(x_1, 0) = x_3, \quad g(x_1, 1) = x_0$$

$$g(x_2, 0) = x_0, \quad g(x_2, 1) = x_3$$
$$g(x_3, 0) = x_1, \quad g(x_3, 1) = x_2$$

Because the state transition graph has cycles, the machine can accept infinite strings. An analysis shows that x_0 represents strings with an even

FIGURE 2-2

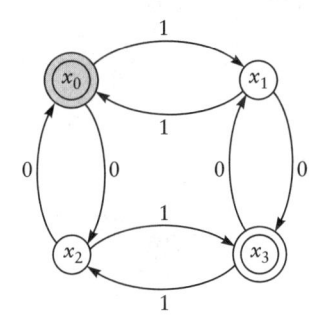

A finite state machine that accepts strings with an even number of inputs.

number of both zeros and ones, x_3 represents strings with an odd number of both zeros and ones, x_1 represents strings with an even number of zeros and an odd number of ones, and x_2 represents strings with an odd number of zeros and an even number of ones. Since $F = \{x_0, x_3\}$, the machine M accepts the language $L(M) = \{r \in \Sigma^* \mid \text{length}(r) \bmod 2 = 0\}$.

Many properties of state machines are not restricted to state machines with a finite state space. Hence we will also define a more general state machine and point to differences between them whenever feasible.

Definition 2.2 A *state machine* is a 5-tuple

$$(\Sigma, X, g, x_0, F)$$

where

Σ is a *countable alphabet*

X is a *countable set of states*

g is a *state transition function*, $g : X \times \Sigma \to X$

x_0 is the *initial state*, $x_0 \in X$

F is the *set of final states*, $F \subseteq X$

Both the input set and the state set can be infinite but countable, such as the natural numbers. We will use the term *state machine* if we mean both finite and infinite state machines, and we will use the terms *finite state machine*, *general state machine*, or *infinite state machine* when we want to be more specific.

2.2.2 Nondeterministic Finite State Machines

So far we have required that, for a given current state and a given input symbol, there is at most one next state. When we lift this restriction, we get a nondeterministic state machine. For instance, the state machine in Figure 2-3 has for state x_0 two possible next states when the input symbol a is received. There are several reasons why we

---

FIGURE 2-3

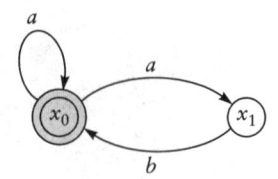

A nondeterministic finite state machine.

want to allow such a model. From a theoretic point of view, nondeterminism is a very interesting and useful concept. Often it is a useful tool for proving theorems. Sometimes it significantly extends the modeling capability; that is, we can model systems that we could not model without nondeterminism. This is not always the case, and it is important to know when it is the case and why. We will see that a nondeterministic finite state machine is not a more powerful modeling concept than a deterministic finite state machine.

From a practical point of view, we can distinguish two different situations where nondeterminism can be convenient. First, we just may not know what exactly the effect of a particular event may be. Suppose for example that we are modeling the braking behavior of a car. When the driver steps on the brake with a given force, it will not always have the same effect and decrease the velocity by the same amount. The precise effect would depend on many factors, such as the age of the car, the humidity of the air, the condition of the road, the type and condition of the tires, and so on. If we still want to develop a useful model, we have to allow for a range of possible effects. If we include all possibilities, we could try to design the car in a way that it would safely stop under all conditions. Or we could prove that a specific behavior of the car—for example, it rolls over—can be ruled out for all conditions we have taken into account.

A second situation, where nondeterminism may be useful arises when we do not care what the precise effect is even if we know it. In the car example, we might know exactly how humidity and the road condition affects the braking behavior, but we don't want to include this in our model because it would overly complicate the model without being useful. Since we have to deal with all possibilities anyway when designing or analyzing the car, it may be superfluous to include all the details of cause and effect in the model.

Due to its central importance in the theory and practice of modeling, we will meet the phenomenon of nondeterminism several more times. But for now we will continue to develop a nondeterministic variant of a finite state machine.

Definition 2.3 A *nondeterministic finite state machine* is a 5-tuple

$$(\Sigma, X, g, x_0, F)$$

where

- Σ is a finite *alphabet*
- X is a finite *set of states*
- g is a *state transition mapping*, $g : X \times \Sigma \to \wp(X)$
- x_0 is the *initial state*, $x_0 \in X$
- F is the *set of final states*, $F \subseteq X$

$\wp(X)$ denotes the power set of set X, that is, the set of all subsets. For example, if $X = \{a, b\}$, we have $\wp(X) = \{\emptyset, \{a\}, \{b\}, \{a, b\}\}$.

The only difference from the deterministic finite state machine is that g evaluates to a set of states rather than a single state.

Example 2.2   For Figure 2-3, the state machine $M = (\Sigma, X, x_0, g, F)$ is defined as follows:

$$\Sigma = \{a, b\}$$
$$X = \{x_0, x_1\}$$
$$F = \{x_0\}$$

$$g(x_0, a) = \{x_0, x_1\}, \qquad g(x_0, b) = \emptyset$$
$$g(x_1, a) = \emptyset, \qquad g(x_1, b) = \{x_0\}$$

The idea is that the machine in state x_0 can either enter state x_1 or remain in state x_0 when it receives an input event a. There is no information available that tells us which alternative the machine will select. We model this situation by carrying the set of possible states around. In a particular experiment, whether it be by simulation or by observation of a real-world entity, we will observe that only one of the possible states is entered.

Again, we extend the state transition function to cover arbitrary sequences of input events.

Let $g : X \times \Sigma \to \wp(X)$ be a state transition function for a nondeterministic finite state machine. Then $g^* : X \times \Sigma^* \to \wp(X)$ is defined as follows, $\forall x \in X, a \in \Sigma, r \in \Sigma^*$:

$$g^*(x, \epsilon) = \{x\}$$
$$g^*(x, a) = g(x, a)$$
$$g^*(x, r + \text{``}a\text{''}) = \{z \mid z \in g(y, r) \text{ for some } y \in g^*(x, r)\}$$

Example 2.3   Consider again the machine in Example 2.2 and confront it with the input sequence "*abaab*".

$$\begin{aligned}
& g^*(\{x_0\}, \text{``}abaab\text{''}) && \text{step 1} \\
& = g^*(\{x_0, x_1\}, \text{``}baab\text{''}) && \text{step 2} \\
& = g^*(\{x_0\}, \text{``}aab\text{''}) && \text{step 3} \\
& = g^*(\{x_0, x_1\}, \text{``}ab\text{''}) && \text{step 4} \\
& = g^*(\{x_0, x_1\}, \text{``}b\text{''}) && \text{step 5} \\
& = g^*(\{x_0\}, \epsilon) \\
& = \{x_0\}
\end{aligned}$$

The string is accepted because x_0 is a final state. In step 1, a is consumed, and we have two possible successor states, x_0 and x_1. In step 2, b is consumed. We must compute both $g(x_0, b)$ and $g(x_1, b)$ and take the union of the result, that is, $g(x_0, b) \cup g(x_1, b) = \emptyset \cup \{x_0\} = \{x_0\}$. In step 3, the machine consumes another a, which yields again two possible successor states. But consuming another a in step 4 allows us to continue only from x_0 (because $g(x_1, a) = \emptyset$), which in turn results again in two potential successor states. Consuming b in step 5 disambiguates the situation, and we end up in state x_0.

In this example, the number of potential states does not grow because a is consumed only in state x_0 and b is consumed only in x_1. The evaluation of $g^*(x_0, \text{"}abaab\text{"})$ shows us that there is a path of consecutive states from the initial to a final state such that the input string is accepted. However, when we consume a particular symbol, for example, a in state x_0, we do not know what is the "right" next state. We know it only when we consume the next symbol.

It is obvious that for every deterministic finite state machine there is a nondeterministic finite state machine that accepts exactly the same language. So the question arises naturally if there are languages that can be accepted by a nondeterministic machine but that cannot be accepted by a deterministic machine. Somewhat counterintuitively, the following theorem gives a negative answer.

Theorem 2.1 For every nondeterministic finite state machine M, there is a deterministic finite state machine M_D with $L(M) = L(M_D)$.

Note that this theorem does not generalize to infinite state machines.

We shall not give a proof for this theorem, which can be found, for instance, in Hopcroft and Ullman (1979, page 22), but we convey the main idea by giving an example.

Example 2.4 We now construct a deterministic machine $M^D = (\Sigma^D, X^D, x_0^D, g^D, F^D)$ that accepts the same language as the machine $M = (\Sigma, X, x_0, g, F)$ from Example 2.2. First we set $\Sigma^D = \Sigma$. Then we use all subsets of X ($\emptyset$, $\{x_0\}$, $\{x_1\}$, $\{x_0, x_1\}$) to construct X^D such that each element of X^D corresponds to a subset of X, that is,

$$X^D = \{y_\emptyset, y_0, y_1, y_{01}\}$$

Intuitively, we have the following correspondence:

$y_\emptyset$: "no possible state of M"
y_0 : x_0
y_1 : x_1
y_{01} : "either state x_0 or state x_1"

We decide the initial state to be $x_0^D = y_0$, and we define the state transition function as follows:

$$g^D(y_\emptyset, a) = y_\emptyset, \quad g^D(y_\emptyset, b) = y_\emptyset$$
$$g^D(y_0, a) = y_{01}, \quad g^D(y_0, b) = y_\emptyset$$
$$g^D(y_1, a) = y_\emptyset, \quad g^D(y_1, b) = y_0$$
$$g^D(y_{01}, a) = y_{01}, \quad g^D(y_1, b) = y_0$$

FIGURE 2-4

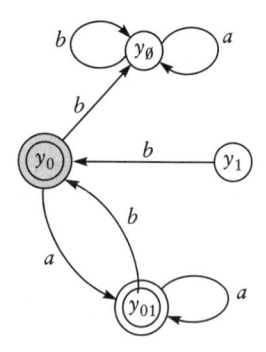

A deterministic counterpart to the machine in Figure 2-3.

Further we set $F^D = \{y_0, y_{01}\}$, and we get the deterministic state machine illustrated in Figure 2-4. It is easy to check that M^D indeed accepts the same set of strings as M. Both machines accept the empty string, and any string containing "*ab*" if *b* is immediately followed by *a* or is at the end of the string.

Note that M^D can be simplified because state y_1 can never be reached and thus could be removed. Further, state $y_\emptyset$ is superfluous because no string leading to it will be accepted. This idea of simplifying machines to equivalent machines with a smaller number of states is explored in the next section.

2.2.3 Finite State Machines with ε-Moves

Once more we extend the basic state machine to include also the possibility of spontaneous transitions, where the only input consumed is the empty string ϵ.

Definition 2.4   A *nondeterministic finite state machine with ε-moves* is a 5-tuple

$$(\Sigma, X, g, x_0, F)$$

where

- X is a finite *set of states*
- g is a *state transition mapping*, $g : X \times (\Sigma \cup \{\epsilon\}) \to \wp(X)$
- x_0 is the *initial state*, $x_0 \in X$
- F is the *set of final states*, $F \subseteq X$

The only difference from the nondeterministic state machine is that here we allow the function g also to have ϵ as a second argument. An example is shown in Figure 2-5.

FIGURE 2-5

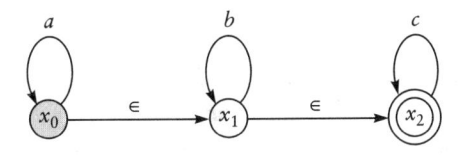

A nondeterministic state machine with ϵ-moves.

Sometimes it is more convenient to use this kind of machine, but again this extension does not fundamentally increase the modeling capability.

Theorem 2.2 For every nondeterministic finite state machine with ϵ-moves M, there is a nondeterministic finite state machine M_D with $L(M) = L(M_D)$.

A proof of this theorem can again be found, for instance, in Hopcroft and Ullman (1979).

2.2.4 State Aggregation

There is no unique way to model a given system as a state machine. Machines with different behaviors may represent a particular system equally well with respect to a given purpose. But even exactly the same behavior could be modeled by different machines, but they may have nonfunctional properties that make one more desirable or convenient to use than another. Such a property may be simulation speed or complexity of analysis. It would be desirable to have a method to transform a state machine into another that is functionally equivalent but better in some other respect.

We will now develop one example of such a method, but first we will give the terms "functionally equivalent" and "better in some respect" more precise meanings. A very important property is the number of states of a machine. The run time of many analysis techniques is very sensitive to the number of states, and most are usually only applicable to problems with a sufficiently small state space. Hence, a machine with a smaller number of states is usually preferable to another if both are functionally equivalent.

Two states are equivalent if the machine will accept exactly the same input strings from both states. This defines an equivalence relation and subsets of the state space with states that are all equivalent to each other.

Definition 2.5 Let $M = (\Sigma, X, g, x_0, F)$ be a finite state machine. Two states x and y are *equivalent*, denoted by $x \sim y$, iff

$$\forall r \in \Sigma^* : g^*(x, r) \in F \Leftrightarrow g^*(y, r) \in F$$

A set $R \subseteq X$ is a set of equivalent states iff

$$\forall x, y \in R, r \in \Sigma^* : x \sim y$$

Note that our notion of equivalence is with respect to a set of final states. Hence, the state machine may in fact distinguish between two equivalent states by accepting the same string with a different final state. Note also that the set of final states F is not a set of equivalent states.

Our idea in the following algorithm to simplify a state machine is that if two states accept the same input strings, they can be merged into a single state, which results in a state machine with fewer states. First, we will make a few observations that we will then use in the algorithm.

1. Two states $x \in F$ and $y \notin F$ are not equivalent because $g(x, \epsilon) \in F$ but $g(y, \epsilon) \notin F$.

2. Two states $x, y \in X$, of which either both are in F or neither is in F, are equivalent if $\forall e \in \Sigma : g(x, e) = g(y, e)$.

3. Two states $x, y \in X$, of which either both are in F or neither is in F, are equivalent if for some $e \in \Sigma : g(x, e) = y$ and $g(y, e) = x$ and $\forall e' \in \Sigma, e' \neq e : g(x, e') = g(y, e')$. Thus, for event e the states x and y are simply interchanged, and for all other events the successor states for x and y are identical.

4. We can generalize the previous observation. A set $R \subseteq X$, for which either $R \subseteq F$ or $R \cap F = \emptyset$, is a set of equivalent states if $\forall x, y \in R, e \in \Sigma : g(x, e) = z \notin R \Rightarrow g(y, e) = z$. Intuitively, if for an event e the set R is left, it is left for all states in R and it leads to the same state outside R.

In the following algorithm, we maintain a data structure L, that holds for each pair of states (x, y) a set of states that are equivalent iff $x \sim y$. The algorithm marks all pairs of states that are not equivalent. At the end, exactly the equivalent state pairs are unmarked. The procedure markall($L, (x, y)$) marks all state pairs (x', y') in the list $L(x, y)$ and, recursively, all state pairs in the lists $L(x', y')$.

Algorithm 2.1 *State Aggregation*

Step 1. **for** all pairs of states (x, y) **do** $L(x, y) := \emptyset$;

mark all pairs of states (x, y) with $x \in F$ and $y \notin F$;

Step 2. **for** every pair (x, y) not marked

do flag := 0;

Step 2.1. **for** every $e \in \Sigma$;

do if $(g(x, e), g(y, e))$ is marked

then markall($L, (x, y)$);

flag := 1;

Step 2.2. **if** (flag = 0)

then for every $e \in \Sigma$

do if $(g(x, e) \neq g(y, e))$

then add (x, y) to $L(g(x, e), g(y, e))$

TABLE 2-1: *Marking after step 1.*

	x_0	x_1	x_2	x_3
x_0	×	•	•	
x_1	×	×		•
x_2	×	×	×	•
x_3	×	×	×	×

Example 2.5 We will now apply Algorithm 2.1 to Example 2.1, shown in Figure 2-2.

To this end we build a table where we mark the state pairs. We do not need to consider pairs designated by × because they are either already covered or they denote pairs (x, x), which are trivially equivalent. Table 2-1 shows the marking after step 1 of the algorithm.

Step 2: According to Table 2-1, we have to consider the pairs $\{(x_1, x_2), (x_0, x_3)\}$. We begin the first iteration through the loop of step 2 by setting $(x, y) := (x_1, x_2)$ and flag := 0.

Step 2.1: The list of events to consider is $\{0, 1\}$.

$e := 0$: $g(x_1, 0) = x_3$, $g(x_2, 0) = x_0$ and (x_0, x_3) are not marked.

$e := 1$: $g(x_1, 1) = x_0$, $g(x_2, 1) = x_3$ and (x_0, x_3) are not marked.

Step 2.2: Since flag has not been set, we enter step 2.2. Again we have to go through the set of events.

$e := 0$: $g(x_1, 0) = x_3 \neq g(x_2, 0) = x_0$; consequently we add (x_1, x_2) to the list associated with (x_0, x_3): $L(x_0, x_3) := \{(x_1, x_2)\}$.

$e := 1$: $g(x_1, 1) = x_0 \neq g(x_2, 1) = x_3$; consequently we should also add (x_1, x_2) to the list associated with (x_0, x_3): $L(x_0, x_3)$, but since we have already done it, $L(x_0, x_3)$ is unchanged.

Step 2: Now we go through the loop a second time and set $(x, y) := (x_0, x_3)$ and flag := 0.

Step 2.1: The list of events to consider is $\{0, 1\}$.

$e := 0$: $g(x_0, 0) = x_2$, $g(x_3, 0) = x_1$ and (x_1, x_2) is not marked.

$e := 1$: $g(x_0, 1) = x_1$, $g(x_3, 1) = x_2$ and (x_1, x_2) is not marked.

Step 2.2: Since flag has not been set, we enter step 2.2. Again we have to go through the set of events.

$e := 0$: $g(x_0, 0) = x_2 \neq g(x_3, 0) = x_1$; consequently we add (x_0, x_3) to the list associated with (x_1, x_2): $L(x_1, x_2) := \{(x_0, x_3)\}$.

$e := 1$: $g(x_0, 1) = x_1 \neq g(x_3, 1) = x_2$; consequently we should also add (x_0, x_3) to the list associated with (x_1, x_2): $L(x_1, x_2)$, but since we have already done it, $L(x_1, x_2)$ is unchanged.

FIGURE 2-6

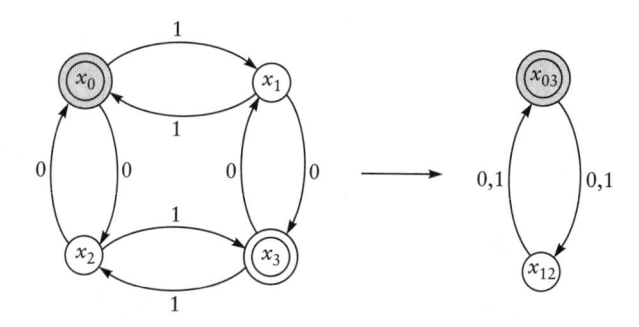

The state machine of Figure 2-2 can be reduced and will still accept the same set of input strings.

We are at the end of the algorithm with two state pairs unmarked. Consequently we can merge x_0 with x_3 and x_1 with x_2. The resulting states get all the input and output arcs of all the originating states, as illustrated in Figure 2-6.

2.2.5 Regular Sets and Expressions

So far we have informally talked about strings and languages accepted by state machines. These languages can be defined more formally and classified into different types that have very close relations to different types of state machines. We now proceed to introduce regular expressions, which define regular sets. Regular sets are precisely those languages that are accepted by finite state machines.

Definition 2.6   Let Σ be a set of symbols and let L, L_1, and L_2 be sets of strings from Σ^*. The *concatenation of two languages* L_1 and L_2, denoted $L_1 L_2$, is defined as follows:

$$L_1 L_2 = \{xy \mid x \in L_1 \text{ and } y \in L_2\}$$

That is, the strings in $L_1 L_2$ are formed by concatenating a string from L_1 with a string from L_2.

We use the notation L^n to designate the concatenation of a language n times with itself; that is,

$$L^0 = \{\epsilon\}, L^i = LL^{i-1}, \forall i > 0$$

The *Kleene closure* of L, denoted L^*, is the set

$$L^* = \bigcup_{i=0}^{\infty} L^i$$

That means L^* contains any number of strings from L including ϵ. The *positive closure*, L^+, is the set

$$L^+ = \bigcup_{i=1}^{\infty} L^i$$

L^+ contains ϵ only if L contains it.

Example 2.6 Let $\Sigma = \{a, b, c\}$, $L_1 = \{\epsilon, a, abb\}$, and $L_2 = \{c\}$. Then we have $L_1^2 = \{\epsilon, a, abb, aa, aabb, abba, abbabb\}$, $L_2^3 = \{ccc\}$, $L_1 L_2 = \{c, ac, abbc\}$, and

$$L_1^* = \{\epsilon, a, abb, aa, abba, aabb, abbabb, aaa, abbaa, aabba, \ldots\}$$
$$L_2^+ = \{c, cc, ccc, \ldots\}$$

Next we define regular expressions as a convenient way to describe and define languages. We use the convention of writing $L(r)$ for the language that is denoted by the regular expression r.

Definition 2.7 Let Σ be an alphabet. The *regular expressions* and the sets that they denote are defined as follows:

1. $\emptyset$ is a regular expression and denotes the empty set.
2. ϵ is a regular expression and denotes $\{\epsilon\}$.
3. For each $a \in \Sigma$, a is a regular expression and denotes $\{a\}$.
4. If r and s are regular expressions with $L(r) = R, L(s) = S$, then
    (a) $(r + s)$ is a regular expression with $L(r + s) = R \cup S$.
    (b) (rs) is a regular expression with $L(rs) = RS$.
    (c) r^* is a regular expression with $L(r^*) = R^*$.

We assume that * has higher precedence than concatenation, which in turn has higher precedence than +. Thus, we write $a^*b + b^*a$ for $((a^*)b) + ((b^*)a)$.

Example 2.7 We give a few examples of regular expressions and the sets they represent.

- aa: $L(aa) = \{aa\}$.
- a^*: $L(a^*) = \{\epsilon, a, aa, \ldots\}$.
- $(a + ab)^*$: $L((a + ab)^*)$ = all strings starting with a and with at most one consecutive b.
- $aab(a + b)^*baa$: $L(aab(a + b)^*baa)$ = all strings of as and bs beginning with aab and ending with baa.

Definition 2.8 Every set (language) that can be denoted by a regular expression is a *regular set (regular language)*.

Theorem 2.3   A language can be denoted by a regular expression iff there exists a nondeterministic finite state machine with ϵ-transitions that accepts the language.

Thus, the regular languages are precisely those languages that can be denoted by a regular expression and be accepted by a finite state machine. We again omit a proof for this theorem (Hopcroft and Ullman 1979), but we illustrate the construction of a state machine for a given regular expression by way of an example. The main idea is to follow the structure of the regular expression and to gradually build up compound state machines from simpler ones based on the templates shown in Figure 2-7.

Example 2.8   Given is the regular expression $r = ab^* + a$, which we write $r = r_1 + r_2 = r_{11}r_{12} + r_2 = r_{11}r_{121}^* + r_2$. The finite state machines for the atomic regular expressions are simply

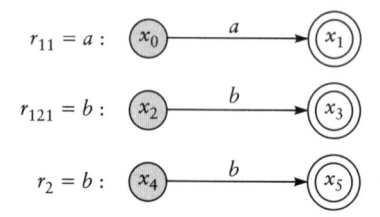

$r_{11} = a$: $x_0 \xrightarrow{a} x_1$

$r_{121} = b$: $x_2 \xrightarrow{b} x_3$

$r_2 = b$: $x_4 \xrightarrow{b} x_5$

To construct $r_{12} = r_{121}^*$, we use the template in Figure 2-7(c) and we get

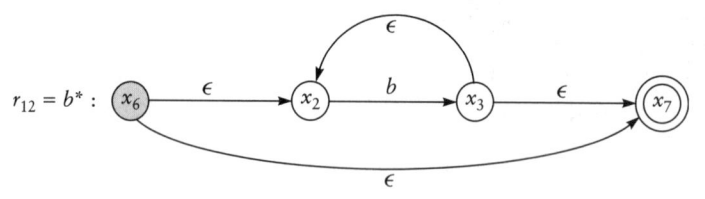

$r_{12} = b^*$: $x_6 \xrightarrow{\epsilon} x_2 \xrightarrow{b} x_3 \xrightarrow{\epsilon} x_7$ (with ϵ loop back from x_3 to x_2 and ϵ bypass from x_6 to x_7)

For expression r_1 we use the template in Figure 2-7(b):

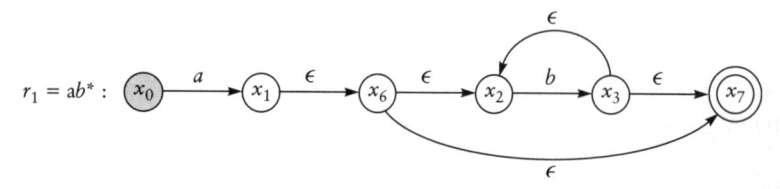

$r_1 = ab^*$: $x_0 \xrightarrow{a} x_1 \xrightarrow{\epsilon} x_6 \xrightarrow{\epsilon} x_2 \xrightarrow{b} x_3 \xrightarrow{\epsilon} x_7$ (with ϵ loop from x_3 to x_2 and ϵ bypass below)

And finally we can construct the machine for r by using the template in Figure 2-7(a):

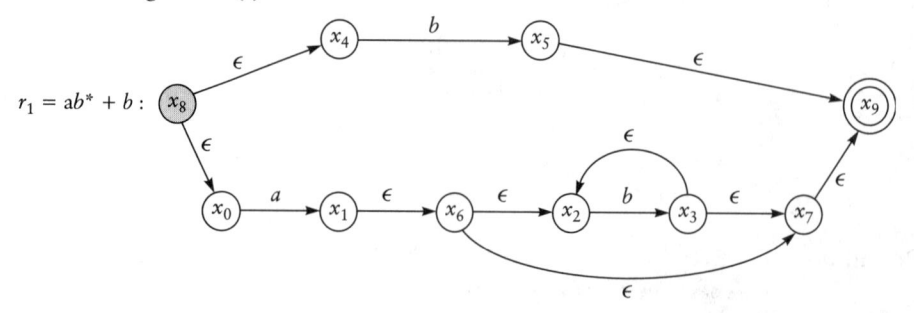

$r_1 = ab^* + b$: (machine starting at x_8, branching via ϵ to upper path $x_4 \xrightarrow{b} x_5$ and lower path $x_0 \xrightarrow{a} x_1 \xrightarrow{\epsilon} x_6 \xrightarrow{\epsilon} x_2 \xrightarrow{b} x_3 \xrightarrow{\epsilon} x_7$, both joining at x_9 via ϵ)

FIGURE 2-7

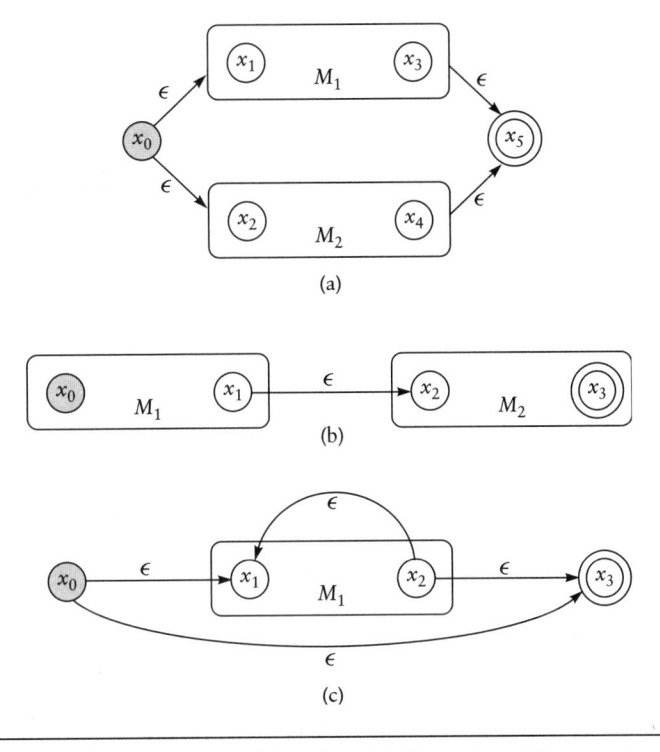

Templates to construct state machines for (a) the union, (b) concatenation, and (c) closure of regular expressions.

2.2.6 Finite State Machines with Output

We are not always content with a machine that merely accepts strings; we also need machines that produce some output as a reaction to the inputs. There are two distinct approaches. The first associates the emission of an output event with a state (Moore machine), and the second with a transition (Mealy machine). Both are equivalent in their modeling capabilities, but they can differ significantly in the number of states that they may require for a given modeling task.

Definition 2.9 A deterministic *Moore machine* is a 6-tuple

$$(\Sigma, \Delta, X, g, f, x_0)$$

where

Σ is a finite alphabet

Δ is a finite *output alphabet*

FIGURE 2-8

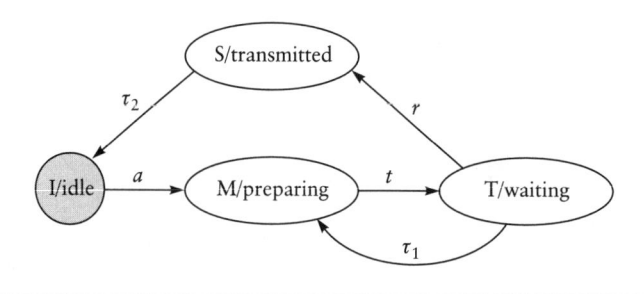

A simple protocol as a Moore machine.

X is a finite set of states

x_0 is the *initial state*, $x_0 \in X$

g is a *state transition function*, $g : X \times \Sigma \to X$

f is an *output function*, $f : X \to \Delta$

Example 2.9 As an example, consider a simple protocol for transmitting a message (Figure 2-8). The machine starts in the idle state I. Upon arrival of a message (a), it enters state M. After transmitting the message (t), it enters state T, where it waits for an acknowledgment. If the acknowledgment arrives (r) before a time-out occurs (τ_1), it enters state S to indicate the successful transmission. Then it again enters state I to wait for new messages. If no acknowledgment arrives before a time-out (τ_1) occurs, the machine re-enters state M to retransmit the message until it is successful. We use the convention of writing the output in the state circle separated from the state name by a "/".

Our machine $M = (\Sigma, \Delta, X, x_0, g, f)$ is specified as follows:

$$\Sigma = \{a, t, \tau_1, \tau_2, r\}$$
$$X = \{\text{I, M, T, S}\}$$
$$x_0 = \text{I}$$
$$\Delta = \{\text{idle, preparing, waiting, transmitting}\}$$
$$g(\text{I}, a) = \text{M}, \quad g(\text{M}, t) = \text{T}$$
$$g(\text{T}, r) = \text{S}, \quad g(\text{T}, \tau_1) = \text{M}$$
$$g(\text{S}, \tau_2) = \text{I}$$
$$f(\text{I}) = \text{idle}, \quad f(\text{M}) = \text{preparing}$$
$$f(\text{T}) = \text{waiting}, \quad f(\text{S}) = \text{transmitted}$$

The output function is used to provide information about the state of the system. As long as it is in state M, it is "preparing" the message. When the message is actually transmitted, a new state is entered and the system is "waiting." The state S is only necessary to inform about the successful transmission. After it has emitted this information, the machine enters the idle state again, triggered by a time-out event. If we had chosen to use a machine with ϵ-transitions, it would have been natural and convenient to trigger the transition from S to I by ϵ.

The second approach to model the output of a machine is to associate it with a transition rather than a state.

Definition 2.10 A deterministic *Mealy machine* is a 6-tuple

$$(\Sigma, \Delta, X, g, f, x_0)$$

where

Σ is a finite alphabet
Δ is a finite *output alphabet*
X is a finite set of states
x_0 is the *initial state*, $x_0 \in X$
g is a *state transition function*, $g : X \times \Sigma \to X$
f is an *output function*, $f : X \times \Sigma \to \Delta$

The only difference from the Moore machine is that the output function takes an input event as argument and hence emits an output for every input received.

Example 2.10   We model the same message transmission protocol with a Mealy machine as shown in Figure 2-9. Now the output is annotating the arcs written together with the inputs that trigger the transition.

Our machine $M = (\Sigma, \Delta, X, x_0, g, f)$ is now specified as follows:

$$\Sigma = \{a, t, \tau, r\}$$

FIGURE 2-9

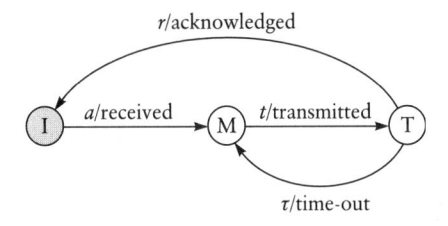

The same protocol of Example 2.9 and Figure 2-8 is modeled as a Mealy machine.

$$X = \{I, M, T\}$$
$$x_0 = I$$
$$\Delta = \{\text{received, transmitted, time-out, acknowledged}\}$$
$$g(I, a) = M, \qquad g(M, t) = T$$
$$g(T, r) = I, \qquad g(T, \tau) = M$$
$$f(I, a) = \text{received}, \qquad f(M, t) = \text{transmitted}$$
$$f(T, \tau) = \text{time-out}, \qquad f(S, r) = \text{acknowledged}$$

The output now provides information about the transitions rather than the states. Note that the Mealy model has one less state than the Moore machine. It is a general phenomenon that Mealy machines often contain fewer states than the corresponding Moore machines, but it greatly depends on the nature of the modeling task which of the two is more appropriate.

Moore and Mealy machines cannot be directly equivalent because the Moore machine produces an output in the initial state before any input has been consumed, but the Mealy machine's first output event occurs during a transition that is triggered by the first input. However, we can define the equivalence of Moore and Mealy machines if we ignore the first initial output of the Moore machine.

Definition 2.11 Let $M = (\Sigma, \Delta, X, x_0, g, f)$ be a deterministic Moore machine and let $M' = (\Sigma', \Delta', X', x'_0, g', f')$ be a deterministic Mealy machine. Let $T_M(s)$ be the output string produced by machine M when given input string s.

M and M' are equivalent if for any input string $s \in \Sigma \cup \Sigma'$ we have $f(x_0)T_M(s) = T_{M'}(s)$.

With this definition of equivalence, we can show that Moore and Mealy machines are equivalent, as the following two theorems disclose.

Theorem 2.4 If $M = (\Sigma, \Delta, X, x_0, g, f)$ is a deterministic Moore machine, then there is a deterministic Mealy machine M' equivalent to M.

Proof Let $M' = (\Sigma, \Delta, X, x_0, g, f')$ be a Mealy machine, and define $f'(x, e) := f(g(x, e))$ for all states x and all input events e. Thus, the output of the Mealy machine during a given transition will be the same as the output of the Moore machine in the target state of the transition. Given the same input string, both machines will go through the same state sequence and produce the same output.

Theorem 2.5 If $M = (\Sigma, \Delta, X, x_0, g, f)$ is a deterministic Mealy machine, then there is a deterministic Moore machine M' equivalent to M.

Proof Let $M' = (\Sigma, \Delta, X', x'_0, g', f')$, where

$$X' = X \times \Delta$$

$$x'_0 = (x_0, d_0), d_0 \in \Delta$$
$$g'((x,d),e) = (g(x,e), f(x,e))$$
$$f'((x,d)) = d$$

Intuitively, we store the output that is produced during a transition by the Mealy machine M in the next state of the Moore machine M'; then we output this stored value in the state. Consequently, the new states of M' are pairs consisting of a state of M and an output symbol. Hence, for every transition $x \to g(x,e), f(x,e)$ of M, the corresponding transition $(x,d) \to (g(x,e), f(x,e)), f'(f(x,e))$ of M' will produce the same output, that is, $f(x,e) = f'(f(x,e))$. In addition, the Moore machine M' will produce an initial output d_0, which we can ignore according to Definition 2.11.

2.2.7 Finite State Machine Extensions

Finite state machines, as we have discussed them so far, have a limited expressiveness. Their theoretical expressiveness is limited since there are input languages that cannot be recognized by any finite state machine. For instance, the language $L(a^n b^n)$—the language consisting of all words with equal numbers of as and bs—cannot be recognized by a finite state machine. The reason is the finiteness of the state space of the state machine. In order to recognize a word $a^n b^n$, a finite state machine has to memorize how many as it has seen before the first b arrives. However, the only way to memorize this is to have a different state for every $n \in \mathbb{N}$. Clearly, whatever the size of the state machine, there will be an $N \in \mathbb{N}$ such that it cannot be represented by the finite state machine.

Push-down automata rectify this limitation. A push-down automaton is a state machine with an unbounded stack. The machine can store tokens on the stack and read them back later. The tokens may not be arbitrary data; in particular, infinite data types such as integers are not allowed. A push-down automaton can recognize $L(a^n b^n)$. Since the stack is unbounded, any $N \in \mathbb{N}$ can be stored in it and the machine will know how many as it has seen when the first b arrives. Even push-down automata are not overly expressive. For instance, they cannot recognize languages such as $L(a^n b^n c^n)$ and $L(a^i b^j c^k)$ with $i \leq j \leq k$ because they cannot recognize arbitrary relations between different parts in the input string. Essentially only one unbounded value can be represented at any time.

The restrictions on push-down automata can be gradually lifted until we obtain machines equivalent to Turing machines, the most expressive models we have. Refer to Hopcroft and Ullman (1979) for a complete account of different automata models, their theoretical expressiveness, and corresponding language classes.

Finite state machines can also be extended for the purpose of modeling convenience. Since it is tedious to represent even the simplest arithmetic operations as finite state machines, it seems natural to annotate state transitions with arithmetic operations in a more compact format. Figure 2-10 shows a *finite state machine with datapath* (FSMD). The variable v can be tested and set during transitions. The transition from state x_0 to x_1 is triggered by the arrival of input event a. To the right of the slash (/)

FIGURE 2-10

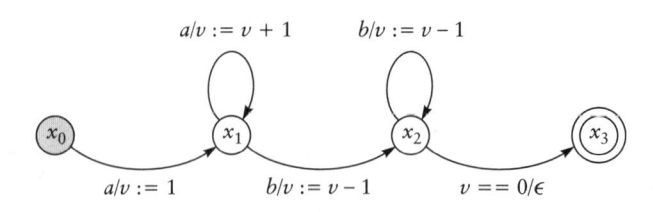

A finite state machine with datapath.

in the transition annotation is the action performed. v is set to 1, indicating that one a has been consumed. For every a, v is incremented in state x_1. With the arrival of the first b, the machine changes to state x_2 and does not accept any more as. However, for each b, v is decremented. Finally, when the condition $v == 0$ becomes true, the transition to x is triggered, no action is taken, and the machine arrives at its final state x_3. In this example the variable v is used to count the number of as and bs seen, which is more compact and natural than to encode this information in states.

If we want the variable values to influence the state sequence, we have to allow for conditions in the transitions where variable values are tested. If no such conditions are allowed, the datapath operations are strictly separated from the state machine behavior. The actions in the transition annotations can be arbitrarily complex, and depending on their complexity and the domains of the variables, we have either a model equivalent to the simple finite state machine or a more expressive one. For instance, if v in Figure 2-10 is drawn from a finite set (e.g., the set $0\ldots 65{,}535$, which can be represented by a 16-bit bit vector), the resulting state machine is not an extension. It can only recognize words of $L(a^n b^n)$ with $n \leq 65{,}535$, which can also be done by ordinary finite state machines. However, the model in Figure 2-10 is still preferable due to its compactness. Also, it reflects the regularity of the behavior much better, which is beneficial for understanding and automatic analysis. If v is drawn from the integers, $v \in \mathbb{Z}$, we have a true extension of the finite state machine model, reflected in the fact that the machine in Figure 2-10 can recognize words of the form $a^n b^n$.

Another extension concerns hierarchy. Since state machines with even a moderate number of states soon become unreadable, and most systems consist of parts repeated several times, hierarchy is a natural and necessary extension to handle complexity. As Figure 2-11 illustrates, in a hierarchical state machine a state can be another state machine. State B is in fact another state machine, and the input sequence accepted by state machine A is *adefbc*.

A further extension introduces concurrency. Like hierarchy it helps to deal with complexity because it allows us to partition the system into smaller parts and deal with them separately and independently. Unlike hierarchy, concurrency is accompanied with a host of difficult issues concerning communication, synchronization, time, and the overall semantics of the model. Different approaches to these problems have led to quite different concurrent finite state machine models.

Statecharts, introduced by Harel (1987), are based on a hierarchical concurrent finite state machine model that has been widely used for hardware modeling

FIGURE 2-11

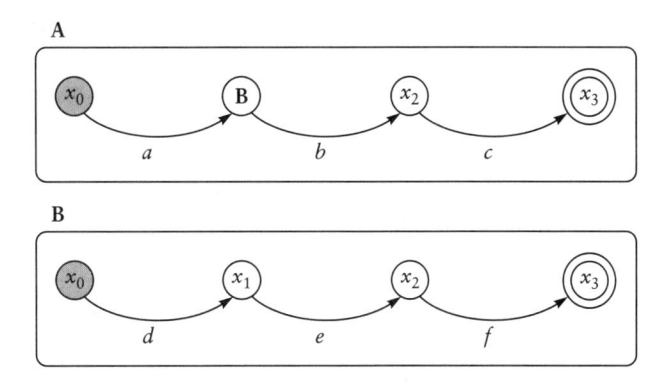

A hierarchical finite state machine.

and design. Statecharts use a synchronous composition model, meaning that all concurrent machines operate in a lockstep mode. During an evaluation cycle, each state machine reads its input and computes the next step and its outputs. The outputs of one evaluation cycle from all machines are available as inputs to all other machines in the next cycle.

An alternative semantics is proposed by Chiodo et al. (1993) as *codesign finite state machines* (CFSMs). In CFSMs individual machines act independently from each other. When two machines exchange data, the receiver blocks until data is available, whereas the sender emits its data and continues its activity without waiting. A buffer of size one holds a token until the receiver consumes it or the sender overwrites it with a new token.

Many other models of interaction have been proposed, and in the following chapters we will discuss several alternatives in great detail. In fact, since we assume in the rest of the book that individual processes are modeled as finite state machines, it can be said that this book essentially is about different approaches to extending finite state machines with concurrency.

2.3 Petri Nets

Models based on finite state machines are inherently focused on the state of a system and the observable input-output behavior. They are not well suited to studying the interaction of concurrently active parts of a system and the combined behavior of distributed parallel systems. To address issues of concurrency, synchronization, and communication, we had to generalize state machines into a model of communicating, concurrent state machines. But before we do so, it is worthwhile to study the phenomena of concurrency isolated from the internal complexities of the subparts of a system.

One of the oldest, most influential, and commonly used in countless tools and methods are *Petri nets*. Petri nets were first proposed by C. A. Petri in the early 1960s.

Since then many variants, specializations, and generalizations have been developed. We will only introduce the basic Petri net model and a few applications.

Petri nets make two principal abstractions that render them so suitable for the study of concurrency. First, they focus only on the act of communication and are not concerned with the data communicated. The means of communication is a *token*, which does not contain any data. Second, all details of the behavior are omitted if they do not contribute to the consumption and emission of tokens. In a sense, concurrency can be studied in a pure form with Petri nets. This is very desirable because concurrency itself quickly leads to a high degree of complexity that can easily be overwhelming even without other sources of complexity interfering.

Definition 2.12 A *Petri net* is a 6-tuple $N = (P, T, A, w, x_0)$, where

- P is a finite set of *places*
- T is a finite set of *transitions*
- A is a set of *arcs*, $A \subseteq (P \times T) \cup (T \times P)$
- w is a weight function, $w : A \to \mathbb{N}$
- $\vec{x}_0$ is an initial marking vector, $\vec{x}_0 \in \mathbb{N}^{|P|}$

Figure 2-12 shows a simple Petri net with two places and one transition. The tokens can move from one place through a transition to another place. The weight function w assigns some weight values to arcs. Weight values are usually small numbers and are represented by several arcs in our diagrams. For instance, in Figure 2-12 the arc between p_1 and t_1 has a weight of 2, and the arc between t_1 and p_2 has a weight of 1. A weight of 0 means there is no arc.

The tokens represent the marking in the above definition. A marking tells us how many tokens are in each place. In Figure 2-12 the marking is $\vec{x} = [1, 0]$. We often denote the marking of a place in state $\vec{x}$ by means of a corresponding marking function $x : P \to \mathbb{N}$. In our example we have $\vec{x} = [x(p_0), x(p_1)] = [1, 0]$. The marking changes when transitions fire. A transition may fire when it is enabled, that is, when there are sufficient tokens on its input places.

Definition 2.13 Let $N = (P, T, A, w, \vec{x}_0)$ be a Petri net. The set $I(t) = \{p \in P | (p, t) \in A\}$ is the set of *input places* of transition t. The set $O(t) = \{p \in P | (t, p) \in A\}$ is the set of *output places* of transition t.

FIGURE 2-12

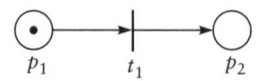

A simple Petri net. Places are designated by circles, transitions by black rules, tokens by black dots, and arcs connect places with transitions.

FIGURE 2-13

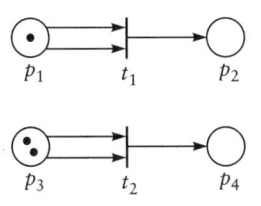

Two Petri nets; one with no enabled transition and one with t_2 enabled.

A transition t is *enabled* in state $\vec{x}$ if

$$x(p) \geq w(p,t) \ \forall p \in I(t)$$

A transition is enabled when there are at least as many tokens on each of its input places as there are arcs (weight) from that input place to the transition.

Transition t_1 of the Petri net in Figure 2-12 is enabled. In Figure 2-13, t_1 is not enabled, but t_2 is enabled. An enabled transition can fire by consuming tokens from its input places and emitting tokens to its output places. The mechanics of this process are defined by the Petri net *transition function G*.

Definition 2.14 Let $N = (P, T, A, w, \vec{x}_0)$ be a Petri net with $P = \{p_0, \ldots, p_{n-1}\}$ and $\vec{x} = [x(p_0), \ldots, x(p_{n-1})]$ be a marking for the n places. Then the transition function $G : (\mathbb{N}^n \times T) \to \mathbb{N}^n$ is defined as follows:

$$G(\vec{x}, t) = \begin{cases} \vec{x}' & \text{if } x(p) \geq w(p,t) \ \forall p \in I(t) \\ \vec{x} & \text{otherwise} \end{cases}$$

with $\quad \vec{x}' = [x'(p_0), \ldots, x'(p_{n-1})]$
$\quad\quad\quad x'(p_i) = x(p_i) - w(p_i, t) + w(t, p_i)$ for $0 \leq i < n$

If the transition is not enabled, the marking is not changed. If it is enabled, the firing of t means that the number of tokens in each $p \in I(t)$ is reduced by $w(p,t)$, and the number of tokens in each $p \in O(t)$ is increased by $w(t,p)$.

Figure 2-14 illustrates this process with a simple example. Transition t_1 is enabled because its input place p_1 contains two tokens, which is sufficient since $w(p_1, t_1) = 2$. After the firing, both tokens are consumed and one token is emitted to place p_2 because $w(t_1, p_2) = 1$.

Example 2.11 As an example consider the Petri net $N = (P, T, A, w, \vec{x}_0)$, with

$$P = \{p_1, p_2, p_3, p_4\}$$
$$T = \{t_1, t_2, t_3\}$$

FIGURE 2-14

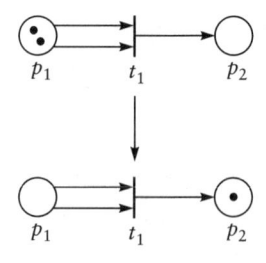

The firing of transition t_1.

FIGURE 2-15

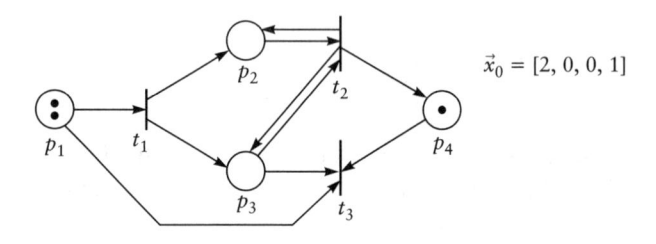

$\vec{x}_0 = [2, 0, 0, 1]$

A Petri net with an initial marking $\vec{x}_0$.

$$A = \{(p_1, t_1), (p_1, t_3), (p_2, t_2), (p_3, t_2), (p_3, t_3), (p_4, t_3),$$
$$(t_1, p_2), (t_1, p_3), (t_2, p_2), (t_2, p_3), (t_2, p_4)\}$$
$$w(a) = 1 \; \forall a \in A$$
$$\vec{x}_0 = [2, 0, 0, 1]$$

which is shown in Figure 2-15. With the initial marking, only transition t_1 is enabled. Firing it results in the marking shown in Figure 2-16. Now all three transitions t_1, t_2, and t_3 are enabled. The Petri net semantics do not define which transition fires when more than one is enabled. Consequently, Petri nets are inherently nondeterministic.

From the marking $\vec{x}_1$, t_2 could fire arbitrarily often with the effect of accumulating tokens in p_4. But, if transition t_3 fires, the net ends up with the marking $\vec{x}_2 = [0, 1, 0, 0]$, which disables all transitions, as shown in Figure 2-17.

In the terminology of state machines, the marking of a Petri net can be considered the *state* of the net, with $\vec{x}_0$ being the initial state. The state of a Petri net is distributed in the sense that the effect of firing a particular transition is local and contained in only

2.3 Petri Nets

FIGURE 2-16

$\vec{x}_0 = [2, 0, 0, 1]$

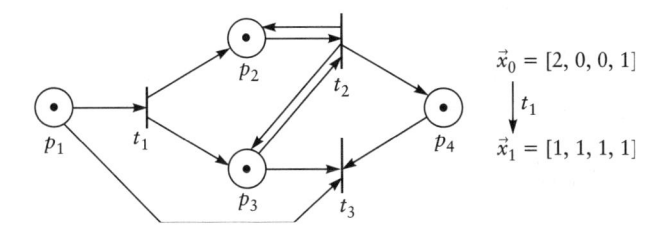

$\downarrow t_1$

$\vec{x}_1 = [1, 1, 1, 1]$

The Petri net of Figure 2-15 after firing t_1.

FIGURE 2-17

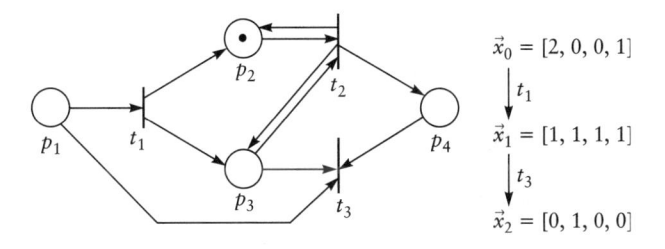

$\vec{x}_0 = [2, 0, 0, 1]$

$\downarrow t_1$

$\vec{x}_1 = [1, 1, 1, 1]$

$\downarrow t_3$

$\vec{x}_2 = [0, 1, 0, 0]$

The Petri net of Figure 2-16 after firing t_3 has no enabled transitions.

part of the state. Obviously, the firing of a transition can change only the marking of places connected to it.

Later it will interest us considerably which states can be reached from a given state.

Definition 2.15 For a Petri net $N = (P, T, A, w, \vec{x}_0)$ and a given state $\vec{x}$, a state $\vec{y}$ is *immediately reachable* from $\vec{x}$ if there exists a transition $t \in T$ such that $G(\vec{x}, t) = \vec{y}$.

The *reachability set* $R(\vec{x})$ is the smallest set of states defined by

1. $\vec{x} \in R(\vec{x})$.

2. If $\vec{y} \in R(\vec{x})$ and $z = G(\vec{y}, t)$ for some $t \in T$, then $\vec{z} \in R(\vec{x})$.

Intuitively, the reachability set of a state includes all the states that can eventually be reached by repeatedly firing transitions. For instance, the reachability set of $\vec{x}_0$ in Example 2.11 is

$$R(\vec{x}_0) = R_1 \cup R_2 \cup R_3 \cup R_4$$

$$R_1 = \{\vec{x}_0\}$$

$$R_2 = \{\vec{y} \mid \vec{y} = [1, 1, 1, n], n \geq 0\}$$

$$R_3 = \{\vec{y} \mid \vec{y} = [0, 2, 2, n], n \geq 0\}$$
$$R_4 = \{\vec{y} \mid \vec{y} = [0, 1, 0, n], n \geq 0\}$$

The set R_2 is generated when t_1 fires once and t_2 n times. R_3 is generated when t_1 fires twice and t_2 n times. And R_4 is generated when t_1 fires once, t_2 fires n times, and finally t_3 fires once.

We can also describe the transition of a Petri net from one state to the next as a vector equation, which will turn out to be convenient.

Definition 2.16 Let $N = (P, T, A, w, \vec{x}_0)$ be a Petri net with $P = \{p_1, \ldots, p_n\}$ and $T = \{t_1, \ldots, t_m\}$. A *firing vector* $\vec{u} = [0, \ldots, 0, 1, 0, \ldots, 0]$ is a vector of length m, where entry j, $1 \geq j \geq m$, corresponds to transition t_j. All entries of the vector are 0 but one, where it has a value of 1. If entry j is 1, transition t_j fires.

The *incidence matrix* $\mathcal{A}$ is an $m \times n$ matrix whose (j, i) entry is

$$a_{j,i} = w(t_j, p_i) - w(p_i, t_j)$$

The (j, i) entry in the incidence matrix contains the information about the net effect of firing transition t_j on place p_i. Hence, the incidence matrix contains all the information about the effect of firing any transition. For instance, the incidence matrix of Example 2.11 is

$$\mathcal{A} = \begin{bmatrix} -1 & 1 & 1 & 0 \\ 0 & 0 & 0 & 1 \\ -1 & 0 & -1 & -1 \end{bmatrix}$$

which perfectly represents the structure of the Petri net without the marking. Together with the marking or state vector and the firing vector, the incidence matrix can be used to compute the evolution of a Petri net. In fact, we can write a state equation in the following way:

$$\vec{x}' = \vec{x} + \vec{u}\mathcal{A} \tag{2.1}$$

Example 2.12 We model the transitions of Example 2.11 by means of state equations. The initial state is $\vec{x}_0 = [2, 0, 0, 1]$, and the firing of transition t_1 is represented by the firing vector $\vec{u}_1 = [1, 0, 0]$. Hence,

$$\vec{x}_1 = \vec{x}_0 + \vec{u}_1 \mathcal{A}$$

$$= [2, 0, 0, 1] + [1, 0, 0] \begin{bmatrix} -1 & 1 & 1 & 0 \\ 0 & 0 & 0 & 1 \\ -1 & 0 & -1 & -1 \end{bmatrix}$$

$$= [2, 0, 0, 1] + [-1 + 0 + 0, 1 + 0 + 0, 1 + 0 + 0, 0 + 0 + 0]$$

$$= [2, 0, 0, 1] + [-1, 1, 1, 0] = [1, 1, 1, 1]$$

which is the same result we got in Figure 2-16 by examination of the diagram. We continue with firing transition t_3.

$$\vec{x}_2 = \vec{x}_1 + \vec{u}_2 \mathcal{A}$$

$$= [1,1,1,1] + [0,0,1] \begin{bmatrix} -1 & 1 & 1 & 0 \\ 0 & 0 & 0 & 1 \\ -1 & 0 & -1 & -1 \end{bmatrix}$$

$$= [1,1,1,1] + [-1,0,-1,-1] = [0,1,0,0]$$

We have to be careful, however, to evaluate the state equation for transitions that are not enabled. If we do this anyway, we get negative numbers in the state vector, which do not represent valid Petri net markings. If we avoid this, we can generalize the state equation.

Let $N = (P, T, A, w, \vec{x}_0)$ be a Petri net and $T' = \langle t_1, t_2, \ldots, t_i, \ldots, t_n \rangle, t_i \in T$, a sequence of n transitions. Further, let $\vec{u}_t$ be the transition vector for t. Then the state after the firing of all transitions in T' is

$$\vec{x}_n = \vec{x}_0 + \left(\sum_{t \in T'} \vec{u}_t \right) \mathcal{A} \qquad (2.2)$$

provided that, for all $t_i \in T'$, t_i is enabled in state

$$\vec{x}_{i-1} = \vec{x}_0 + \left(\sum_{t \in \langle t_1, \ldots, t_{i-1} \rangle} \vec{u}_t \right) \mathcal{A}$$

If all transitions are enabled when they are supposed to fire, the order of firing is not relevant. It is relevant only if some firing orders lead to the situation that not all consecutive transitions can fire.

2.3.1 Inputs and Outputs

So far we have not been concerned with the inputs and outputs of a Petri net. In fact, many useful modeling tasks can be accomplished with isolated nets without interfaces to the outside. However, frequently it is useful to explicitly model inputs and outputs. Fortunately we can do this without extensions of the formalism that we have introduced. We only need to agree on a convention.

There exist two principal equivalent ways to model I/O, as illustrated by Figures 2-18 and 2-19. First, we can model inputs with places with no input arcs and outputs with places with no output arcs. We assume that the environment puts tokens into the input places and consumes them from the output places. Second, we can use transitions. The environment would select the input transitions to fire and would consume the tokens emitted by the output transitions.

In both cases we need some environment process to interact with the Petri net. The connection to and interaction with this process will then define which places or transitions are inputs, outputs, or internal. For instance, place p_1 in Figure 2-15 could be an input if connected properly to the environment.

It can easily be shown that both methods are equivalent by modeling one by the other. We choose to represent I/O with places as used in Figure 2-18 because it seems slightly more intuitive.

FIGURE 2-18

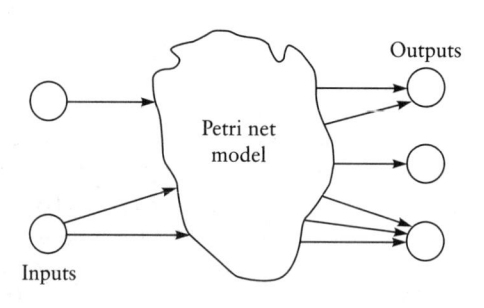

The I/O of a Petri net modeled as places.

FIGURE 2-19

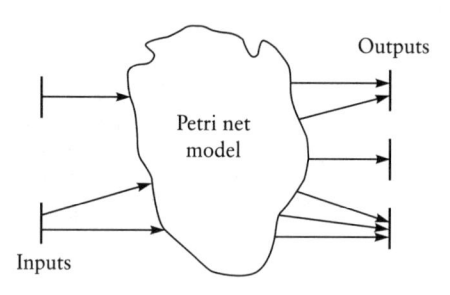

The I/O of a Petri net modeled as transitions.

Example 2.13 We model a simple server that serves arriving customers one by one as illustrated in Figure 2-20. Place p_1 is an input. Whenever a token is put there, we say a customer arrives (Figure 2-20(a)). If the server is idle, represented by a token in p_3, the server accepts the customer by firing transition t_1. That would remove the token in place p_3, putting the server into a "busy" state (Figure 2-20(b)). After the server completes the task, transition t_2 fires, allowing the customer to depart, and the server becomes ready for new requests (Figure 2-20(c)).

2.3.2 Petri Nets and Finite State Machines

As we have mentioned, Petri nets and finite state machines emphasize different aspects. Finite state machines represent states and state transitions explicitly, while Petri nets put the focus on concurrency and parallel behavior. Therefore, the composition of subsystems into a complete system is much easier with Petri nets. Given two nets with input and output places, we can easily connect them together by merging the appropriate output places of one net with the corresponding input places of the other net.

FIGURE 2-20

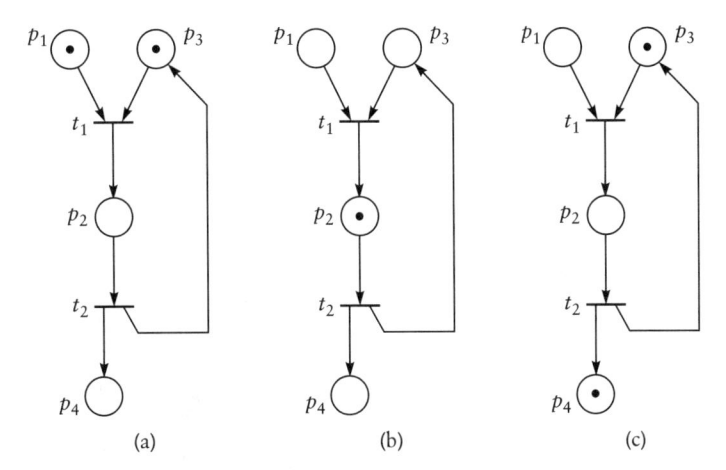

A server model where customers arrive at input p_1 (a), are processed by the server (b), and depart at output p_4 (c).

FIGURE 2-21

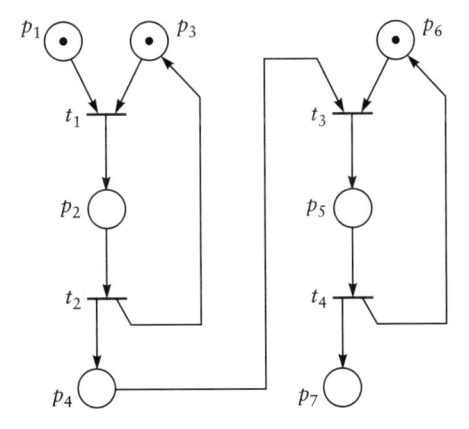

Two server models are sequentially connected together.

Example 2.14 Consider again Example 2.13 and assume we want every customer to go through two servers. We can simply copy the net of Figure 2-20, rename places and transitions, and connect the output of the first to the other net's input. The result is shown in Figure 2-21. We call this sequential composition, where the output of one component feeds the input of another component.

FIGURE 2-22

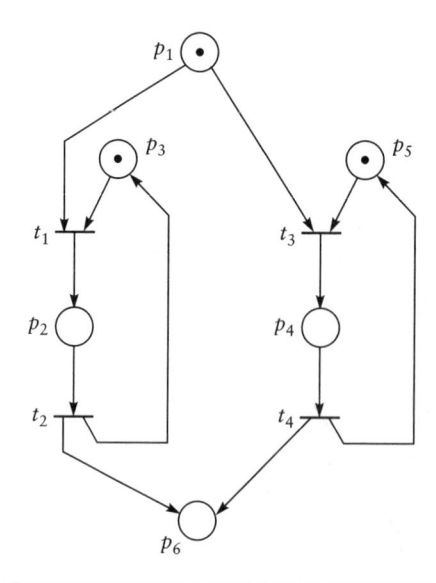

Two servers connected in parallel.

An alternative is parallel composition, as illustrated in Figure 2-22. A common input is feeding both servers, and both servers contribute to the same output. The two servers can operate in parallel, and a customer can be served by the server that is idle.

The composition of state machines is more complicated because the new state space is a product of the two component state spaces with a potential explosion of the number of states.

But apart from issues of convenience and efficiency, we can ask if both techniques are equally powerful. The answer is "no" because Petri nets are more general than finite state machines in the sense that every finite state machine can be modeled by a Petri net but not vice versa. To see this, we construct a Petri net for an arbitrary finite state machine.

Let $M = (\Sigma, \Delta, X, x_0, g, f)$, with mutually exclusive sets Σ and Δ, be a deterministic finite state machine. Then we define the Petri net $N = (P, T, A, \vec{y}_0)$ with

$$P = X \cup \Sigma \cup \Delta$$

$$T = \{t_{x,a} | x \in X, a \in \Sigma\}$$

$$A = I(t_{x,a}) \cup O(t_{x,a}) \quad \forall t_{x,a} \in T$$

$$I(t_{x,a}) = \{x, a\}$$

$$O(t_{x,a}) = \{g(x, a), f(x, a)\}$$

$$\vec{y}_0 = [1, 0, \ldots, 0]$$

2.3 Petri Nets

Each element of Σ becomes an input place of N, each element of Δ becomes an output place, and each state in X becomes an internal place in the Petri net. For each (state, input) pair of M, we have a transition in N, so that each possible state transition of the finite state machine has a corresponding transition in the Petri net. The arcs are defined by specifying all input and output places of all transitions. Input places of a transition $t_{q,a}$ are both the internal place q and the input place a. Its output places correspond to the next state of the state machine ($g(x,a)$) and to the output symbol $f(x,a)$.

The initial marking is 0 for all places but the one corresponding to x_0 because we expect the environment to provide tokens through the input places.

If sets Σ and Δ are not mutually exclusive, they have to be transformed such that they are, for instance, by simple renaming. This is necessary to make the input and output places distinct and to avoid having the Petri net consume the very same tokens it has produced as outputs before.

Example 2.15 This procedure is best illustrated by way of an example. Consider a state machine $M = (\Sigma, \Delta, X, x_0, g, f)$ with

$$\Sigma = \{0, 1\}$$
$$\Delta = \{0, 1\}$$
$$X = \{x_0, x_1\}$$
$$g(x_0, 0) = x_0, \quad g(x_0, 1) = x_1$$
$$g(x_1, 0) = x_1, \quad g(x_1, 1) = x_1$$
$$f(x_0, 0) = 0, \quad f(x_0, 1) = 1$$
$$f(x_1, 0) = 1, \quad f(x_1, 1) = 0$$

M produces the two's complement of a binary number (Figure 2-23). The corresponding Petri net is $N = (P, T, A, w, \vec{y}_0)$ with

$$P = \{x_0, x_1, 0_I, 1_I, 0_O, 1_O\}$$
$$\vec{y}_0 = [0, 0, 0, 0, 0, 0, 0, 0]$$

FIGURE 2-23

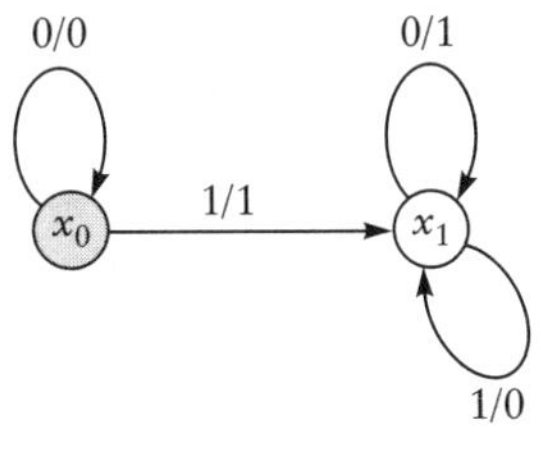

This finite state machine computes the two's complement of a binary number. Both input and output are represented with the least significant bit first.

FIGURE 2-24

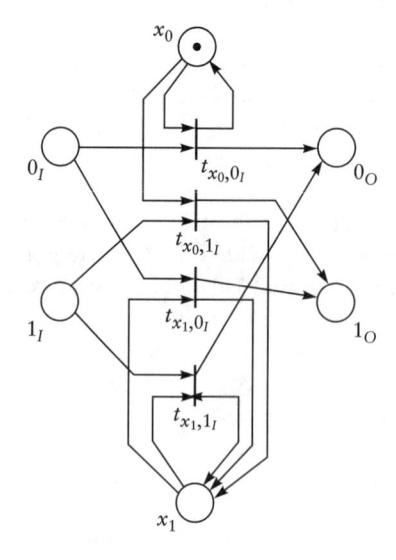

This Petri net is derived from the finite state machine in Figure 2-23 and also computes the two's complement of a binary number.

$$T = \{t_{x_0,0_I}, t_{x_0,1_I}, t_{x_1,0_I}, t_{x_1,1_I}\}$$
$$A = I(t) \cup O(t) \quad \forall t \in T$$

$$I(t_{x_0,0_I}) = \{x_0, 0_I\}, \quad I(t_{x_0,1_I}) = \{x_0, 1_I\}$$
$$I(t_{x_1,0_I}) = \{x_1, 0_I\}, \quad I(t_{x_1,1_I}) = \{x_1, 1_I\}$$
$$O(t_{x_0,0_I}) = \{x_0, 0_O\}, \quad O(t_{x_0,1_I}) = \{x_1, 1_O\}$$
$$O(t_{x_1,0_I}) = \{x_1, 1_O\}, \quad O(t_{x_1,1_I}) = \{x_1, 0_O\}$$

It is illustrated in Figure 2-24.

2.3.3 Modeling Templates

To illustrate the usage of Petri nets, we will briefly introduce a few practical templates for standard modeling tasks.

Sequence and Concurrency

In Example 2.14 we saw how to combine two nets in a sequential or parallel manner. In principle, all transitions can act in parallel if not restricted by the availability of tokens. Consequently, if we just place two nets beside each other, we have a parallel composition (Figure 2-25(b)), and if we connect them such that one produces a token that is required by the other net, we realize a sequential composition (Figure 2-25(a)).

FIGURE 2-25

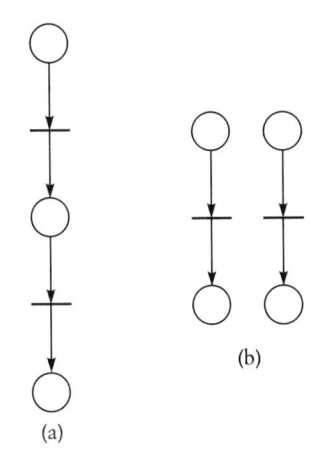

(a) A sequential and (b) a parallel composition of two Petri nets.

Fork and Join

In software development, control over parallel and sequential execution is often exercised via fork and join operations. A fork operation doubles the control flow, resulting in two parallel threads. The inverse operation is the join, which merges two control threads into one. This can be modeled with the Petri nets shown in Figure 2-26(a). With the nets in Figure 2-26(b), which we have used in Example 2.14, we can model the availability of parallel resources. Input tokens can choose one of the branches in the fork and are merged again in the corresponding join.

Conflict

Interacting concurrent activities can come in conflict with each other when they compete for data or resources. Figure 2-27 shows how to model a conflict situation. Transitions t_1 and t_2 are competing for the token in place p. Only one of the two transitions can fire.

Mutual Exclusion

The conflict pattern can be used to model the mutually exclusive access to a shared resource. Mutual exclusion is a general problem that emanates whenever two or more concurrent activities interact with an object that allows only exclusive interaction with one of the activities. Examples are shared resources, such as printers, which should only print one file at a time, and memory locations and files, which should be updated in a disciplined way, implying that arbitrary interleaving of reads and writes from different processes is not allowed. This situation is often cumbersome to analyze because all possible combinations of the states and actions of the participating processes have to be taken into account. For instance, consider two processes A and B (Figure 2-28), that modify a shared variable x. Both simply increment it once. After both processes

FIGURE 2-26

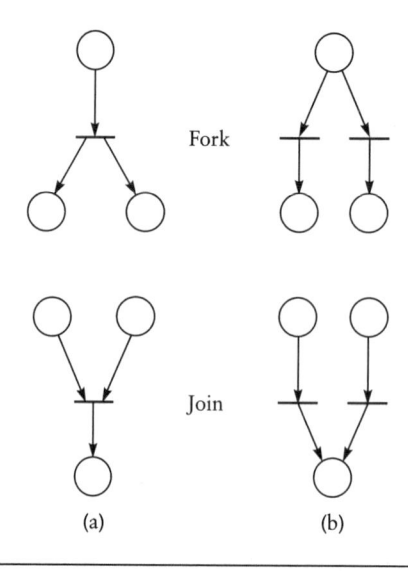

Two alternative versions of fork and join operations. In (a) a control thread is doubled and re-united. In (b) dataflow is doubled and re-united but individual data tokens are not doubled.

FIGURE 2-27

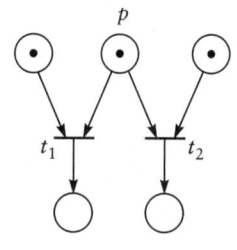

The two transitions t_1 and t_2 are in conflict and only one of them can fire.

have finished, the value of x depends on the order in which the statements in the two processes have been executed. It may be either 1 or 2, as shown in Figure 2-29.

The solution is to make the sequence of `read(x); x <- x + 1; write(x);` indivisible. Once a process has started this sequence, another process cannot start the same sequence before the first process has finished it. We call this sequence of statements a *critical region* to which processes have mutually exclusive access rights.

Figure 2-30 shows two parallel processes that cannot enter their respective critical regions at the same time. If one of the two enters its critical region, the token from place m is removed until the process leaves the region again. Hence place m is a guard for

FIGURE 2-28

```
read(x);

set x <- x + 1;

write(x);

       Process A
```

```
read(x);

set x <- x + 1;

write(x);

       Process B
```

Two parallel processes modify a common data object.

FIGURE 2-29

```
x <- 0;
A.read(x);
A.set x <- x + 1;
A.write(x);
B.read(x);
B.set x <- x +1;
B.write(x);
x == 2
```

```
x <- 0;
A.read(x);
B.read(x)
A.set x <- x + 1;
A.write(x);
B.set x <- x +1;
B.write(x);
x == 1
```

Different execution orders of statements in concurrent processes lead to different results.

the critical region. This scheme can easily be generalized to more than two processes. We can also allow more than one process to enter the critical region with some upper limit on the number of processes. If the initial marking puts n tokens into place m, up to n processes are allowed to be in the critical region at the same time.

Producer/Consumer Relationship

Another common pattern of interaction between parallel processes is the producer-consumer relationship. One process (the producer) produces data that is used by another process (the consumer). The processes operate independently from one another. Hence, the producer does not wait until the consumer reads the data. Consequently a buffer is needed between the two processes. If the producer emits data continuously at a higher rate than the consumer reads it, any finite buffer will eventually overflow. Since both the production and the consumption rate are usually not constant, it is an interesting question of how big a buffer must be to avoid a loss of data.

In Figure 2-31 the place B represents the buffer. Whenever there is a token in place p_1, the producer is ready to initiate the data generation process, which ends by

FIGURE 2-30

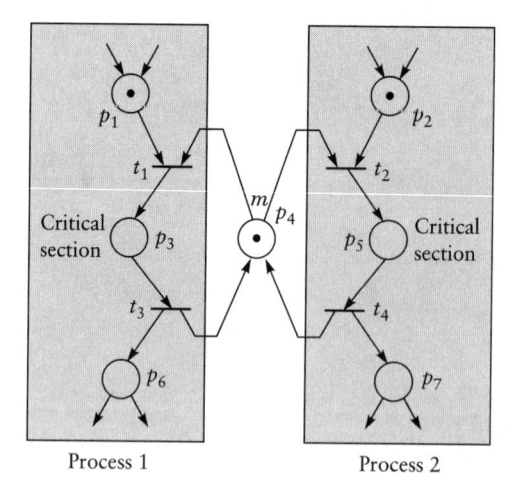

The two processes cannot enter their critical region at the same time. A present token in place m means the critical region is empty and can be entered.

FIGURE 2-31

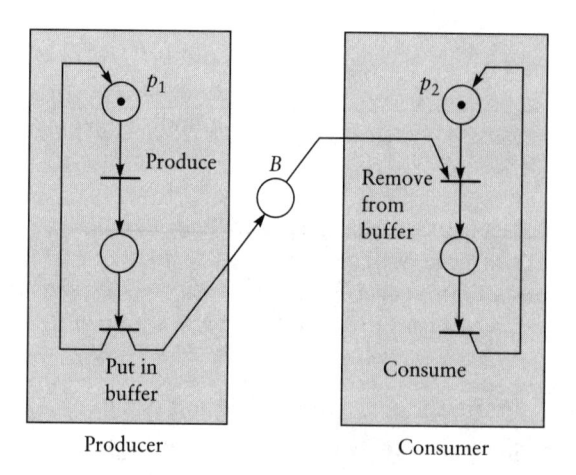

A producer process emits data into a buffer, and a consumer process reads its input from the buffer.

putting a token into B. Whenever place p_2 has a token, the consumer is ready to read data.

The Petri net in Figure 2-31 has an unbounded buffer B, which can be used to analyze the question of the smallest required buffer size. If we want to examine the system behavior under the assumption of a fixed buffer of size n, we can use the Petri

FIGURE 2-32

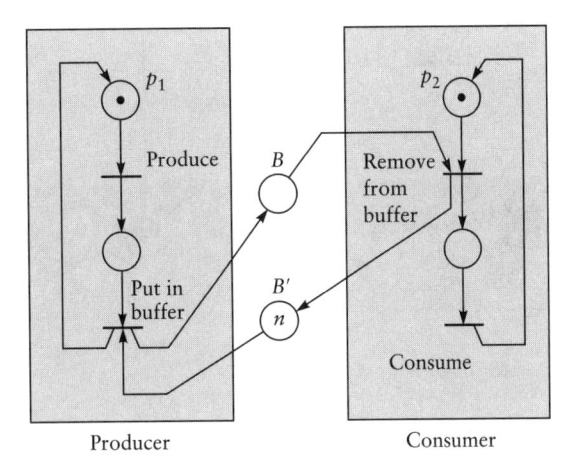

A producer process connected to a consumer via a buffer of size n.

net of Figure 2-32. The place B' is initialized with n tokens. When B' becomes empty, the producer cannot place new data into the buffer.

Dining Philosophers

A famous problem involving parallel processes is Dijkstra's table of dining philosophers (Di-jkstra 1968). A number of philosophers sit around a table with Chinese food. In front of each philosopher is a plate, and between every two plates there is one chopstick. All the philosophers do is alternate eating and contemplating. When a philosopher wants to eat, he needs two chopsticks. They all follow the convention that they first pick up the chopstick on their left and, if successful, then take the chopstick on their right and start eating. When they are finished, they put the chopsticks back. In addition to being unhygienic, this convention faces the potential problem of deadlock. If all the philosophers start eating at the same time, they will pick up the chopstick on their left and then wait until the other chopstick becomes available. But this never happens because each philosopher's right neighbor is also waiting for a chopstick while holding one.

Figure 2-33 shows a Petri net that implements a different convention. $C_1, \ldots, C_6$ represent the chopsticks. If there is a token in that place, then the corresponding chopstick is free. $M_1, \ldots, M_6$ represent the philosophers in meditation, while $E_1, \ldots, E_6$ designate the philosophers eating. A meditating philosopher can only switch to the eating state if both chopsticks are free. When a philosopher stops eating, he will switch to the meditating state and release both chopsticks.

2.3.4 Analysis Methods for Petri Nets

We will now examine some analysis problems that can be addressed and often solved very neatly for Petri nets. At the same time they resemble questions and situations that

FIGURE 2-33

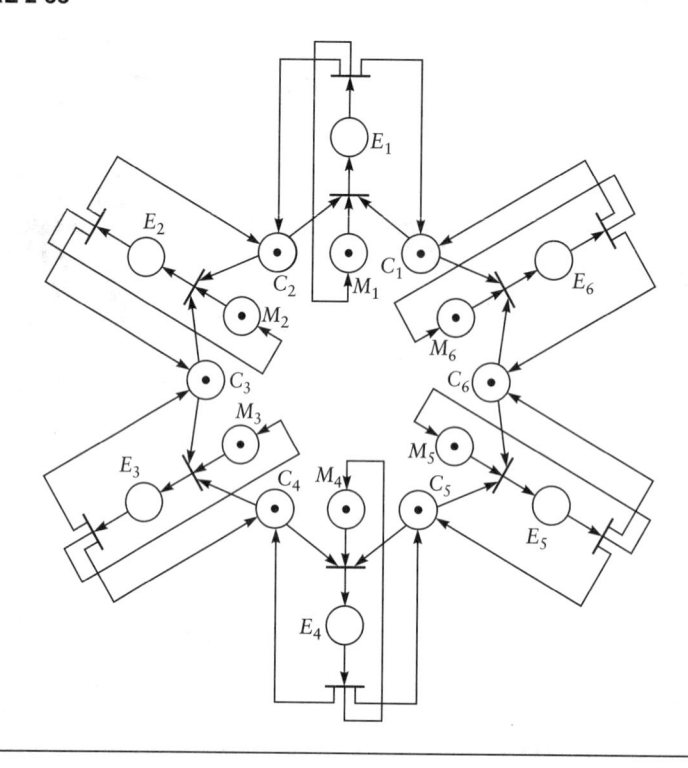

Dining philosophers without the danger of deadlock.

you frequently encounter with system models in general. Hence, you can commonly translate the problem into a Petri net and solve it there.

Typical general questions include the following: Will the system ever enter a particular state? Will the system always be able to avoid a particular dangerous state? Will the system always be able to react to inputs? Will the system always eventually reach some some desired state? In the following section, we will precisely define questions like these in the context of Petri nets, and then discuss if and how they can be answered. But we should keep in mind that these are indeed very general problems that can be formulated in most modeling frameworks and that are of utmost importance to many real systems. As an example, consider a robot controller. It is obviously desirable to show that it will always be able to react to new inputs and not end up in a nonresponsive deadlock situation. Similarly we could identify undesirable states of the controller that might be hazardous for the mechanics of the robot. Or we would like to be able to guarantee that, given a particular task like moving an object from A to B, the robot will eventually end up in a state representing the successful completion of the task.

Problems of these kinds can often easily be formulated for Petri nets. We can classify the problems in the following ways.

Boundedness

The places are often used to represent a buffer, a memory, or a queue. For instance, if the tokens represent arriving and emitted messages, the number of tokens in a particular place may represent the number of messages that have arrived but not yet been processed. Consequently they have to be stored in a buffer. If this is the case, we would like to guarantee that the number of tokens in that place never exceeds a particular limit. In general, when tokens accumulate in a place without an upper limit, we often have a defective situation that requires special attention.

Definition 2.17 A place $p \in P$ in a Petri net $N = (P, T, A, w, \vec{x}_0)$ is *k-bounded* or *k-safe* if for all

$$\vec{y} \in R(\vec{x}_0) : y(p) \leq k$$

The Petri net is called *k-bounded* or *k-safe* if all places $p \in P$ are *k*-bounded.

If a place or a net is *k*-bounded, we often say simply it is "bounded" if we are not interested in the particular value of *k*.

Boundedness is an important and usually desirable property. If a net is not bounded, we have to fear a pathological situation. The net of Example 2.11 is not bounded because p_4 is not bounded.

Note that this definition is only useful for nets in isolation with no inputs because an input place may always be unbounded, depending of course on the environment's behavior. To handle Petri nets with inputs, we assume that the environment has delivered all tokens in the beginning such that all input tokens are part of the initial state. With this assumption, the server model in Example 2.13 is also bounded for any given initial state. Place p_1 is bounded because it can never have more tokens than in the initial state, and p_4 is also bounded because it can never have more tokens than p_1.

Conservation

Tokens may represent quite different things, such as data, requests, customers, services, resources, and so on. In some cases it is interesting to ask if the number of tokens for all reachable states is constant. For instance, consider Figure 2-30. Assume the tokens in place m represent available printers and the tokens in p_1 and p_2 represent incoming printing requests. The initial marking for m would represent the total number of printers, say, n. A reasonable requirement for our model would then be that the number of printers is always a constant n, whatever happens in the system.

Definition 2.18 A Petri net $N = (P, T, A, w, \vec{x}_0)$ is *strictly conservative* if for all $\vec{y} \in R(\vec{x}_0)$,

$$\sum_{p \in P} y(p) = \sum_{p \in P} x_0(p)$$

A Petri net is strictly conservative if and only if all transitions conserve the number of tokens; no transition can produce more or less tokens than it consumes.

Strict conservation is very strong and severely restricts the expressiveness of Petri nets. The Petri net in Figure 2-30 is not strictly conservative because transition t_1 consumes two tokens and produces only one. However, transition t_3 reverses this effect and produces one token more than it consumes. Even though the net is not strictly conservative, the number of printer resources are conserved in some way. To capture this weaker conservation property, we provide a weight for each place, which determines how much a place contributes to the overall conservation property.

Definition 2.19 A Petri net $N = (P, T, A, w, \vec{x}_0)$ with n places is *conservative* with respect to a *weighting vector* $\vec{\gamma} = [\gamma_1, \gamma_2, \ldots, \gamma_n]$, $\gamma_i \in \mathbb{N}$, if

$$\sum_{i=1}^{n} \gamma_i x(p) = \text{constant for all } p \in P \text{ and } \vec{x} \in R(\vec{x}_0)$$

The Petri net is conservative if it is conservative with respect to a weighting vector that has a positive nonzero weight for all places.

For example, the Petri net in Figure 2-30 is conservative with respect to $\vec{\gamma} = [0, 0, 1, 1, 1, 0, 0]$. This captures the intuition that the input and output places p_1, p_2, p_6, and p_7 are not relevant for the number of printers, but the places p_3, p_4, and p_5 are. If we consider the places p_1, p_2, p_6, and p_7 in Figure 2-30 as inputs and outputs, the Petri net is conservative because it is conservative with respect to the vector $\vec{\gamma} = [1, 1, 2, 1, 2, 1, 1]$. Places p_3 and p_5 get a weight of 2 because their respective input transitions t_1 and t_2 decrease the number of tokens by one.

The Petri net of Example 2.11 is not conservative because it is not conservative with respect to any weighting vector with all nonzero weights.

A Petri net is strictly conservative if it is conservative with respect to the weighting vector $\vec{\gamma} = [1, 1, \ldots, 1]$.

Every Petri net is conservative with respect to the null vector $\vec{\gamma} = [0, 0, \ldots, 0]$.

Liveness

Concurrent processes that interact with each other can come into a situation of deadlock—all processes waiting for some action to be taken by another process. We have seen this for the table of dining philosophers, but the phenomenon is widespread and potentially very harmful. Moreover it is difficult to predict and to avoid because it requires the holistic analysis of all involved processes. Deadlock cannot be predicted by analyzing each process separately.

Consider the situation in Figure 2-34. Two processes need two resources, designated by the places m_1 and m_2, to complete their respective tasks. However, if they allocate these resources in different orders they may deadlock. Process A first allocates resource m_1, then m_2. After it completes its task, it releases both resources when transition t_3 fires. Process B allocates m_2 first and then m_1. If Process A allocates m_1 and then B allocates m_2, both processors end up waiting until the other resource becomes available, which never happens. Clearly, this is an undesirable situation, and we would like to be able to guarantee that a particular Petri net is deadlock free.

FIGURE 2-34

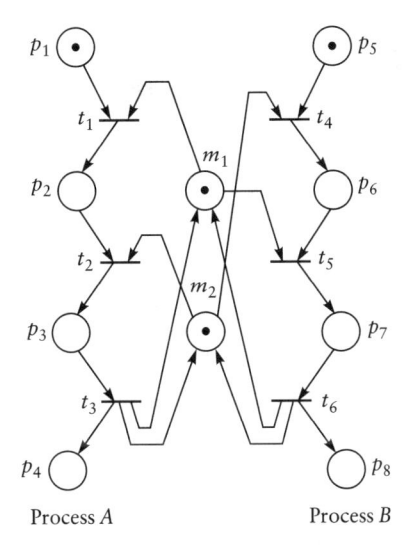

Two processes can deadlock when they allocate the same resources in different sequences.

Deadlock is defined for transitions. A transition is deadlocked in a given state $\vec{x}$ if it cannot fire in all states reachable from $\vec{x}$. A transition is live if it is not deadlocked. We distinguish between several levels of liveness.

Definition 2.20 Let $N = (P, T, A, w, \vec{x}_0)$ be a Petri net and $\vec{x}$ a state reachable from $\vec{x}_0$.

L0-live: A transition t is *live at level 0* in state $\vec{x}$ if it cannot fire in any state reachable from $\vec{x}$; that is, it is deadlocked.

L1-live: A transition t is *live at level 1* in state $\vec{x}$ if it is potentially firable; that is, if there exists a $\vec{y} \in R(\vec{x})$ such that t is enabled in $\vec{y}$.

L2-live: A transition t is *live at level 2* in state $\vec{x}$ if for every integer n there exists a firing sequence in which t occurs at least n times.

L3-live: A transition t is *live at level 3* in state $\vec{x}$ if there is an infinite firing sequence in which t occurs infinitely often.

L4-live: A transition t is *live at level 4* in state $\vec{x}$ if it is L1-live in every $\vec{y} \in R(\vec{x})$.

A Petri net is live at level i if every transition is live at level i.

Example 2.16 Consider the Petri net in Figure 2-35. Transition t_0 is dead because it can never fire. Transition t_1 can only fire once because it removes the token in place p_0, which cannot be replaced after that. Hence, t_1 is L1-live. Transition t_2 is L2-live because for every given n there exists a transition

FIGURE 2-35

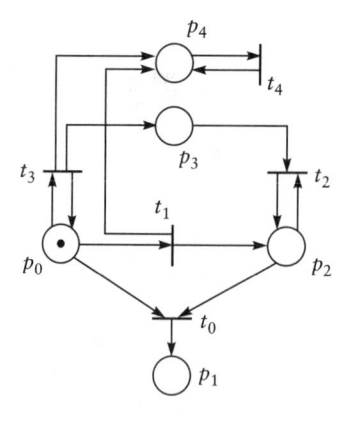

Transition t_i is Li-live.

sequence such that t_2 fires n times, for example, the sequence t_3^n, t_1, t_2^n. Thus, t_2 is L2-live, but it is not L3-live because there is no sequence such that it fires infinitely often. To be able to fire, t_1 must fire first, but then t_2 can only fire a finite number of times.

Transition t_3 is L3-live because it can fire infinitely often. However, it cannot do this for any arbitrary sequence. In particular it cannot do this if t_1 fires. Thus, it is not L4-live. Only transition t_4 is L4-live because it is L1-live in every state no matter what transition sequence we choose.

Reachability and Coverability

Many other problems, such as deadlock avoidance, can be viewed as a special case of the state reachability problem, which asks if a particular state can be reached from a given state. In other words, given a state $\vec{x}$, is a state $\vec{y} \in R(\vec{x})$?

If we can answer this question for arbitrary $\vec{x}$ and $\vec{y}$, we can also answer all the questions about boundedness, conservation, deadlock, and liveness. Regrettably, we do not have an efficient method to answer this question for Petri nets with infinite state space. Fortunately, we can still solve some of the problems by analyzing a weaker property than reachability: coverability.

Definition 2.21 Let $N = (P, T, A, w, \vec{x}_0)$ be a Petri net and let $\vec{x}$ and $\vec{y}$ be arbitrary states. State $\vec{x}$ *covers* state $\vec{y}$ if at least all transitions are enabled in $\vec{x}$ that are enabled in $\vec{y}$. This is equivalent to the condition that

$$x(p) \geq y(p) \forall p \in P$$

State $\vec{x}$ *strictly covers* state $\vec{y}$ if $\vec{x}$ covers $\vec{y}$ and, in addition,

$$\exists p \in P : x(p) > y(p)$$

Coverability now asks the question if, from a given state $\vec{x}_0$, we can reach a state that covers another state $\vec{y}$. The property is weaker than reachability because we are not asking for one specific state but only if we can reach any state out of a potentially infinite set of states that all share a common property. Answering this question will allow us to solve all the problems for which it does not matter which particular state in this set we can reach.

Definition 2.22 Let $N = (P, T, A, w, \vec{x}_0)$ be a Petri net and let $\vec{x}$ be an arbitrary state in $R(\vec{x}_0)$. A state $\vec{y}$ is *coverable by* $\vec{x}$ iff there exists a state $\vec{x}' \in R(\vec{x})$ such that $x'(p) \geq y(p)$ for all $p \in P$.

Example 2.17 Consider again the Petri net in Figure 2-35. Observe that state $\vec{y}_0 = [y_0(p_0), y_0(p_1), y_0(p_2), y_0(p_3), y_0(p_4)] = [1, 0, 1, 0, 0]$ is the state with the minimal number of tokens required to enable transition t_0. The same holds for state $\vec{y}_1 = [0, 0, 1, 1, 0]$ with respect to transition t_2.

Observe further that state $\vec{y}_1$ can be covered by the initial state $\vec{x}_0$, but state $\vec{y}_0$ cannot. Consequently, transition t_0 is dead while transition t_2 is at least L1-live.

Hence, the question whether a particular transition is dead or L1-live can be formulated as a coverability problem.

Persistence

When analyzing concurrent activities, the understanding of the impact of one activity on another activity is of particular interest. Can the occurrence of an interrupt cause some tasks to miss their deadlines?

The property of persistence captures some aspects of the relationship between transitions.

Definition 2.23 Two transitions are *persistent with respect to each other* if when both are enabled, the firing of one does not disable the other.

A Petri net is *persistent* if any two transitions are persistent with respect to each other.

The Petri net in Figure 2-35 is not persistent because, if both transitions t_1 and t_3 are enabled, the firing of t_3 will disable t_1. But t_2 and t_3 are persistent with respect to each other.

2.3.5 The Coverability Tree

The coverability tree is an efficient technique for addressing some of the problems discussed in the previous section. Intuitively, the coverability tree is a tree with the arcs representing transitions and the nodes denoting sets of states that can be covered by a sequence of transitions. The root of the tree is the initial state of the net.

FIGURE 2-36

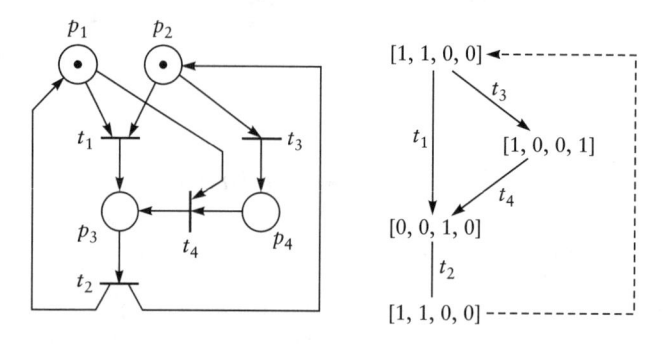

The coverability tree for a Petri net with a finite state space.

FIGURE 2-37

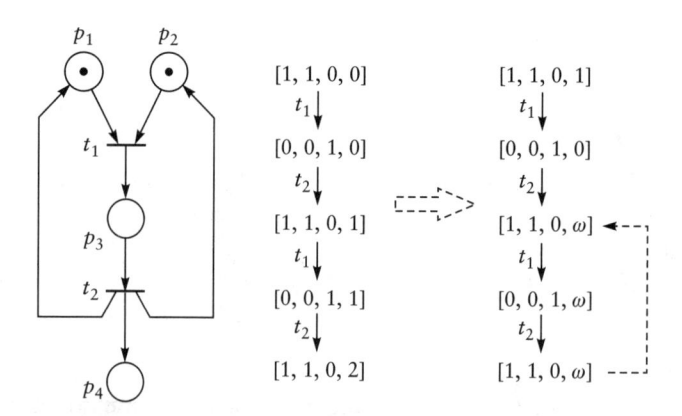

The coverability tree for a Petri net with an infinite state space uses the ω symbol to represent an unbounded number of tokens.

For a Petri net with a finite state space, that is, a bounded Petri net, the coverability tree simply contains all the reachable states. An example of this special case is shown in Figure 2-36. The initial state is $[1, 1, 0, 0]$, where transitions t_1 and t_3 are enabled. Firing t_1 leads to state $[0, 0, 1, 0]$, which disables t_1 and t_3 and enables t_2. Alternatively, t_3 can fire followed by t_4, which lands us in the same state $[0, 1, 0, 0]$. From there, firing t_4 returns the net to the initial state, which is already in the coverability tree. This is indicated by the dotted line. In this case the coverability tree contains all states reachable from the initial state. Furthermore, it gives the sequence of transitions leading to each reachable state, where a branching in the tree designates alternative transitions.

This procedure works well for bounded Petri nets. However, for unbounded nets, like the one shown in Figure 2-37, it would lead to an infinite coverability tree. In order to avoid this, we represent two states, where one covers the other, by the same node in the tree. Consider the Petri net in Figure 2-37. When transition t_4 fires, we return "almost" to the initial state, with the only difference being that place p_4 has one token more. Since state $[1, 1, 0, 1]$ covers the initial state $[1, 1, 0, 0]$, all transitions enabled in state $[1, 1, 0, 0]$ are also enabled in state $[1, 1, 0, 1]$. Consequently, we can again fire the same transition sequence that led us from $[1, 1, 0, 0]$ to $[1, 1, 0, 1]$. Applying the same transition sequence again and again will only increase the tokens in place p_4 but will not affect the other places. This means that we can accumulate an arbitrary number of tokens in place p_4, with the other places unaffected. This is expressed by the notation $[1, 1, 0, \omega]$, where the ω represents an unbounded place.

The key observation is that if, during the construction of the coverability tree, we find a state $\vec{y}$ that covers another state $\vec{x}$ already contained in the tree, we can set an ω for all places, where the marking of $\vec{y}$ is strictly greater than the marking of $\vec{x}$. We have detected a quasi-cyclic transition sequence, which we denote by a dashed line from $\vec{y}$ to $\vec{x}$.

Essentially we have gained a final representation for infinite coverability trees. Unfortunately, we lose some information on the way. If we have an ω on a place, we only know that that place is unbounded, but we do not know how many tokens it can potentially contain. For the net of Figure 2-37, place p_4 may contain an arbitrary number $n \in \mathbb{N}$ of tokens. But assume that the arc (t_2, p_4) had a weight 2; that is, for each firing of transition t_2 two tokens would be added, and p_4 could contain only an even number of tokens. However, this situation would be designated by the same symbol $[1, 1, 0, \omega]$ in the coverability tree. Such a distinction cannot be made by the coverability tree.

We proceed by first formally defining the coverability tree in order to introduce then an algorithm for the construction of a coverability tree.

Definition 2.24 Let $N = (P, T, A, w, \vec{x}_0)$ be a Petri net.

> A coverability tree is a tree where the arcs denote transitions $t \in T$ and the nodes represent ω-enhanced states of the Petri net.
>
> An ω-enhanced state $\vec{x}$ is a marking vector $\vec{x} \in (\mathbb{N} \times \omega)^{|P|}$ that represents all the states in the Petri net, which can be derived by replacing ω by a natural number $n \in \mathbb{N}$. In the following we sometimes use the term "ω-enhanced state $\vec{x}$" when we actually mean the set of states represented by $\vec{x}$.
>
> We extend the notion of coverage to ω-enhanced states. $\vec{x}$ covers $\vec{y}$ if for all $p \in P : x(p) = \omega$ or $(y(p) \neq \omega$ and $x(p) \geq y(p))$.
>
> The root node of the tree is $\vec{x}_0$.
>
> A *terminal node* is an ω-enhanced state in which no transition is enabled.
>
> A *duplicate node* is an ω-enhanced state that already exists somewhere else in the coverability tree.
>
> An arc t connects two nodes $\vec{x}$ and $\vec{y}$ in the tree iff the firing of t in state $\vec{x}$ leads to state $\vec{y}$.

The following algorithm constructs the coverability tree for the Petri net $N = (P, T, A, w, \vec{x}_0)$.

94 chapter two *Behavior and Concurrency*

Algorithm 2.2 *Coverability Tree*

Step 1. $L := \{\vec{x}_0\}$ (L is the list of open nodes);

Step 2. Take one node from L, named $\vec{x}$, and remove it from L;

Step 2.1. **if** $(G(\vec{x}, t) = \vec{x} \;\; \forall t \in T)$

then $\vec{x}$ is a terminal node|; **goto** Step 3;

Step 2.2. **forall** $(\vec{x}' \in G(\vec{x}, t), t \in T, \vec{x} \neq \vec{x}')$

Step 2.2.1. **do if** $(x(p) = \omega)$ then $x'(p) := \omega$;

Step 2.2.2. **if** (there is a node $\vec{y}$ already in the tree, such that $\vec{x}$ covers $\vec{y}$, and there is a path from $\vec{y}$ to $\vec{x}$)

then $x(p) := \omega$ for all p for which $x(p) > y(p)$;

Step 2.2.3. **if** ($\vec{x}$ is not a duplicate node) then $L := L \cup \{\vec{x}\}$;

Step 3. **if** (L is not empty)

then goto Step 2

Example 2.18 As an example we construct the coverability tree of the Petri net from Example 2.11, which is again drawn in Figure 2-38. In the initial state only transition t_1 is enabled, which leads to state $[1, 1, 1, 1]$. Now, all three transitions are enabled. We follow the leftmost branch first. Firing t_1 leads to $[0, 2, 2, 1]$, where only transition t_2 is enabled. Firing it we come to state $[0, 2, 2, 2]$, which strictly covers the previous state. Consequently we have to denote the new node in the coverability tree by $[0, 2, 2, \omega]$. Here again only t_2 is enabled, which when fired leads to the same ω-enhanced state. Thus we have a duplicate node and stop with this branch.

Now we follow the next branch from state $[1, 1, 1, 1]$, which is introduced by firing t_2. The result is state $[1, 1, 1, 2]$, which strictly covers its predecessor state. Hence, we denote the new node by $[1, 1, 1, \omega]$. Here, all three

FIGURE 2-38

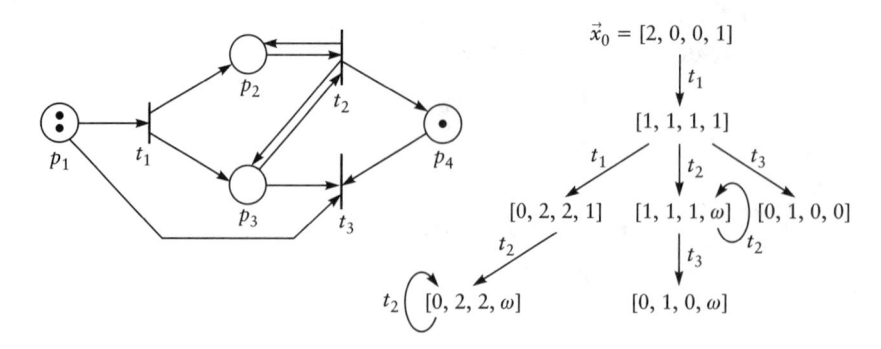

The coverability tree for the Petri net of Example 2.11.

transitions are enabled, but firing two of them leads to duplicate nodes; t_1 leads to $[0, 2, 2, \omega]$, and t_2 leads to $[1, 1, 1, \omega]$. Only t_3 leads to a new state, namely $[0, 1, 0, \omega]$, which is a terminal node.

Finally, we follow the third branch from state $[1, 1, 1, 1]$. Firing transition t_3 leads to state $[0, 1, 0, 0]$. Since this is a terminal node, we have completed the coverability tree.

The coverability tree is a very useful tool for the analysis of Petri nets. One important reason for its usefulness is that it is finite. This is not obvious, but fortunately there exists the following theorem.

Theorem 2.6   The coverability tree of a Petri net is finite.

The proof is lengthy and is not repeated here. The interested reader is referred to Peterson (1981).

We now show how some of the problems outlined above can be solved by means of the coverability tree.

Safeness and Boundedness

It is obvious that a Petri net is k-bounded iff the ω symbol never appears in its coverability tree. If it appears in a place, we can easily find a transition cycle that, when traversed sufficiently often, leads to more than k tokens in that place, for any given k.

In Figure 2-38 the ω appears in place p_4. If, starting from state $\vec{x}_0$, we fire the transition sequence $\langle t_1, t_1, t_2^k \rangle$, that is, 2 times transition t_1 and k times transition t_2, we end up with at least k tokens in p_4. The coverability tree does not tell us how many tokens accumulate in p_4 because how many tokens are added for each firing of t_2 is not visible. All we know is that it is at least one.

Conservation

We have defined the conservation of tokens with respect to a weight vector (Definition 2.19):

$$\sum_{i=1}^{n} \gamma_i x(p) = \text{constant for all } p \in P \text{ and } \vec{x} \in R(\vec{x}_0)$$

We can use the coverability tree to check if a Petri net is conservative with respect to a given weighting vector $\vec{\gamma}$.

First note that the ω symbol represents an infinite number of different values. As a consequence, the weighting factor for every place with an ω symbol must be 0 because if it is not we get infinitely many different results for the weighted sum. To avoid this, the number of tokens in any place with an ω must be irrelevant; that is, the weighting factor for that place must be 0.

If this condition is fulfilled, we can check for conservation by evaluating the weighted sum for each node in the coverability tree. If the result is the same for every node, the net is conservative with respect to the weighting vector.

Conversely, we can check if there exists a weighting vector for which a Petri net is conservative, and if yes, we can calculate the vector. First we set $\gamma_i = 0$ for each unbounded place p_i. Then, supposing we have b bounded places and r nodes in the coverability tree, we set up r equations with $b + 1$ unknown variables:

$$\sum_{i=1}^{r} \gamma_i x(p_i) = C$$

These equations constitute a set of linear equations for which we have several standard methods. We can effectively check if there exists a solution and, if so, we can calculate the solution to find the weighting vector and the constant C.

Example 2.19 To determine if the Petri net of Figure 2-38 is conservative with respect to any weighting vector, we set up the equations

$$2\gamma_1 + 0\gamma_2 + 0\gamma_3 = C$$
$$1\gamma_1 + 1\gamma_2 + 1\gamma_3 = C$$
$$0\gamma_1 + 2\gamma_2 + 2\gamma_3 = C$$
$$0\gamma_1 + 1\gamma_2 + 0\gamma_3 = C$$

for the nodes in the coverability tree with a distinct number of tokens in places p_1, p_2 and p_3.

It turns out that the only nonnegative solution is $\vec{\gamma} = [0, 0, 0, 0]$ and $C = 0$. Thus, the Petri net is not conservative with respect to any nontrivial weighting vector.

Coverability

Obviously, we can solve the coverability problem for a Petri net by simple inspection of the coverability tree. However, finding the shortest transition sequence that leads to a covering state is more involved, but efficient algorithms do exist.

Limitations

Unfortunately, the coverability tree does not contain enough information to determine the set of reachable states. Consequently, many problems related to liveness, deadlock avoidance, and possible transition sequences cannot be tackled in general with the coverability tree.

Figures 2-39 through 2-41 illustrate the lack of information in the coverability tree. Figure 2-39 shows two distinct Petri nets with the same coverability tree. The Petri net in Figure 2-39(a) can have any integer number of tokens in place p_3, while the Petri net in Figure 2-39(b) can only have an even number of tokens in that place.

Figures 2-40 and 2-41 show two similar Petri nets with the same coverability tree. However, the net in Figure 2-40 can easily deadlock (e.g., after the transition sequence t_1, t_2, t_3), but the net in Figure 2-41 cannot.

Even though the coverability tree cannot always solve reachability problems, sometimes it can. For instance, if a state without an ω symbol is in the coverability tree,

FIGURE 2-39

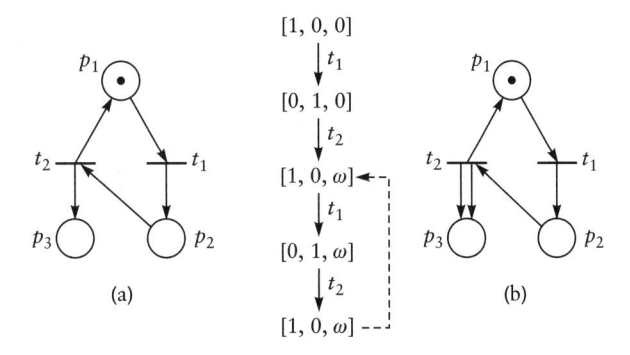

Two similar but distinct Petri nets (a) and (b) have the same coverability tree.

FIGURE 2-40

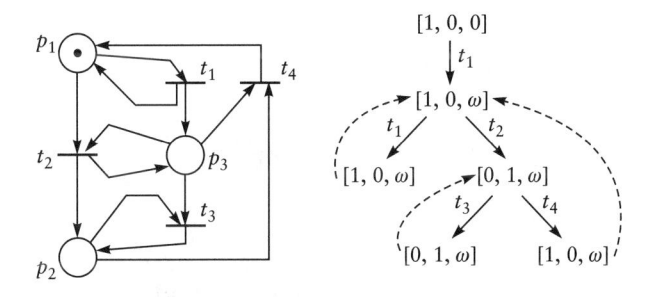

A Petri net that can potentially deadlock.

FIGURE 2-41

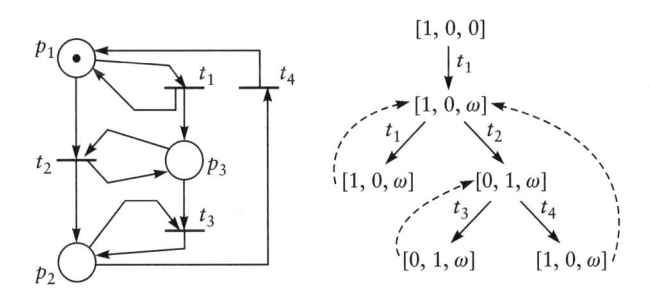

A Petri net that is similar to the net of Figure 2-40 and that has the same coverability tree. However, this net cannot deadlock.

FIGURE 2-42

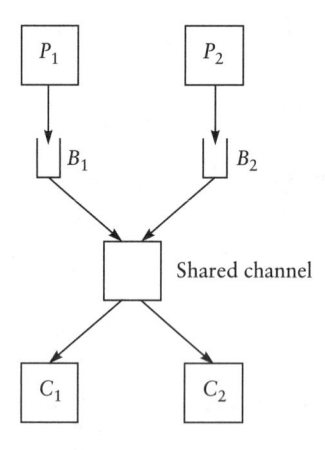

Two producer/consumer pairs communicating over a shared channel (Peterson 1981).

it is also reachable. And, conversely, if a state cannot be covered, it cannot be reached. So the coverability tree may help in many practical situations to solve problems of reachability, liveness, and deadlock.

2.4 Extended and Restricted Petri Nets

Petri nets represent a compromise between expressive power and decision power. Higher expressive power means that more situations can be formulated and expressed. Higher decision power means that more problems can be decided effectively. For instance, Turing machines have very high expressive power but very low decision power; that is, many problems cannot be efficiently decided for Turing machines. It has been shown that Petri nets are not as expressive as Turing machines because they cannot test for the emptiness of a place. Certain situations cannot be represented by a Petri net due to this limitation (Kosaraju 1973; Agerwala and Flynn 1973). Figure 2-42 shows two producer/consumer pairs that communicate over a shared channel. The communication is buffered, represented by buffers B_1 and B_2. P_1 sends tokens to C_1 and P_2 sends tokens to C_2, but the P_1/C_1 path has higher priority. Thus, communication from B_2 to C_2 takes place only when buffer B_1 is empty. However, in Petri nets we cannot test for the emptiness of a place, and hence we cannot model this rather realistic example.

It turns out that extensions of Petri nets that allow a so-called *zero test* are equivalent to Turing machines.

One extension introduces inhibitor arcs. An *inhibitor arc* inhibits a transition when there is a token in its input place. Thus, the transition rule is changed as follows. A transition is enabled when there are tokens in all of its normal inputs and there are zero tokens on all of its inhibitor inputs. Other essentially equivalent extensions introduce constraints (Patil 1970), priorities (Hack 1972), exclusive-or transitions, and switches (Peterson 1981, Chapter 7).

2.4 Extended and Restricted Petri Nets

Most variants for modeling time are true extensions of Petri nets in the sense that they allow zero testing. Time may be associated with places, with transitions, or with arcs.

If time is associated with places, a duration d is assigned to a place; that is, when a token arrives at the place at time τ, it is not available for out transitions before time $\tau + d$ (Sifakis 1977).

If a duration d is associated with a transition, it means that if a transition is enabled at time τ, the token is not available at the output places at time $\tau + d$ (Ramchandani 1973). In another variant a transition is associated with two time values, τ_1 and τ_2. A transition can only fire if it has been enabled for at least time τ_1 and it must fire before τ_2 has elapsed (Merlin 1974).

Time can also be associated with arcs. Two different approaches have been proposed (Palanque and Bastide 1996). *Arc-timed Petri nets* allow us to model varying durations of transitions depending on the output places. *Temporal arcs* constitute another interesting variant inspired by the watchdog concept. An interval $[d_1, d_2]$ is associated with an arc connecting place p and transition t. If a token arrives at place p at time τ, it contributes to enabling the transition τ during the interval $[\tau + d_1, \tau + d_2]$. If the transition has not fired and consumed the token until $\tau + d_2$ (e.g., because another input token was not available), it cannot consume the token from place p.

All these variations on timed Petri nets are equivalent to each other in their expressive power, but they emphasize different aspects and tend to promote different modeling styles. For instance, timed places associate a duration with a system state, while transitions correspond to instantaneous events. In contrast, timed transitions associate durations with activities, while the places represent state variables that change as a result of the activities.

Another direction for extension is to introduce types of tokens. In *colored Petri nets* each token has a value or color. Places are also typed and restrict the types of tokens they can hold. When a transition fires, tokens from the input places are transformed into other types of tokens on the output places. Colored Petri nets are widely used today for a great variety of applications (Kristensen et al. 1998; Jensen 1997a).

Extending Petri nets makes them more expressive but at a price—many interesting decision problems become undecidable. Therefore, researchers have searched for more restricted models with a higher decision power but that are still expressive enough to model important classes of problems.

Two interesting Petri net variants are state machines and marked graphs. *State machines* are Petri nets where each transition has exactly one input and one output. As a consequence, state machines are strictly conservative in preserving the initial number of tokens. Thus, the state space is finite, and state machines are equivalent to the finite state machines we discussed earlier (Section 2.2). *Marked graphs* are Petri nets where each place has exactly one input and one output. State machines and marked graphs (see Figure 2-43) are dual to each other also in terms of the kind of problems they can model. Recall Figure 2.26, which depicted two types of fork-join operations. Figure 2-26(a) can be modeled as a marked graph, and Figure 2-26(b) as a state machine. State machines can easily represent conflicts among several output transitions of a place, but they cannot create and destroy the tokens necessary to model concurrency of activities and synchronization. Marked graphs, in contrast, can easily model concurrency and synchronization, but they have difficulties with conflict and data-dependent decisions.

FIGURE 2-43

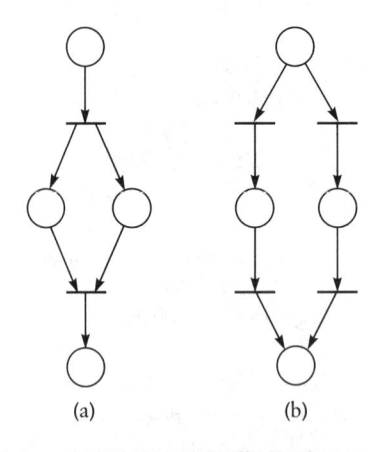

(a)   (b)

Two restricted Petri nets: (a) a marked graph and (b) a state machine.

Hack (1972) proposed a restricted Petri net that combined both these aspects in a controlled way. A *free-choice Petri net* is a Petri net where, for each transition t and each of its input places $p \in I(t)$, either the transition is the only output of p ($O(p) = \{t\}$) or the place is the only input of t ($I(t) = \{p\}$). For instance, both nets in Figure 2-43 are free-choice Petri nets. Although free-choice Petri nets are fairly expressive, they are restricted Petri nets, and Hack (1972) could give necessary and sufficient conditions for a marked free-choice Petri net to be live and safe.

2.5 Further Reading

The standard textbook on the theory of automata and languages was written by Hopcroft and Ullman (1979). But there are numerous other good books on automata, languages, and the countless applications of these theories. For instance, the classic book on the decomposition of automata is Hartmanis and Stearns (1966). A detailed introduction of their application to parsers and compilers is presented by Aho, Sethi, and Ullman (1988). A well-written and comprehensive introduction into finite automata can be found in the *Handbook of Theoretical Computer Science* (Perrin 1994), which is an excellent source for systematic, introductory material on various subjects.

A short and comprehensive introduction to Petri nets is given by Cassandras (1993). Excellent and complete accounts are presented by Reisig (1985) and Peterson (1981). Murata (1989) provides good introduction and summary of application areas. Jensen (1997a, 1997b) develops colored Petri nets and their applications in two books. A practical introduction is given by Kristensen et al. (1998). Palanque and Bastide (1996) provides a short and good overview of timed Petri nets. A selection of recent applications of a number of Petri net variants in the area of hardware design and hardware/software co-design is collected by Yakovlev et al. (2000). The University of Aarhus in Denmark maintains a Web page about Petri nets with an extensive bibliography, links to Petri-net-based tools, and applications (*www.daimi.au.dk/PetriNets*).

2.6 Exercises

2.1 Consider the English language alphabet $\Sigma = \{a, \ldots, z\}$. Construct a finite state machine that recognizes the words "man" and "woman".

2.2 Let $\Sigma = \{0, 1\}$ and consider a 1-bit parity check code for 8-bit words. Construct a deterministic finite state machine that recognizes only those 9-bit words with an even number of 1s.

2.3 Let $\Sigma = \{a, b\}$.

   a. Construct a deterministic finite state machine that recognizes all strings with a p number of as, where p is a prime number less than 10.

   b. Construct a deterministic finite state machine that recognizes all strings with a p number of as and bs together, where p is a prime number less than 10 and there is always one more b than a.

   c. Construct a deterministic finite state machine that recognizes all strings with an initial sequence "*aab*" and a final sequence "*baa*".

2.4 Let $\Sigma = \{a, b, c\}$.

   a. Construct a nondeterministic finite state machine that recognizes all strings with a p number of as where p is a prime number less than 10 and where each a is followed either by one b or two cs.

   b. Construct a nondeterministic finite state machine that recognizes all strings with a p number of as and bs together, where p is a prime number less than 10 and there is always one more or less b than a.

   c. Construct a nondeterministic finite state machine that recognizes all strings with an initial sequence "*aab*" or "*abb*" and a final sequence "*baa*".

2.5 For each nondeterministic state machine of Exercise 2.4, construct a deterministic finite state machine following the method described in Example 2.4.

2.6 Construct a finite state machine corresponding to the circuit in Figure 2-44, such that the final states correspond to a logic 1 at output O. Assume there is sufficient time for the output to become stable before the next input arrives.

2.7 Construct a finite state machine corresponding to the neural network in Figure 2-45, such that the final states correspond to a value of 1 at output O. Each neuron has excitory (open circles) and inhibitory (shaded circles) synapses. It produces a 1 at its output if the number of excitory synapses with a 1 as input exceeds the number of inhibitory synapses with a 1 as input by at least the threshold value of the neuron. All initial values on the internal connections are 0. Assume there is sufficient time for the output to become stable before the next input arrives.

102   chapter two *Behavior and Concurrency*

FIGURE 2-44

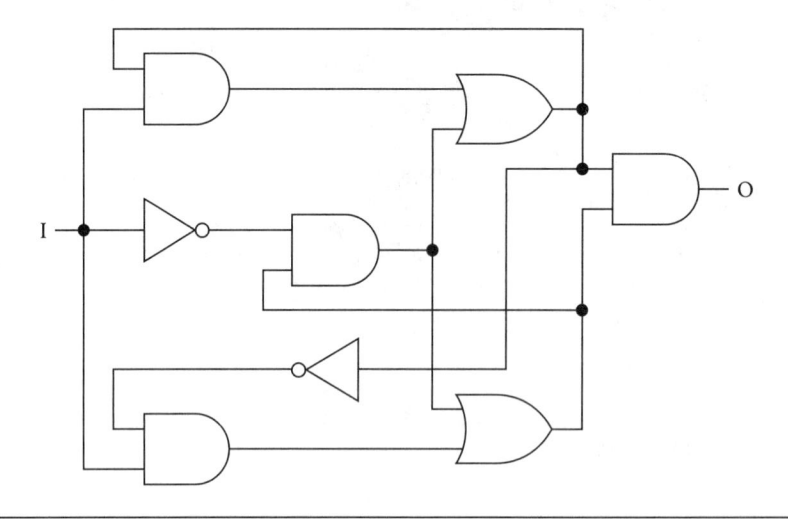

A logic circuit.

FIGURE 2-45

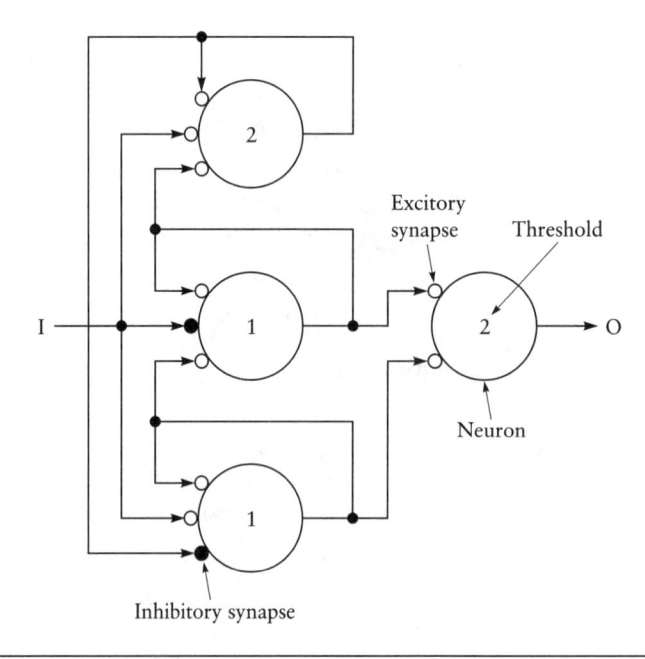

A neural network.

FIGURE 2-46

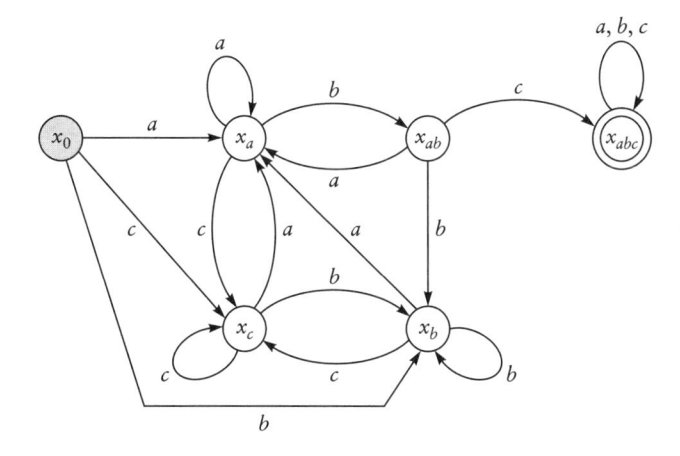

A state machine that recognizes strings with the sequence "abc".

2.8 The state machine depicted in Figure 2-46 recognizes all strings that contain the sequence "*abc*". Construct an equivalent state machine with fewer states by applying Algorithm 2.1.

2.9 Consider the alphabet $\Sigma = \{a, b, c\}$. Find the following:

   a. a regular expression for the language where each string contains at least one *b*

   b. a regular expression for the language where each string contains exactly two *b*s

   c. a regular expression for the language where each string contains at least two *b*s

   d. a regular expression for the language where each string contains, at least once, the sequence "*aba*" or "*bab*"

   e. a regular expression for the language where each string contains the sequence "*aa*" at most once, and the sequence "*bb*" at most once

   f. a regular expression for the language where the sequence "*aa*" appears before any sequence "*bb*" in each string

   g. a regular expression for the language where no string contains the sequence "*aba*"

2.10 Describe in English the languages defined by the following regular expressions:

   a. $(bb + a)^*(aa + b)^*$

   b. $(b + ab + aab)^*(\epsilon + a + aa)$

   c. $(aa + bb + (ab + ba)(aa + bb)^*(ab + ba))^*$

FIGURE 2-47

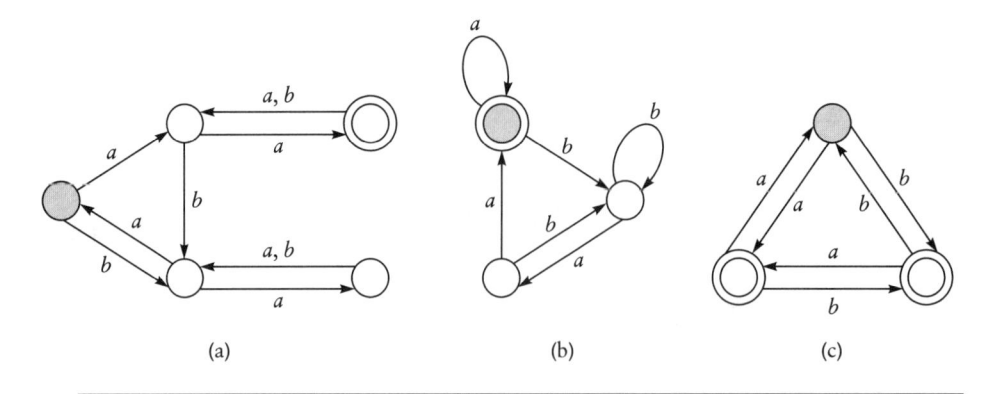

State transition diagrams.

2.11 Construct finite state machines equivalent to the following regular expressions:

a. $(abc) + (abb) + (bcc)$

b. $(ab)^*c$

c. $a(ba)^*a^*$

d. $ab + (b + aa)b^*a$

e. $ab(((ab)^* + aaa)^* + b)^*a$

f. $((a+b)(a+b))^* + ((a+b)(a+b)(a+b))^*$

2.12 Find regular expressions for the languages recognized by the finite state machines in Figure 2-47.

2.13 Consider an elevator that serves four stories, 1...4. Model the elevator controller as a deterministic finite state machine. The possible input events are "move to story n." The possible output events are "move upwards," "move downwards," and "stop." Input events are only accepted after the processing of the previous input has been completed.

2.14 Consider the Petri net defined by

$$P = \{p_1, p_2, p_3\}$$
$$T = \{t_1, t_2, t_3\}$$
$$A = \{(p_1, t_1), (p_1, t_3), (p_2, t_1), (p_2, t_2), (p_3, t_3),$$
$$(t_1, p_2), (t_1, p_3), (t_2, p_3), (t_3, p_1), (t_3, p_2)\}$$

with all weights being one except $w(p_1, t_1) = 2$.

a. Draw the corresponding graph.

b. Let $\vec{x}_0 = [1, 0, 1]$ be the initial state. Show that, in any subsequent operation of the Petri net, transition t_1 can never be enabled.

c. Let $\vec{x}_0 = [2, 1, 1]$ be another initial state. Show that, in any subsequent operation of the Petri net, either a deadlock occurs or a return to $\vec{x}_0$ results.

2.15 Petri nets are not well suited to modeling dataflow transformations and arithmetic operations because data and operations on data cannot be conveniently represented. However, it can be done.

a. Model an adder for arbitrary positive integers.

b. Model a subtractor for arbitrary integers.

c. Model a multiplier for arbitrary positive integers.

2.16 The operation of a Petri net is not coordinated, and any transition can fire whenever it is enabled independently of the status of other parts of the net. However, different parts of the net can be synchronized as is desired.

a. Develop a scheme to model a clock signal that triggers the execution of subnets.

b. Construct a Petri net that models the expression $y = (x_1 + x_2) \times (x_3 + x_4)$. Use two adders and one multiplier from the previous exercise and connect them such that the two adders operate in parallel and, when they have finished, they feed their results to the multiplier. Make sure that the model works as a pipeline, that is, while the multiplier is active, the adders accept the next input numbers and work in parallel.

c. The adders are much faster than the multiplier. Eliminate one adder without sacrificing performance by using the same adder for both additions.

2.17 Petri nets are much better at modeling dataflow at a more abstract level without considering the values of the operations. One technique is to use the symbolic expressions for the names of places. For example, the place representing the result of $(x_1 + x_2)$ would be named "$(x_1 + x_2)$". The flow of tokens represents the activation of operations.

a. Using this technique and abstraction levels, construct Petri nets corresponding to Exercises 2.16(b) and 2.16(c).

b. Construct a Petri net model with three pipeline stages for the expression

$$y = (x_1 \times x_2) + ((x_3 + c_1) \times (x_4 + c_2)) + (x_5 \times x_6)$$

where $x_1, \ldots, x_6$ are inputs and c_1 and c_2 are constants.

2.18 Consider the Petri net defined by

$$P = \{p_1, p_2, p_3, p_4\}$$
$$T = \{t_1, t_2, t_3, t_4\}$$

$$A = \{(p_1, t_1), (p_1, t_2), (p_2, t_3), (p_2, t_4), (p_3, t_2), (p_4, t_3),$$
$$(t_1, p_1), (t_1, p_2), (t_1, p_3), (t_2, p_4), (t_3, p_1), (t_4, p_3)\}$$

with all weights being 1 except $w(t_1, p_2) = 2$. Let the initial state $\vec{x}_0 = [1, 1, 0, 2]$.

a. Let the Petri net fire twice such that then all transitions are dead.

b. Show that it is not possible to apply the infinite firing sequence $\langle t_3, t_1, t_3, t_1, \ldots \rangle$.

c. Find the state $\vec{x}$ resulting from $\langle t_1, t_2, t_3, t_3, t_3 \rangle$.

d. Construct the coverability tree.

2.19 Consider the Petri net defined by

$$P = \{p_1, p_2, p_3, p_4\}$$
$$T = \{t_1, t_2, t_3\}$$
$$A = \{(p_1, t_1), (p_2, t_1), (p_3, t_2), (p_4, t_3),$$
$$(t_1, p_3), (t_2, p_1), (t_2, p_2), (t_2, p_4), (t_3, p_2)\}$$

with all weights being 1. Let the initial state $\vec{x}_0 = [1, 1, 0, 0]$. Construct the coverability tree and use it to show that the state $\vec{x} = [0, 0, 0, 0]$ cannot be reached.

2.20 Petri nets without closed loops are called *acyclic*. For acyclic Petri nets, a necessary and sufficient condition for state $\vec{x}$ to be reachable from state $\vec{x}_0$ is that there exists a nonnegative solution $\vec{z}$ to the equation

$$\vec{x} = \vec{x}_0 + \vec{z}\mathcal{A}$$

where $\mathcal{A}$ is the incidence matrix of the Petri net.

a. Use this result to show that $\vec{x} = [0, 0, 1, 2]$ is reachable from $\vec{x}_0 = [3, 1, 0, 0]$ for the Petri net defined by

$$P = \{p_1, p_2, p_3, p_4\}$$
$$T = \{t_1, t_2\}$$
$$A = \{(p_1, t_1), (p_2, t_2), (p_3, t_2),$$
$$(t_1, p_2), (t_1, p_3), (t_2, p_4)\}$$

with all weights being 1 except $w(p_2, t_2) = 2$.

b. What is the interpretation of vector $\vec{z}$? Why is this condition not sufficient for Petri nets with cycles?

2.21 Determine if there exists a weight vector with respect to which the Petri net of Figure 2-34 is conservative, and determine such a weight vector.

chapter three
The Untimed Model of Computation

We now introduce the formal framework for representing different MoCs. It is based on processes, events, and signals. Events are the elementary units of information exchange between processes. Processes receive or consume events, and they send or emit events. The medium through which events are communicated from one process to another is called a signal. One and only one process can emit into a signal, but one or several can receive events from a signal. Signals preserve the order in which events enter them. Thus, they are represented as potentially infinite sequences of events.

The activity of processes is divided into evaluation cycles. In each evaluation cycle a process consumes inputs, computes its new internal state, and emits outputs. Processes can be modeled in any formalism as long as they comply with the rules of a particular MoC. We use a notation loosely based on finite state machines with an initial state, an output encoding function, and a next-state function as principal components. This leads directly to the concept of signal partitioning. A process partitions its input and output signals into subsequences corresponding to its evaluation cycles. During each evaluation cycle a process consumes exactly one subsequence of each of its input signals and emits exactly one subsequence of each of its output signals.

Process constructors are parameterizable templates that instantiate processes. We define a number of process constructors: some with no internal state, some with internal state, some with one input and one output, some with several inputs and outputs. The parameters of the process constructors determine the next-state function, the output encoding function, and the partitionings of input and output signals. In addition, we define three operators, called combinators, to compose process networks out of simpler networks and elementary processes. These combinators are parallel composition, sequential composition, and a feedback operator. This allows us to build arbitrary process networks.

We define a MoC to be the set of processes and process networks that can be constructed by a given set of process constructors and combinators. Then we define the untimed MoC by a particular set of process constructors and combinators. It is characterized by the way its processes communicate and synchronize with other processes and, in particular, by the absence of timing information available to and used by processes. Only the order of events and cause and effect of events are relevant. Thus, the untimed MoC operates on the causality abstraction level of time.

The characteristic function and the signature of a process are devices to represent the functional behavior and the interaction pattern of processes. The characteristic function determines the behavior in each evaluation cycle. The process signature describes how a process partitions its input and output signals. The transformations up-rating and down-rating change the signature of a process and, as a consequence, the characteristic function, to keep the overall relation between input and output signals unmodified. For instance, to up-rate a process by a factor 2 means that the process consumes and produces twice as many events in each evaluation cycle and that the characteristic function of the up-rated process accomplishes what two consecutive invocations of the original characteristic function attain. Based on process signatures we investigate the conditions to merge processes. Up-rating sometimes facilitates the merging of processes.

By way of the relationship between the untimed MoC and Petri nets we identify the synchronous dataflow (SDF), a practically important and influential restricted untimed MoC. As an application example we present the scheduling and buffer minimization for single- and multi-processor architectures for SDF graphs.

3.1 The MoC Framework

In Section 2.2 we discussed using state machines to model the input-output behavior of an entity with minimal assumptions about the internals of the entity. State machines are in a way purely sequential, and they have a simple evaluation cycle. They receive an input and then, as reaction to that input, they compute an output that depends on the input and their own internal state. In the following, we will use state machines as the primary concept to model components and "atomic" entities—entities that we treat as indivisible because we either have no knowledge about their internal structure or we do not want to bother.

In Section 2.3 we have introduced the notion of concurrency with Petri nets. Petri nets are ideally suited for a study of concurrency because various issues and problems of concurrency appear in a "pure" form; they are not mixed up with the behavior of individual components or with the values of data and effects of computations on data.

However, we cannot be content with modeling component behavior and concurrency issues separately. Eventually we must be able to study their combined effects. Our objective in this and the following chapters is to investigate three different modes of interaction between component behavior in a network of concurrently active components. The main distinguishing feature between the three *models of computation* is the abstraction level of time. First we investigate the *untimed model of computation*, which adopts the simplest timing model, corresponding to the Causality abstraction level in the Rugby metamodel. Processes, modeled as state machines, are connected to one another via signals. Signals transport data values from a sending process to a receiving process. The data values do not carry time information, but the signals preserve the order of emission. Values that are emitted first by the sender are received first by the receiving process.

Then, in Chapter 4 we introduce the synchronous model of computation in two variants. Both partition time into time slots or clock cycles. The clock synchronous model assumes that every evaluation of a process takes one cycle. Hence, the reaction of a process to an input becomes effective in the next cycle. In contrast, the perfectly synchronous model assumes that no time advances during the evaluation of a process. Consequently, the results of a computation on input values is already available in the same cycle. This leads to interesting situations in feedback loops.

Finally, in Chapter 5 we deal with the timed model of computation, which assigns a time stamp to each value communicated between two processes. This allows us to model timing-related issues in great detail, but it complicates the model and the task of analysis, simulation, and synthesis significantly.

One effect of increased complexity in system design is heterogeneity in various forms, which necessitates different modeling concepts for different system parts and in different design steps. Consequently, we observe that different models of computation appear together, as illustrated in Figure 3-1. Sometimes one part of a system is modeled as a dataflow graph based on an untimed model, while the system control part is modeled as a discrete event system based on a synchronous or a fully timed model. Hardware systems are frequently specified as clocked synchronous models and later refined into timed models. Sometimes, as in Figure 3-1, we say a design model or part of a model is in a particular *MoC domain* if it is modeled solely based on the principles of that MoC. For instance, the system in Figure 3-1 consists first of two domains, an untimed domain and a synchronous domain. The untimed part is refined into two different domains, untimed and synchronous, while the synchronous part is refined into a timed model.

Due to this entanglement, in practice we cannot be content with a good but isolated understanding of the different models of computation; in addition we have to study the possibilities of the interaction and integration of different models and the transformation of one model into another. This is the objective of Chapter 6.

Before we take up the issues of an untimed model of computation, we introduce some basic notation and definitions to determine our modeling framework (Section 3.2).

FIGURE 3-1

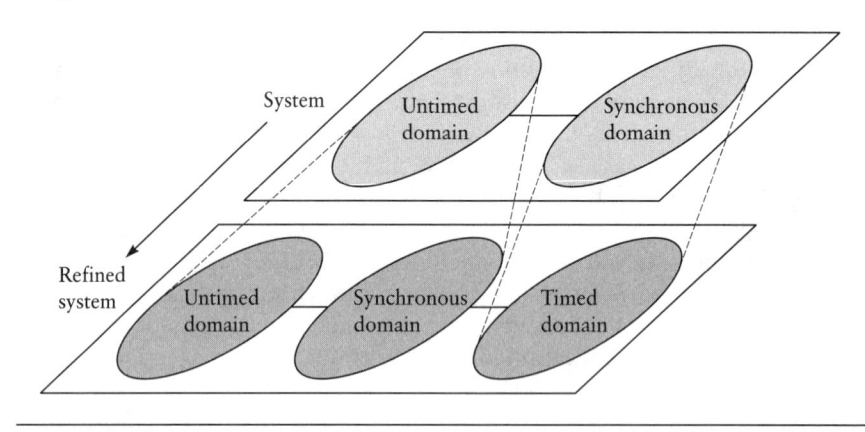

Different MoC domains coexist and interact.

Throughout this chapter we introduce concepts that we also use in later chapters. For instance, here we first describe process constructors, characteristic functions, up-rating and down-rating, and so on, which will be applied to the other computational models as well. We discuss them here in more detail because they are simpler in the untimed context.

In the rest of this book we mostly focus on models of concurrency. Hence, we deal with concurrently active processes and how they communicate with each other. We use a state machine approach to model the individual processes, and we require that they comply with the restrictions of a particular model of concurrency. In practice, different processes may be modeled in different ways with different languages. This is allowed as long as they agree on how to communicate with each other and they adhere to the rules of the overall model of concurrency. If we want to simulate the system, the simulation environment must provide simulation engines for each of the languages involved. For instance, if some processes are modeled in Matlab and others in VHDL, the simulation environment must include a VHDL simulator and a Matlab interpreter. In addition it must facilitate communication between the different simulation engines.

Example 3.1   **Amplifier** We will illustrate the discussion in this chapter with one pedagogic example, which we call the Amplifier. The example is simplified to illustrate the main concepts.

Figure 3-2 shows the example, which consists of three processes forming a feedback loop. A_2 multiplies each element in the input signal by the value of the control signal. A_3 compares the average of the amplified signal to a preferred range. If it is too high, the control signal is lowered; if the amplified signal is too low, the control signal is increased. We will use this example to illustrate how processes and signals of all three computational models are represented in our framework, how processes can be merged, and how processes can migrate from one computational domain to another.

FIGURE 3-2

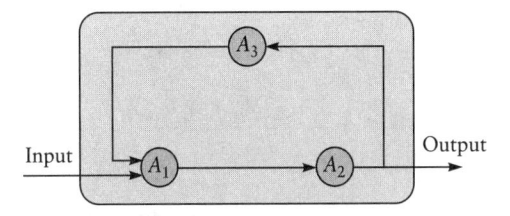

Process A_1 merges the primary input signal with a control signal that contains the amplifying factor. A_2 multiplies the control signal with each element in the primary input signal. A_3 analyzes the amplified signal and adapts the control signal accordingly. All signals carry only integer values.

3.2 Processes and Signals

Processes communicate with each other by writing to and reading from signals. Given is a set of values V, which represents the data communicated over the signals. *Events*, which are the basic elements of signals, are or contain values. We distinguish between three different kinds of events.

Untimed events $\dot{E}$ are just values without further information, $\dot{E} = V$. *Synchronous events* $\bar{E}$ include a pseudovalue ⊔ (pronounced "absent") in addition to the normal values; hence $\bar{E} = V \cup \{⊔\}$. *Timed events* $\hat{E}$ are identical to synchronous events, $\hat{E} = \bar{E}$. However, since it is often useful to distinguish them, we use different symbols. Intuitively, timed events occur at a much finer granularity than synchronous events, and they would usually represent physical time units such as a nanosecond. In contrast, synchronous events represent abstract time slots or clock cycles. This model of events and time can only accommodate discrete time models. Continuous time would require a different representation of time and events.

An alternative for modeling timed events is to annotate each event with a time tag and avoid the explicit modeling of absent events. This approach, as taken by Lee and Sangiovanni-Vincentelli (1998), very elegantly covers a variety of different models such as untimed, discrete time, and continuous time models. We discuss some of its differences to the approach based on absent events after the treatment of the timed MoC in Section 5.2.3.

We use the symbols $\dot{e}, \bar{e}$, and $\hat{e}$ to denote individual untimed, synchronous, and timed events, respectively. We use $E = \dot{E} \cup \bar{E} \cup \hat{E}$ and $e \in E$ to denote any kind of event.

Signals are sequences of events. Sequences are ordered, and we use subscripts as in e_i to denote the ith event in a signal. For example, a signal may be written as $\langle e_0, e_1, e_2 \rangle$. In general, signals can be finite or infinite sequences of events, and S is the set of all signals. We also distinguish between three kinds of signals. $\dot{S}, \bar{S}$ and $\hat{S}$ denote the untimed, synchronous, and timed signal sets, respectively, and $\dot{s}, \bar{s}$, and $\hat{s}$ designate individual untimed, synchronous, and timed signals (Figure 3-3).

FIGURE 3-3

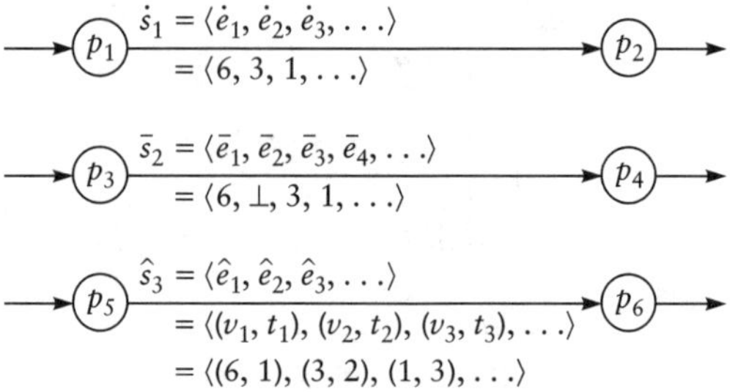

$$\dot{s}_1 = \langle \dot{e}_1, \dot{e}_2, \dot{e}_3, \ldots \rangle$$
$$= \langle 6, 3, 1, \ldots \rangle$$

$$\bar{s}_2 = \langle \bar{e}_1, \bar{e}_2, \bar{e}_3, \bar{e}_4, \ldots \rangle$$
$$= \langle 6, \bot, 3, 1, \ldots \rangle$$

$$\hat{s}_3 = \langle \hat{e}_1, \hat{e}_2, \hat{e}_3, \ldots \rangle$$
$$= \langle (v_1, t_1), (v_2, t_2), (v_3, t_3), \ldots \rangle$$
$$= \langle (6, 1), (3, 2), (1, 3), \ldots \rangle$$

Processes p_1 and p_2 are connected by the untimed signal $\dot{s}_1$. Processes p_3 and p_4 are connected by the synchronous signal $\bar{s}_2$. Processes p_5 and p_6 are connected by the timed signal $\hat{s}_3$.

$\langle\rangle$ is the empty signal, and $\oplus$ concatenates two signals. Concatenation is associative and has the empty signal as its neutral element: $s_1 \oplus (s_2 \oplus s_3) = (s_1 \oplus s_2) \oplus s_3$, $\langle\rangle \oplus s = s \oplus \langle\rangle = s$. To keep the notation simple we often treat individual events as one-event sequences; for example, we may write $e \oplus s$ to denote $\langle e \rangle \oplus s$.

We use angle brackets, "$\langle$" and "$\rangle$", to denote ordered sets or sequences of events, but also for sequences of signals if we impose an order on a set of signals.

length(s) gives the length of signal s. Infinite signals have infinite length, and length($\langle\rangle$) = 0.

[ ] is an index operation to extract an event on a particular position from a signal. For example, $s[2] = e_2$ if $s = \langle e_1, e_2, e_3 \rangle$.

Function take(n, s) takes the first n elements of signal s:

$$\text{take}(n, s) = \begin{cases} \langle e_0, \ldots, e_{n-1} \rangle & \text{if length}(s) \geq n \\ s & \text{otherwise} \end{cases}$$

Function drop(n, s) drops the first n elements of signal s:

$$\text{drop}(n, s) = \begin{cases} \langle e_n, \ldots, e_{\text{length}(s)-1} \rangle & \text{if length}(s) \geq n \\ \langle\rangle & \text{otherwise} \end{cases}$$

Function head(s) takes the first element of a signal s:

$$\text{head}(s) = \begin{cases} e_0 & \text{if } s \neq \langle\rangle \\ \text{undefined} & \text{otherwise} \end{cases}$$

Function tail(s) drops the first event from a signal s:

$$\text{tail}(s) = \text{drop}(1, s)$$

Processes are defined as functions on signals:

$$p : S \rightarrow S$$

Processes are functions in the sense that for a given input signal we always get the same output signal; that is, $s = s' \Rightarrow p(s) = p(s')$. Note that this still allows processes to have an internal state. Thus, a process does not necessarily react identically to the same event applied at different times. But it will only have the same (possibly infinite) output signal when confronted with identical (possibly infinite) input signals at different times provided it starts from the same initial state.

3.3 Signal Partitioning

A *partition* $\pi(v, s)$ of a signal s defines an ordered set of signals, $\langle r_i \rangle$, that, when concatenated together, form "almost" the original signal s. The function $v : \mathbb{N}_0 \to \mathbb{N}_0$ defines the lengths of all elements in the partition. $v(0) = \text{length}(r_0)$ gives the length of the first element in the partition, $v(1) = \text{length}(r_1)$ gives the length of the second element, and so on.

Example 3.2 Let $s_1 = \langle 1, 2, 3, 4, 5, 6, 7, 8, 9, 10 \rangle$ and $v_1(0) = v_1(1) = 3, v_1(2) = 4$. Then we get the partition $\pi(v_1, s_1) = \langle \langle 1, 2, 3 \rangle, \langle 4, 5, 6 \rangle, \langle 7, 8, 9, 10 \rangle \rangle$.

Let $s_2 = \langle 1, 2, 3, \ldots \rangle$ be the infinite signal with ascending integers. Let $v_2(i) = 2$ for all $i \geq 0$. The resulting partition is infinite: $\pi(v_2, s_2) = \langle \langle 1, 2 \rangle, \langle 3, 4 \rangle, \ldots \rangle$.

To formally define partitions we use the recursive function parts, which constructs this sequence of sequences:

$$\text{parts}(v, i, s) = \begin{cases} \langle \langle \text{take}(v(i), s) \rangle \rangle \oplus \text{parts}(v, i+1, \text{drop}(v(i), s)) & \text{if length}(s) \geq v(i) \\ \langle \rangle & \text{otherwise} \end{cases}$$

for all $s \in S, i \in \mathbb{N}_0$.

For infinite signals, parts constructs an infinite number of subsignals, unless $v(i)$ becomes infinite for some i. For finite signals, parts constructs a finite number of subsignals, and it may drop some elements at the end of s. Therefore it is not always possible to completely reconstruct the original signal from the subsignals defined by $\pi(v, s)$.

Definition 3.1 Let $v : \mathbb{N}_0 \to \mathbb{N}_0$ be a function on natural numbers and $s \in S$ be a signal. The *partition* $\pi(v, s)$ is defined as follows:

$$\pi(v, s) = \text{parts}(v, 0, s) \quad \forall s \in S$$

We usually write $\pi(v, s) = \langle r_i \rangle$ with $i = 0, 1, 2, \ldots$ to designate the individual subsignals of the partitioning with r_i.

The *remainder* $\text{rem}(\pi, v, s)$ of a partitioned signal is defined as

$$s = \left(\bigoplus_{\langle r_i \rangle = \pi(v, s)} r_i \right) \oplus \text{rem}(\pi, v, s), \quad s \in S, i \in \mathbb{N}_0$$

FIGURE 3-4

$$s = \langle e_0, e_1, e_2, e_3, e_4, e_5, \ldots \rangle$$
$$\pi(v, s) = \langle r_i \rangle \text{ for } v(i) = 3 \text{ for all } i$$
$$\langle r_0, r_1, \ldots \rangle = \langle\langle e_0, e_1, e_2 \rangle, \langle e_3, e_4, e_5 \rangle, \ldots \rangle$$

p

$$s' = \langle e'_0, e'_1, e'_2, e'_3, e'_4, e'_5, \ldots \rangle$$
$$\pi(v', s') = \langle r'_i \rangle \text{ for } v'(i) = 2 \text{ for all } i$$
$$\langle r'_0, r'_1, \ldots \rangle = \langle\langle e'_0, e'_1 \rangle, \langle e'_2, e'_3 \rangle\rangle, \ldots$$

The input signal of process p is partitioned into an infinite sequence of subsignals, each of which contains three events, while the output signal is partitioned into subsignals of lengths 2.

FIGURE 3-5

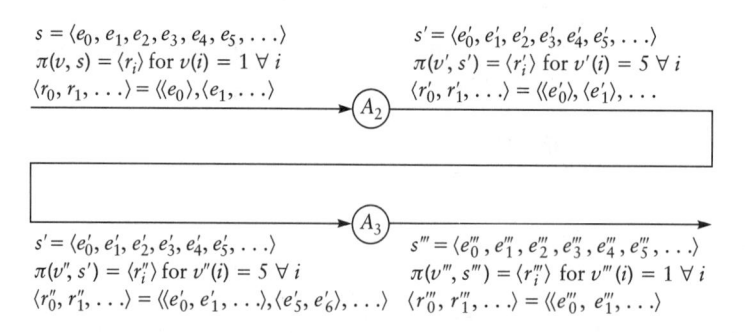

$$s = \langle e_0, e_1, e_2, e_3, e_4, e_5, \ldots \rangle$$
$$\pi(v, s) = \langle r_i \rangle \text{ for } v(i) = 1 \; \forall \; i$$
$$\langle r_0, r_1, \ldots \rangle = \langle\langle e_0 \rangle, \langle e_1 \rangle, \ldots \rangle$$

A_2

$$s' = \langle e'_0, e'_1, e'_2, e'_3, e'_4, e'_5, \ldots \rangle$$
$$\pi(v', s') = \langle r'_i \rangle \text{ for } v'(i) = 5 \; \forall \; i$$
$$\langle r'_0, r'_1, \ldots \rangle = \langle\langle e'_0 \rangle, \langle e'_1 \rangle, \ldots$$

A_3

$$s' = \langle e'_0, e'_1, e'_2, e'_3, e'_4, e'_5, \ldots \rangle$$
$$\pi(v'', s') = \langle r''_i \rangle \text{ for } v''(i) = 5 \; \forall \; i$$
$$\langle r''_0, r''_1, \ldots \rangle = \langle\langle e'_0, e'_1, \ldots \rangle, \langle e'_5, e'_6 \rangle, \ldots \rangle$$

$$s''' = \langle e'''_0, e'''_1, e'''_2, e'''_3, e'''_4, e'''_5, \ldots \rangle$$
$$\pi(v''', s''') = \langle r'''_i \rangle \text{ for } v'''(i) = 1 \; \forall \; i$$
$$\langle r'''_0, r'''_1, \ldots \rangle = \langle\langle e'''_0, e'''_1, \ldots \rangle$$

The signal partitions for the processes A_2 and A_3 of the Amplifier example.

The remainder is the tail of the signal s that has been dropped by the partitioning process.

The function $v(i)$ defines the length of the subsignals r_i. If it is constant for all i, we usually omit the argument and write v. Figure 3-4 illustrates a process with an input signal s and an output signal s'. s is partitioned into subsignals of length 3, and s' into subsignals of length 2.

Example 3.3 **Amplifier** Figure 3-5 shows how the processes A_2 and A_3 partition their input and output signals. Each event that A_2 consumes contains one control value and five values from the input. It multiplies each input value by the control value and outputs them as separate events. Thus, it partitions the output signal into sequences of five events.

A_3 consumes five events every time it is invoked, checks if their average is above or below a certain level, and emits the new control value. Consequently, it partitions its input signal into five-event sequences, and its output signal into single-event sequences.

Note that the signal partitioning is a property of the process and not the signal. The same signal can be partitioned differently by different processes. Although it is not the case in this example, signal s' could be partitioned differently by processes A_2 and A_3.

3.4 Process Constructors

Our aim is to relate functions on events to processes. Therefore we introduce *process constructors*, which essentially are templates for processes. They can be considered as higher-order functions that take functions on events as arguments and return processes. We define only a few basic process constructors that can be used to compose more complex processes and process networks. In particular we define constructors for processes without state and with state, with one or two inputs, with one or two outputs, data sources, and sinks. We distinguish between three families of processes and their constructors, ranging from simple stateless processes to the most general and complex processes.

Table 3-1 enumerates them. A process at a higher level is more general and can be used to express a process at a lower level. Moore- and Mealy-based processes are the most general form. They have arbitrary next-state and output encoding functions, and the output encoding functions depend both on the current state and the current input. As we know from Section 2.2.7, Moore and Mealy machines are equivalent. Therefore we will use either Moore or Mealy machines as the most general case, whatever is more convenient. Usually we denote next-state functions with g and output encoding functions with f.

Because many problems can be easier explained with map-based or scan-based processes, we use them as much as possible to introduce new concepts. Then we

TABLE 3-1: *Process family hierarchy.*

Level	Name	Constructor	Description
1	Map	*mapU*	Processes without internal state.
2	Scan	*scanU*	Processes with an internal state and a next-state function. The state is directly visible at the output.
3	Moore	*mooreU*	Processes with a state. The output is a function of the state, but not directly of the input.
3	Mealy	*mealyU*	Processes with a state. The output is a function of the state and the current input.

move on to develop them also for Moore- or Mealy-based processes to cover the general case.

A *map* process constructor creates a process that takes one input signal, generates one output signal, and has no internal state:

$$mapU(c,f) = p \tag{3.1}$$

where

$$p(s) = s'$$
$$f(a_i) = a'_i$$
$$\pi(v,s) = \langle a_i \rangle, \ v(i) = c$$
$$\pi(v',s') = \langle a'_i \rangle, \text{rem}(\pi, \gamma', s') = \langle \rangle, v'(i) = \text{length}(f(a_i))$$
$$s, s', a_i, a'_i \in S, i \in \mathbb{N}$$

We use the suffixes U, S, and T to designate constructors for the untimed, the synchronous, and the timed MoCs, respectively. Hence, $mapU$ is a map-based process constructor that takes the constant c and the function f as arguments and returns the process p. p in turn is a process with one input signal (s) and one output signal (s'). c is a positive integer constant and defines the partitioning of the input signal s. This partitioning determines how many events are consumed by the process during each evaluation cycle. The partitioned signal s is denoted by the sequence of subsequences a_i; the partitioned output signal s' is denoted by the sequence a'_i. The length of each a_i is c, while the length of the a'_i can vary.

The function f defines the functionality of the process. It takes c input events as arguments and produces the output events. Thus, the process p is defined by f, which is repeatedly applied on parts of the input signal and produces parts of the output signal. We call an application of f an *activation cycle*, *firing cycle*, or *evaluation cycle* of the process. The function v, which is constant c in the case of $mapU$, defines how the input signal is partitioned and on how many events f is applied in each evaluation cycle. In evaluation cycle i it takes c events a_i and generates a'_i.

Note that in contrast to v, v'—and hence the partitioning of the output signal s'—depends on f.

Example 3.4 **Amplifier** Process A_2 of our example can be defined by means of the $mapU$ constructor:

$$A_2 = mapU(c,f)$$

where

$$c = 1$$
$$f((\langle x \rangle, \langle y_1, y_2, y_3, y_4, y_5 \rangle)) = \langle xy_1, xy_2, xy_3, xy_4, xy_5 \rangle$$

The process consumes one event containing a pair of sequences that are the result of process A_1, as we will see below. The first sequence in the pair contains one value, the control value originating from A_3. The second sequence contains five values from the input. Function f produces a sequence of five values that are emitted by the process as five separate events.

For instance,

$$A_2(\langle(\langle 10\rangle, \langle 1,2,3,4,5\rangle), (\langle 10\rangle, \langle 6,7,8,9,10\rangle)\rangle)$$
$$= \langle 10, 20, 30, 40, 50, 60, 70, 80, 90, 100\rangle$$

scanU instantiates processes with an internal state, which is directly visible at the output as a single event:

$$\text{scanU}(\gamma, g, w_0) = p \qquad (3.2)$$

where

$$p(s) = s'$$
$$g(a_i, w_i) = w_{i+1}$$
$$\langle w_{i+1}\rangle = a'_i$$
$$\pi(v,s) = \langle a_i \rangle, \ v(i) = \gamma(w'_i)$$
$$\pi(v',s') = \langle a'_i \rangle, \text{rem}(\pi, \gamma', s') = \langle\rangle, v'(i) = \text{length}(g(a_i)) = 1$$
$$s, s', a_i, a'_i \in S, w_i \in E, i \in \mathbb{N}_0$$

Therefore, processes constructed by scanU generate only one event in each activation cycle. $v(i)$—and hence the partitioning of the input signal—depends on the internal state w_i via the function γ. Thus, the current state of the process determines how many events are consumed for the next evaluation cycle. Note further that the initial state w_0 is never visible at the output. The first output event is only generated after the first input has been consumed.

Example 3.5 **Amplifier** We model A_3 as a scan-based process. The control value emitted is at the same time the internal state of the process. It reads in five values, sums them up, and compares the result with the range [400, 500]. If it falls within the range, the state is not changed. If it is greater than the upper limit, the state is decreased by one; otherwise it is increased by one. Thus, the amplifier tries to keep the output values between 80 and 100. The initial state is 10.

$$A_3 = \text{scanU}(\gamma, g, w_0)$$

where

$$w_0 = 10$$
$$\gamma(w_i) = 5 \quad \forall i \in \mathbb{N}_0$$
$$g(w_i, \langle x_1, x_2, x_3, x_4, x_5\rangle) = \begin{cases} w_i - 1 & \text{if } x_1 + x_2 + x_3 + x_4 + x_5 > 500 \\ w_i + 1 & \text{if } x_1 + x_2 + x_3 + x_4 + x_5 < 400 \\ w_i & \text{otherwise} \end{cases}$$

For instance,

$$A_3(\langle 10, 20, 30, 40, 50, 60, 70, 80, 90, 100\rangle) = \langle 11, 11\rangle$$

`scanU`-constructed processes are the simplest among processes with state. `mealyU`-based processes resemble Mealy state machines in that they have, in addition to a next-state function, an output encoding function f that depends on both the input and the current state.

$$\mathtt{mealyU}(\gamma, g, f, w_0) = p \qquad (3.3)$$

where

$$p(s) = s'$$
$$f(w_i, a_i) = a'_i$$
$$g(w_i, a_i) = w_{i+1}$$
$$\pi(\nu, s) = \langle a_i \rangle, \ \nu(i) = \gamma(w_i)$$
$$\pi(\nu', s') = \langle a'_i \rangle, \nu'(i) = \mathsf{length}(f(w_i, a_i)), \mathsf{rem}(\pi, \gamma', s') = \langle\rangle$$
$$s, s', a_i, a'_i \in S, w_i \in E, i \in \mathbb{N}_0$$

Consequently, more than one event may be generated during each evaluation cycle. A Moore state machine will also be convenient to have available.

$$\mathtt{mooreU}(\gamma, g, f, w_0) = p \qquad (3.4)$$

where

$$p(s) = s'$$
$$f(w_i) = a'_i$$
$$g(w_i, a_i) = w_{i+1}$$
$$\pi(\nu, s) = \langle a_i \rangle, \ \nu(i) = \gamma(w_i)$$
$$\pi(\nu', s') = \langle a'_i \rangle, \nu'(i) = \mathsf{length}(f(w_i)), \mathsf{rem}(\pi, \gamma', s') = \langle\rangle$$
$$s, s', a_i, a'_i \in S, w_i \in E, i \in \mathbb{N}_0$$

All processes we can create so far have only one input and one output signal. But of course we also want processes with several inputs and outputs. To be able to do this, we define constructors `zipU` and `unzipU`, which allow us, together with the already defined constructors, to compose processes with arbitrarily many input and output signals.

`zipU` takes two input signals, zips them together, and produces a signal consisting of 2-tuples of events. The first component of this event contains the event sequence from input signal s_a, and the second component contains the event sequence from s_b. s_c is a control signal. The two functions γ_a and γ_b take their arguments from s_c to determine how many events are taken in each evaluation cycle from each of the two

3.4 Process Constructors

input signals s_a and s_b, respectively.

$$zipU(\gamma_a, \gamma_b) = p \tag{3.5}$$

where

$$p(s_a, s_b, s_c) = s'$$
$$\langle a_i, b_i \rangle = e'_i$$
$$\pi(v_a, s_a) = \langle a_i \rangle, \ v_a(i) = \gamma_a(c_i)$$
$$\pi(v_b, s_b) = \langle b_i \rangle, \ v_b(i) = \gamma_b(c_i)$$
$$\pi(v_c, s_c) = \langle c_i \rangle, \ v_c(i) = 1$$
$$\pi(v', s') = \langle \langle e'_i \rangle \rangle, \ v'(i) = 1, \text{rem}(\pi, \gamma', s') = \langle \rangle$$
$$s_a, s_b, s_c, s', a_i, b_i, c_i \in S, e' \in E, i \in \mathbb{N}_0$$

Because we will need it later, we also define simplified versions that have no control input but two constants as parameters that define the partitioning of the input signals. The input sequences are packed into a pair.

$$zipUs(c_1, c_2) = p \tag{3.6}$$

where

$$p(s_a, s_b) = s'$$
$$(a_i, b_i) = e'_i$$
$$\pi(v_a, s_a) = \langle a_i \rangle, \ v_a(i) = c_1$$
$$\pi(v_b, s_b) = \langle b_i \rangle, \ v_b(i) = c_2$$
$$\pi(v', s') = \langle \langle e'_i \rangle \rangle, \ v'(i) = 1, \text{rem}(\pi, \gamma', s') = \langle \rangle$$
$$s_a, s_b, s', a_i, b_i \in S, e' \in E, i \in \mathbb{N}_0$$

Finally, we also define a process constructor that allows the zipping of two signals with an arbitrary function.

$$zipWithU(c_1, c_2, f) = p \tag{3.7}$$

where

$$p(s_a, s_b) = s'$$
$$f((a_i, b_i)) = c_i$$
$$\pi(v_a, s_a) = \langle a_i \rangle, \ v_a(i) = c_1$$
$$\pi(v_b, s_b) = \langle b_i \rangle, \ v_b(i) = c_2$$
$$\pi(v', s') = \langle c_i \rangle, \ v'(i) = \text{length}(c_i), \text{rem}(\pi, \gamma', s') = \langle \rangle$$
$$s_a, s_b, s', a_i, b_i, c_i \in S, c_1, c_2, i \in \mathbb{N}_0$$

Note that zip-based processes perform a synchronization between their input signals. For instance, if a zip-based process was expecting one event from both signals s_a and s_b and only an event from s_a was available at some particular time instant, it would wait for a matching event from s_b before it re-emits both events at its output.

Example 3.6 **Amplifier** Process A_1 is based on a $\mathtt{zipUs}$ constructor. For each control input value from A_3, it consumes five events from the primary input:

$$A_1 = \mathtt{zipUs}(1,5)$$

For instance,

$$A_1(\langle 10, 11\rangle, \langle 1,2,3,4,5,6,7,8,9,10\rangle)$$
$$= \langle(\langle 10\rangle, \langle 1,2,3,4,5\rangle), (\langle 11\rangle, \langle 6,7,8,9,10\rangle)\rangle$$

where
$$A(s_{in}) = s_{out}$$
$$s_{out} = A_2(s_1)$$
$$s_1 = A_1(s_2, s_{in})$$
$$s_2 = A_3(s_{out})$$

Now we can put together the amplifier (Figure 3-6). However, the model A cannot be simulated due to an initialization problem. Assuming we receive an input signal s_{in}, A_1 cannot fire since s_2 is empty. But A_3 cannot generate output before it received input from A_2, which in turn requires A_3 to fire first. There are two possible remedies. We can define A_3 such that it outputs its initial state before any input is consumed. This would require a variant of the $\mathtt{scanU}$ constructor. The second remedy is to initialize the signal s_2, which also requires a new process constructor.

We must introduce a few more process constructors before we can complete the adaptive amplifier model. A $\mathtt{scandU}$-based process behaves identically to a $\mathtt{scanU}$ with the addition that it also emits its initial state.

$$\mathtt{scandU}(\gamma, g, w_0) = p \qquad (3.8)$$

FIGURE 3-6

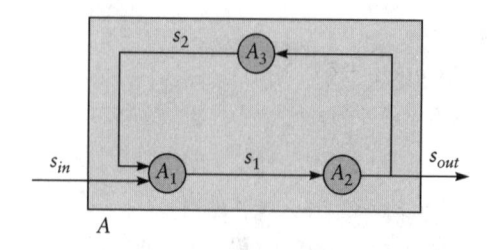

The amplifier process A as a composition of processes.

where
$$p(s) = \langle w_0 \rangle \oplus scanU(\gamma, g, w_0)(s)$$

Unzip-based processes perform the reverse operation of zip processes.

$$unzipU() = p \tag{3.9}$$

where
$$p(s) = \langle s', s'' \rangle$$
$$e_i = \langle a'_i, a''_i \rangle$$
$$\pi(v, s) = \langle e_i \rangle, \ v(i) = 1$$
$$\pi(v', s') = \langle a'_i \rangle, \ v'(i) = \text{length}(a'_i), \text{rem}(\pi, \gamma', s') = \langle \rangle$$
$$\pi(v'', s'') = \langle a''_i \rangle, \ v''(i) = \text{length}(a''_i), \text{rem}(\pi, \gamma'', s'') = \langle \rangle$$
$$s, s', s'', a'_i, a''_i \in S, e_i \in E, i \in \mathbb{N}_0$$

Finally, we define source and sink process constructors and constructors to initialize a signal:

$$sourceU(g, w_0) = p \tag{3.10}$$

where
$$p() = s'$$
$$w_i = e'_i$$
$$g(w_i) = w_{i+1}$$
$$\pi(v', s') = \langle \langle e'_i \rangle \rangle, \ v'(i) = \text{length}(g(w_i)) = 1, \text{rem}(\pi, \gamma', s') = \langle \rangle$$
$$s' \in S, w_i, e'_i \in E, i \in \mathbb{N}_0$$

$$sinkU(\gamma, g, w_0) = p \tag{3.11}$$

where
$$p(s) = \langle \rangle$$
$$g(w_i) = w_{i+1}$$
$$\pi(v, s) = \langle a_i \rangle, \ v(i) = \gamma(w_i)$$
$$s, a_i \in S, w_i \in E, i \in \mathbb{N}_0$$

$$initU(r) = p \tag{3.12}$$

where
$$p(s) = r \oplus s$$
$$v(0) = 0$$
$$v'(0) = \text{length}(r)$$

FIGURE 3-7

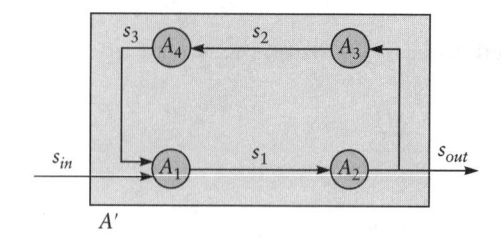

The amplifier process A' with an initialization of s_3.

$$v(i) = v'(i) = 1 \quad \text{for } i \geq 1$$

$$r, s \in S$$

Note that $initU$-based processes always partition their input and output signals into single-event sequences after having emitted the initial sequence.

Example 3.7 **Amplifier** Let us now complete our adaptive amplifier model. We solve the initialization problem by adding another process $A_4 = initU(\langle 10 \rangle)$. The resulting model A' (Figure 3-7) is defined as follows:

$$A'(s_{in}) = s_{out}$$

where

$$s_{out} = A_2(s_1)$$
$$s_1 = A_1(s_3, s_{in})$$
$$s_3 = A_4(s_2)$$
$$s_2 = A_3(s_{out})$$

If we feed A' with signal $s_{in} = \langle 1, 2, 3, 4, 5, 6, 7, 8, 9, 11, 12, 13, 14, 15, 16, 17, 18, 18, 20 \rangle$, we observe

$$s_{out} = \langle 10, 20, 30, 40, 50, 66, 77, 88, 99, 110,$$
$$121, 132, 143, 154, 165, 160, 170, 180, 190, 200 \rangle$$
$$s_3 = \langle 10, 11, 11, 10, 9 \rangle$$

After the initial value 10, the control signal s_3 is raised to 11 due to low values on s_{in}. But then, as the values on s_{in} increase, the control signal drops rapidly.

For practical convenience many more process constructors should be provided. For example, we would like to instantiate directly processes that take several input signals and produce events on several output signals. However, for the sake of conciseness we are content with the few process constructors we have defined and assume that we can in principle instantiate arbitrary processes.

FIGURE 3-8

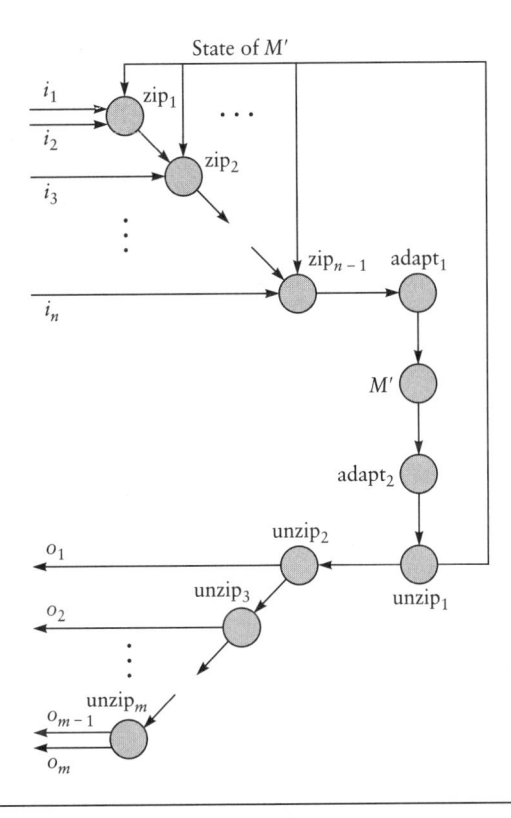

A n-input m-output Mealy machine constructed from simpler processes. M' is a Mealy-based process, while $adapt_1$ and $adapt_2$ are map-based processes.

To see why, consider a Mealy state machine M with n input signals and m output signals. Both the next-state function g_M and the output encoding function f_M depend on all inputs. The number of input events consumed from input i is determined by a function γ_i that takes only the current state of the machine as an argument. This is the most general machine we can model, and it can be constructed from the process types we have introduced above. First we use $n-1$ `zipU` processes, denoted by $zip_1 \ldots zip_{n-1}$ in Figure 3-8, to assemble all the input events from the input signals that are required for a single evaluation cycle of M. All these processes have as a third input signal the sequence of states of M, and they use the respective function γ_i to determine how many events to consume from input signal i. Then we need a function that formats the assembled input events into a form expected by the functions g_M and f_M ($adapt_1$). M' is a Mealy-based process applying g_M and f_M. The result of f_M together with the state of M' is formatted again ($adapt_2$). Process $unzip_1$ extracts the state of M' required by the zip processes. The result of f_M is distributed by $m-1$ `unzipU` processes ($unzip_2 \ldots unzip_m$) into the m output signals $o_1,\ldots,o_m$. Altogether we need $n-1$ `zipU` processes, one `mealyU` process, two `mapU` processes, and m `unzipU` processes to emulate the process M.

For simplicity, from now on we will restrict the discussion to processes with at most two input and output signals based on the assumption that we can generalize all arguments to arbitrary processes.

These process constructors can be considered to define a very general model of concurrency with very few restrictions on what a process can do. One important restriction is that all processes must be *determinate*—for a given set of input signals, they will always generate the same set of output signals.

This restriction is lifted if we introduce a nondeterministic merge process:

$$ndmergeU() = p$$

where

$$p(s, s') = \text{head}(s) \oplus p(\text{tail}(s), s')$$
$$\text{or } \text{head}(s') \oplus p(s, \text{tail}(s'))$$

p takes the first event of either s or s', emits it, and processes the rest of the input signals. The order of both s and s' is preserved in the output signal, but the way they are merged is nondeterministic. In fact, one possible behavior is that p never consumes an event from one of its inputs by always favoring the other signal. This means that p is not necessarily *fair* because a particular event at one of the inputs is not guaranteed to appear at the output.

In the following discussion we do not allow $ndmergeU$-based processes because they lead to nondeterministic models and have undesirable theoretical properties. Nondeterminism will be addressed in more detail in Chapter 8.

3.5 Process Properties

To base our processes on a sound foundation, we relate them to the well-established theory of denotational semantics. We will do this only briefly, and the exposition may appear terse if you are not familiar with this theory. However, if you are not interested in the formal underpinnings you may safely skip this and the following sections as well as Sections 3.10.5 and 3.11. They are not necessary to obtain an intuitive understanding of the computational models and the presented application examples.

3.5.1 Monotonicity

Important properties of processes are *monotonicity* and *continuity*. We define these in the standard way—based on the prefix ordering of signals. We start with monotonicity.

The relation $\sqsubseteq : (S \times S)$ is defined by

$$s_1 \sqsubseteq s_2 \text{ iff } s_1 \text{ is an initial segment of } s_2$$

A process $p : S \to S$ is *monotonic* if

$$s_1 \sqsubseteq s_2 \Rightarrow p(s_1) \sqsubseteq p(s_2)$$

Monotonicity means that receiving more input can only provoke a process to generate more output, but not change the already emitted output. This means that a process can start computing before all the input is available because new input will only add to the previously created output but not change it. For instance, a sorting process p is not monotonic since

$$p(\langle 5, 4, 3\rangle) = \langle 3, 4, 5\rangle$$
$$p(\langle 5, 4, 3, 2\rangle) = \langle 2, 3, 4, 5\rangle$$

Appending 2 to the input signal changes the entire output signal.

Theorem 3.1 All processes constructed with the process constructors defined by (3.1) through (3.12) are monotonic.

Proof Both `sourceU`- and `sinkU`- based processes are monotonic because they all produce constants as outputs. `initU` simply copies the input at the output, prepended by a constant sequence. Zip- and unzip-based processes also just copy their inputs to their outputs without changing values or order.

Next we turn to map-, scan-, Mealy-, and Moore-based processes. In fact we only need to deal with Mealy processes since they are the most general. Recall the basic computation mechanism. Whenever a new event sequence r_i, as defined by the input signal partitioning, is consumed, the process is activated and will emit a new event sequence r'_i at the output as defined by the output signal partitioning. As long as the entire input sequence r_i is not seen, the process is not activated and no output is generated.

Can we create all monotonic processes? The answer is negative due to two restrictions. First, a process cannot detect if input signals are infinite, since it always only sees finite prefixes of infinite input signals. For instance, consider process $p : S \to S$:

$$p(s) = \begin{cases} \langle 0 \rangle & \text{if } s \text{ is a finite sequence} \\ \langle 0, 1 \rangle & \text{otherwise} \end{cases}$$

Clearly, p is monotonic, but none of our process constructors can create it. This restriction is related to continuity, which we take up in the next section.

The second restriction concerns processes with several inputs. Our zip constructors link the different inputs strongly together; they cannot be processed independently from each other. Consider a process $p : S^2 \to S^2$:

$$p(s_1, s_2) = (s_1, s_2)$$

Each of the two outputs only depends on one of the two inputs. However, a zip-based process would stop producing further output altogether if one of the inputs does not provide sufficient data. This restriction is called the *sequentiality* of processes, which we discuss below.

3.5.2 Continuity

While monotonicity ensures that processes can produce partial results based on partial inputs, continuity guarantees that infinite output signals can be gradually approximated.

The prefix relation $\sqsubseteq$ defines a partial order on S with $\langle\rangle$ as a minimal element. Any increasing chain $C = c_1 \sqsubseteq c_2 \sqsubseteq \ldots \sqsubseteq c_n$ has a *least upper bound*, which we designate $\sqcup C$. Hence, S is a *complete partial order* (CPO).

A monotonic process p is also continuous if

$$p(\sqcup C) = \sqcup p(C) \qquad (:= \sqcup\{p(s) \mid s \in C\})$$

for every chain $C \subseteq S$.

Continuity means that infinite signals can be approximated with arbitrary accuracy by finite signal prefixes. Thus, a process never needs to consume all its infinite input signal before starting to produce output.

Theorem 3.2 All processes constructed with the process constructors defined by (3.1) through (3.12) are continuous.

Proof We take the simple cases first. Both $sourceU$- and $sinkU$-based processes are continuous because they all produce constants as outputs. $initU$ simply copies the input at the output, prepended by a constant sequence. Zip- and unzip-based processes also just copy their inputs to their outputs without changing values or order.

It suffices again to deal with Mealy-based processes to cover map-, scan-, and Moore-based processes also.

To see that Mealy processes are continuous, we have to show that $p(\sqcup C) = \sqcup p(C)$. For finite chains C or for chains with a finite least upper bound, this is obviously true due to the monotonicity of p. For infinite chains with least upper bounds of infinite length, we first develop $\sqcup p(C)$. C is a set of sequences, $C = \{c_i \mid i \geq 0, c_i \sqsubseteq c_{i+1}\}$. The process p partitions the input signal $\dot{s}$ according to $\pi(v, \dot{s})$. Hence, all c_i can be expressed as

$$c_i = \left(\bigoplus_{\langle r_j \rangle = \mathsf{parts}(v, 0, c_i)} r_j\right) \oplus \mathsf{rem}(\pi, v, c_i)$$

which for convenience can be written as

$$c_i = \bigoplus_j r_j \oplus \mathsf{rem}(\pi, v, c_i)$$

for some appropriate j. Because p does not evaluate the remainder, we get

$$p(c_i) = \bigoplus_j f(w_j, r_j)$$

for the output encoding function f and the evolving internal state w_i of process p. For $i \to \infty$, we also get $j \to \infty$, and hence

$$\sqcup p(C) = \bigoplus_{j=0}^{\infty} f(w_j, r_j)$$

On the other hand, since $c_i = \bigoplus_j r_j \oplus \text{rem}(\pi, v, c_i)$, we get

$$\sqcup C = \bigoplus_{j=0}^{\infty} r_j$$

because the partitioning of infinite sequences never gives a remainder. Applying p to this sequence results in

$$p(\sqcup C) = p\left(\bigoplus_{j=0}^{\infty} r_j\right) = \bigoplus_{j=0}^{\infty} f(w_j, r_j)$$

which is the same as we got for $\sqcup p(C)$ above.

We noted that not all monotonic processes can be created by our constructors. But can at least all continuous processes be created? The answer is again negative due to the second restriction mentioned in the previous section, which couples different inputs tightly together. However, the answer is positive for single-input processes.

Theorem 3.3 All continuous processes with one input signal can be created by means of the process constructors defined by (3.1) through (3.11).

Proof First we need to define the difference operator for two signals. $\ominus$ subtracts one sequence from another; that is, $s_1 \sqsubseteq s_2 \Rightarrow s_2 = s_1 \oplus (s_2 \ominus s_1)$.

Let $p: S \to S$ be a continuous process. Because it is continuous, it is also monotonic; that is, $s_1 \sqsubseteq s_2 \to p(s_1) \sqsubseteq p(s_2)$. That means for $s_1 = \langle e_0, e_1, \ldots \rangle$ there exists a chain $C = c_0, c_1, \ldots, c_n, \ldots$ such that

$$c_0 = \langle \rangle$$
$$c_{i+1} = c_i \oplus e_i$$

and we have

$$c_i \sqsubseteq c_{i+1} \sqsubseteq s_1 \quad \forall i \geq 0$$
$$p(c_i) \sqsubseteq p(c_{i+1}) \sqsubseteq p(s_1) \quad \forall i \geq 0$$

Due to the continuity of p, that is, $p(\sqcup C) = \sqcup p(C)$, we know that these chains also exist for infinite input and output sequences.

Now we can construct a process q as follows:

$$\text{mealyU}(1, g, f, w_0) = q$$

where

$$w_0 = \langle \rangle$$
$$w_{i+1} = g(w_i, a_i) = w_i \oplus a_i$$
$$f(w_i, a_i) = p(w_i \oplus a_i) \ominus p(w_i)$$

The input signal is partitioned into single events. The state simply accumulates the input events and $w_i = c_i$; thus functions γ and g are simple. f is more complex, but it exists due to the existence of p, which is functional. Furthermore, it will approximate infinite output signals in exactly the same way as p. Consequently, the process q is equivalent to p.

3.5.3 Sequential Processes

A process with two or more inputs is sequential if the inputs are not independently processed. First we extend the prefix relation to tuples of signals. If $\vec{s}, \vec{s}\,' \in S^n$ are tuples of n signals, $\vec{s} = (s_1, \ldots, s_n)$, $\vec{s}\,' = (s'_1, \ldots, s'_n)$, then the relation $\sqsubseteq_n : (S^n \times S^m)$, $n, m \in \mathbb{N}_0$, is defined by

$$\vec{s} \sqsubseteq_n \vec{s}\,' \text{ iff } s_i \sqsubseteq s'_i \; \forall i, 1 \leq i \leq n$$

It defines a partial order on tuples of n signals with $(\langle\rangle, \ldots, \langle\rangle)$ as its minimal element.

Definition 3.2 A process $p : S^n \to S^m$ with n input signals and m output signals is *sequential* if it is continuous and if, for any $\vec{s} = (s_1, s_2, \ldots, s_n)$, there exists an i, $1 \leq i \leq n$, such that for any $\vec{s}\,'$, where $\vec{s} \sqsubseteq_n \vec{s}\,'$ and $s_i = s'_i$, we have $p(\vec{s}) = p(\vec{s}\,')$.

For an arbitrary input constellation, there is one particular input i that, if not extended, will prevent the extension of the output, independently of whether other inputs are extended arbitrarily. Note that i may be different for different input constellations; it is only required that one such input exists for any input tuple.

Theorem 3.4 All processes constructed with the process constructors defined by (3.1) through (3.12) are sequential.

Proof Apparently, processes with one input are sequential, and we can concentrate on `zipU` processes. A `zipU`-based process has three inputs: one control input, which is partitioned into single events, and two data inputs, which are partitioned as determined by the values of the control input. At any time if the process cannot produce more output, it is waiting for more input data on a particular input signal. This blocking input signal may be any of the three, but providing the other two input signals with more data will not cause the process to fire.

The same argument goes for the `zipUs` processes.

By now we have established the properties monotonicity, continuity, and sequentiality for the processes we can construct. This is useful for understanding their behavior precisely and for formally analyzing them. However, have we been too restrictive? Are there processes that we would like to build in practical situations but that we cannot? Intuitively, nonmonotonic and noncontinuous processes have undesirable properties that we clearly want to avoid. They require that we read in the entire input streams and determine if the input is infinite or not before we can produce a single output token. This is infeasible for long or infinite input streams.

However, there are obviously nonsequential processes that are useful. There is no fundamental reason why we shouldn't define the process $p(s_1, s_2) = (s_1, s_2)$. The sequentiality property ties the processing of inputs together, which is only justified if the produced outputs depend on all the inputs. It cannot be motivated if one output depends only on some but not all inputs. In that case sequentiality imposes additional nonfunctional dependencies. This may or may not be intended by the user. Fortunately it is easy to rectify this situation if we allow processes to consist of other, simpler processes. For instance, $p(s_1, s_2) = (s_1, s_2)$ could be defined as

$$p(s_1, s_2) = (p'(s_1), p'(s_2))$$

where

$$p' = mapU(1, f)$$
$$f(x) = x$$

Therefore in the following section we define composition operators that allow us to build complex processes from simpler parts. Anyway this is an obvious practical necessity to be able to model complex systems.

3.6 Composition Operators

We consider only three basic composition operators: sequential composition, parallel composition, and feedback. As is well known, these three composition operators suffice to model all kinds of networks of reactive information processing components, provided we have simple components available for permuting and copying input and output signals.

We give the definitions only for processes with one or two input and output signals because the generalization to arbitrary numbers of inputs and outputs is straightforward.

3.6.1 Parallel Composition

Let p_1 and p_2 be two processes with one input and one output each, and let $s_1, s_2 \in S$ be two signals. Their parallel composition (Figure 3-9), denoted as $p_1 \parallel p_2$, is defined as follows:

$$(p_1 \parallel p_2)(\langle s_1, s_2 \rangle) = \langle p_1(s_1), p_2(s_2) \rangle \qquad (3.13)$$

It is straightforward to show that $(p_1 \parallel p_2)$ is a continuous process, provided that p_1 and p_2 are continuous processes.

FIGURE 3-9

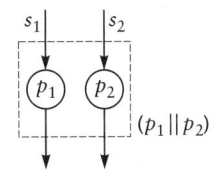

Parallel composition of processes.

FIGURE 3-10

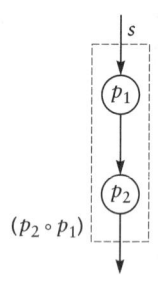

Sequential composition of processes.

3.6.2 Sequential Composition

Since processes are functions, we can easily define sequential composition in terms of functional composition. Again let p_1 and p_2 be two processes and let $s \in S$ be a signal. The sequential composition (Figure 3-10), denoted as $p_1 \circ p_2$, is defined as follows[1]:

$$(p_2 \circ p_1)(s) = p_2(p_1(s)) \qquad (3.14)$$

It is again straightforward to show that $(p_2 \circ p_1)$ is a continuous process, provided that p_1 and p_2 are continuous processes.

3.6.3 Feedback Operator

Given a process $p : (S \times S) \to (S \times S)$ with two input signals and two output signals, we define the process $\mathbf{FB_P}(p) : S \to S$ (Figure 3-11) by the equation

$$\mathbf{FB_P}(p)(s_1) = s_2 \qquad (3.15)$$

where

$$p(s_1, s_3) = (s_2, s_3)$$

The behavior of the process $\mathbf{FB_P}(p)$ is defined by the least fixed-point semantics. The well-known fixed-point theorem states that a continuous function f over a complete partial order always has a least fixed point that is the unambiguous solution to the recursive equation $f(x) = x$ (Stoy 1989, Kahn 1974). Without delving into the theoretical foundations that justify this approach and guarantee the existence of an unambiguous solution to recursive definitions, we illustrate the implications by giving the solution to a few interesting examples.

---

[1] Although it may appear more natural to define $(p_1 \circ p_2)(s) = p_2(p_1(s))$, historically the notation has been established as we give it in the text. To be consistent with the literature, we follow this convention.

FIGURE 3-11

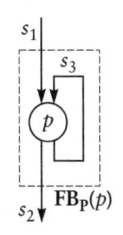

Feedback composition of a process.

FIGURE 3-12

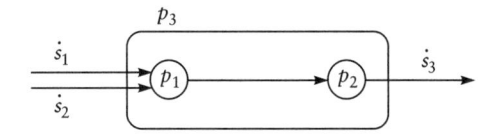

Processes $p_1 = \mathtt{zipWithU}(1,g)$ and $p_2 = \mathtt{initU}(\langle 0 \rangle)$ form process p_3.

FIGURE 3-13

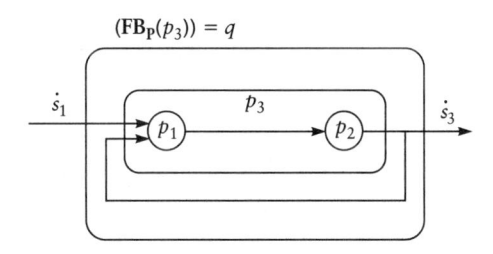

Process $q = \mathbf{FB_P}(p_3)$ is equivalent to a $\mathtt{scanU}$-based process $\mathtt{scanU}(1,g,0)$.

Example 3.8 Let process p_3 (Figure 3-12) be defined as follows:

$$p_3(\dot{s}_1, \dot{s}_2) = p_2(p_1(\dot{s}_1, \dot{s}_2))$$
$$p_1 = \mathtt{zipWithU}(1,1,g)$$
$$p_2 = \mathtt{initU}(\langle 0 \rangle)$$
$$g(x,y) = x+y$$

If we use the feedback operator to form a new process $q = \mathbf{FB_P}(p_3)$ (Figure 3-13), the least fixed point of p_3 defines the behavior of q. To find this fixed point, we start evaluating p_3 with the empty sequence for its input $\dot{s}_2$. The result is used as the new input in the next evaluation cycle. We repeat this procedure until the input equals the output and we have reached a fixed point.

$$\langle\rangle = \dot{r}_0$$
$$p_3(\dot{s}_1, \dot{r}_0) = \dot{r}_1$$
$$p_3(\dot{s}_1, \dot{r}_1) = \dot{r}_2$$
$$\vdots$$
$$p_3(\dot{s}_1, \dot{r}_n) = \dot{r}_f$$
$$p_3(\dot{s}_1, \dot{r}_f) = \dot{r}_f$$

We set $\dot{s}_1 = \langle 1, 2, 3 \rangle$ to determine the behavior of $q(\dot{s}_1) = (\mathbf{FB_P}(p_3))(\langle 1, 2, 3 \rangle)$.

$$\dot{s}_1 = \langle 1, 2, 3 \rangle$$
$$\langle\rangle = \dot{r}_0$$
$$p_3(\langle 1, 2, 3 \rangle, \langle\rangle) = p_2(p_1(\langle 1, 2, 3 \rangle, \langle\rangle))$$
$$= p_2(\langle\rangle) = \langle 0 \rangle = \dot{r}_1$$
$$p_3(\langle 1, 2, 3 \rangle, \langle 0 \rangle) = p_2(p_1(\langle 1, 2, 3 \rangle, \langle 0 \rangle))$$
$$= p_2(\langle 1 \rangle) = \langle 0, 1 \rangle = \dot{r}_2$$
$$p_3(\langle 1, 2, 3 \rangle, \langle 0, 1 \rangle) = p_2(p_1(\langle 1, 2, 3 \rangle, \langle 0, 1 \rangle))$$
$$= p_2(\langle 1, 3 \rangle) = \langle 0, 1, 3 \rangle = \dot{r}_3$$
$$p_3(\langle 1, 2, 3 \rangle, \langle 0, 1, 3 \rangle) = p_2(p_1(\langle 1, 2, 3 \rangle, \langle 0, 1, 3 \rangle))$$
$$= p_2(\langle 1, 3, 6 \rangle) = \langle 0, 1, 3, 6 \rangle = \dot{r}_4$$
$$p_3(\langle 1, 2, 3 \rangle, \langle 0, 1, 3, 6 \rangle) = p_2(p_1(\langle 1, 2, 3 \rangle, \langle 0, 1, 3, 6 \rangle))$$
$$= p_2(\langle 1, 3, 6 \rangle) = \langle 0, 1, 3, 6 \rangle = \dot{r}_4$$

With $\dot{r}_4 = \langle 0, 1, 3, 6 \rangle$ we have found a fixed point that defines the behavior of $(\mathbf{FB_P}(p_3))(\langle 1, 2, 3 \rangle) = \langle 0, 1, 3, 6 \rangle$. There may exist other fixed points, but with the procedure we applied we have found the least of them since p_3 is a monotonic function in both arguments.

Example 3.9 The possible existence of several fixed points is illustrated by the following example (Figure 3-14).

$$p_1 = \mathtt{zipWithU}(1, 1, g)$$
$$g(x, y) = \text{if } (y == 3) \text{ then } 2 \text{ else } 1$$

g tests its second argument and evaluates to 2 if the test succeeds, and to 1 otherwise. When we form a feedback loop from the output $\dot{s}_3$ of p_1 to

FIGURE 3-14

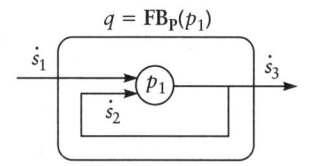

Process $p_1 = \text{zipWithU}(1, 1, g)$ has several fixed points, but the behavior of $q = \textbf{FB}_\textbf{P}(p_1)$ is determined by the least fixed-point solution.

its second input $\dot{s}_2$, the second argument of g becomes equal to its result. Hence, the only consistent result of g is $g(x, 1) = 1$.

Process p_1, however, has several fixed points for a given argument on its first input. For $\dot{s}_1 = \langle 1, 2, 3 \rangle$, p has four fixed points:

$$p_1(\langle 1, 2, 3 \rangle, \langle \rangle) = \langle \rangle$$
$$p_1(\langle 1, 2, 3 \rangle, \langle 1 \rangle) = \langle 1 \rangle$$
$$p_1(\langle 1, 2, 3 \rangle, \langle 1, 1 \rangle) = \langle 1, 1 \rangle$$
$$p_1(\langle 1, 2, 3 \rangle, \langle 1, 1, 1 \rangle) = \langle 1, 1, 1 \rangle$$

All four solutions represent a logically consistent behavior of process q. However, operator **FB**$_\textbf{P}$ requires that we select the *least* fixed point as its behavior. Obviously, the least fixed point in this case is $\langle \rangle$ since it is a prefix to all other fixed points.

We do not need to compute all fixed points to determine the least one, but we can apply the same procedure as above. Starting the evaluation with the empty sequence for $\dot{s}_2$, we immediately obtain

$$p_1(\langle 1, 2, 3 \rangle, \langle \rangle) = \langle \rangle$$

3.7 Definition of the Untimed MoC

Now we are in a position to define precisely what we mean by a "model of computation."

Definition 3.3 A *model of computation* (MoC) is a 2-tuple MoC $= (C, O)$, where C is a set of process constructors, each of which, when given constructor-specific parameters, instantiates a process. O is a set of process composition operators, each of which, when given processes as arguments, instantiates a new process.

An MoC is defined by the set of processes that can be instantiated or constructed. By carefully selecting the process constructors, we can restrict the way processes interact

FIGURE 3-15

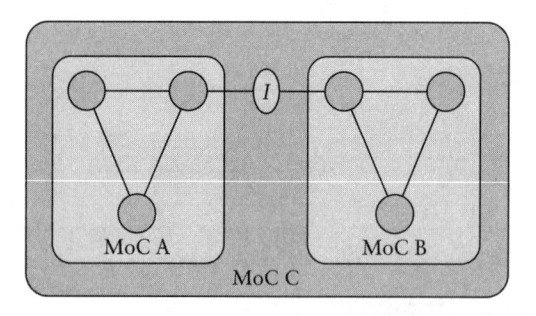

Two parts of the system are isolated in specialized models of computation MoC A and MoC B. Together with an interface process I, they are integrated into the more general model of computation MoC C.

with their environment and thus impose specific properties on all the processes and process networks. This mechanism can also be used for establishing subdomains in a system with specific, more specialized properties. More general MoCs can then be defined to integrate several specialized subdomains.

Figure 3-15 illustrates how two specialized MoCs are integrated, together with the appropriate interface process, into a more general MoC. For instance, the datapath of a system can conveniently be described in an untimed MoC, while the control part requires some timing information and can be modeled in a timed MoC. Tools and techniques specific to these two MoCs can be applied to the two different parts of the system to exploit specific properties. In order to simulate, design, and analyze the entire system, a more general MoC can be used that has to be content with much weaker properties valid for the entire system.

In this and the next two chapters, we define specialized MoCs with different amounts of timing information. In Chapter 6 we develop processes for interfacing the different MoCs and a more general MoC.

Definition 3.4 The *untimed model of computation (untimed MoC)* is defined as untimed MoC $= (C, O)$, where

$$C = \{\texttt{mapU}, \texttt{scanU}, \texttt{scandU}, \texttt{mealyU}, \texttt{mooreU},$$
$$\texttt{zipU}, \texttt{zipUs}, \texttt{zipWithU}, \texttt{unzipU},$$
$$\texttt{sourceU}, \texttt{sinkU}, \texttt{initU}\}$$
$$O = \{\,\|, \circ, \mathbf{FB_P}\}$$

In other words, a process or a process network belongs to the *untimed MoC domain* iff all its processes and process compositions are constructed either by one of the named process constructors or by one of the composition operators. We call such processes *U-MoC processes*.

This definition is a bit redundant because, for instance, map- and scan-based processes could also be instantiated by the `mealyU` constructor.

3.8 Characteristic Functions

The characteristic function is the reverse of a process constructor. Our objective is to reason about the composition of processes, which is, in general, a delicate problem due to the concurrency, synchronization, and communication aspects involved. We can avoid these difficulties in some cases by reducing the problem to the composition of the elementary functions contained by the processes.

A characteristic function of a process is defined by a *function extractor*. One function extractor, which corresponds to `mapU`, is `pamU`. Given a process p and two signal partitionings $\pi(v,s)$ and $\pi(v',s)$, it defines a characteristic function f:

$$pamU(p, v, v') = f$$

where

$$p(\dot{s}) = \dot{s}'$$
$$\pi(v, \dot{s}) = \langle \dot{a}_i \rangle$$
$$\pi(v', \dot{s}') = \langle \dot{a}'_i \rangle$$
$$f(\dot{a}_i) = \dot{a}'_i \quad \forall i$$

f, when applied to all subsignals of $\dot{s}$ in proper order, will completely generate the output signal $\dot{s}'$.

For instance, for a process generated by the `mapU` constructor, $mapU(c,f) = p$, one possible characteristic function is $f' = pamU(p, v, v')$. Then we have that $f = f'$ if $v(i) = c \ \forall i$. As required by `mapU`, $v'(i) = \text{length}(f(\dot{a}'_i))$ for all i due to the definition of `pamU`.

Note that a characteristic function is not unique for a given process and a process may have infinitely many different characteristic functions. As an example, consider process $p = mapU(1, f)$ with $f(x) = 2x$. Every input value (we assume integers as input and output values) is doubled. With partitions defined by $v(i) = v'(i) = 1$, f is a characteristic function because $f = pamU(p, v, v')$. However, with different partitionings $v(i) = v'(i) = 2$, we derive $f' = pamU(p, v, v')$ as another characteristic function. Apparently, $f'(\langle x, y \rangle) = \langle 2x, 2y \rangle$. Obviously, we can identify a large number of characteristic functions.

If a process p has an internal state, one characteristic function does not suffice. We need one function for computing the next state and one for the output, and we have to find a sequence of process states w_i. Given are again nonoverlapping partitions $\pi(v,s) = \langle \dot{a}_i \rangle$ and $\pi(v',s') = \langle \dot{a}'_i \rangle$ of the input and output signals s and s', respectively. Then the next-state function, the output encoding function, and the sequence of internal state values together are characteristic functions of p and are defined by the function extractor `ylaemU`:

$$\texttt{ylaemU}(p, v, v') = (g, f, \langle w_i \rangle)$$

where

$$p(\dot{s}) = \dot{s}'$$
$$\pi(v, \dot{s}) = \langle \dot{a}_i \rangle$$
$$\pi(v', \dot{s}') = \langle \dot{a}'_i \rangle$$
$$g(\dot{a}_i, w_i) = w_{i+1} \quad \forall i$$
$$f(\dot{a}_i, w_i) = \dot{a}'_i \quad \forall i$$

As you would expect, there is an intimate relation between `ylaemU` and `mealyU`. Let $p = \texttt{mealyU}(\gamma, g, f, w_0)$ and $(g', f', \langle w'_i \rangle) = \texttt{ylaemU}(p, v, v')$. Then we have that $f = f'$ and $g = g'$ if $w_i = w'_i$ and $v(i) = \gamma(w_i), \forall i \geq 0$. As above, we can infer that $v'(i) = \texttt{length}(f(\dot{a}_i, w_i))$ for all $i \geq 0$.

Does every process have a characteristic function? The following theorem gives a positive answer for continuous processes with one input signal and a negative answer for discontinuous processes.

Theorem 3.5 A process with one input and one output is continuous iff it has a characteristic function.

Proof On the one hand, Theorem 3.3 tells us that every continuous process with one input indeed must have a characteristic function. On the other hand, every characteristic function can be used to create a process by means of a process constructor. Since processes derived from process constructors are always continuous, as Theorem 3.2 states, noncontinuous processes cannot have characteristic functions.

Unfortunately, Theorem 3.3 does not provide a constructive method to build a characteristic function. It is in fact unlikely that a general method exists that does not take more specific information about process p into account.

We conclude our discussion of characteristic functions with Table 3-2, which summarizes the process families, their process constructors, and characteristic functions.

3.9 Process Signatures

We have seen that when input and output signals of a process are partitioned into many subsequences, the process can be characterized by its characteristic function. Conversely, given a partitioning of the input and output signals and functions operating on these subsequences, process constructors completely define processes. However, it is justified to view the signal partitioning functions as a property of the process. We consider them as part of the type of a process.

TABLE 3-2: *Process family hierarchy.*

Level	Name	Constructor	Extractor	Description
1	Map	mapU	pamU	Processes without internal state.
2	Scan	scanU	nacsU	Processes with an internal state and a next-state function. The state is directly visible at the output.
3	Moore	mooreU	eroomU	Processes with a state; the output is a function of the state, but not directly of the input.
3	Mealy	mealyU	ylaemU	Processes with a state; the output is a function of the state and the current input.

The *type of a process TYPE* is a four tuple $\langle TI, TO, NI, NO \rangle$. *TI* and *TO* are sets of signal types for the input and output signals, respectively. The type of a signal is the type of its events. Since we are not particularly interested in the types of events and signals, we will not elaborate this aspect here. $NI = \{vi_1, \ldots, vi_n\}$ is the set of partitioning functions for the n input signals, and $NO = \{vo_1, \ldots, vo_m\}$ is the set of partitioning functions for the m output signals. The tuple $\langle NI, NO \rangle$ is also called the *process signature*. In the following we only consider processes with one input and one output signal, and for a process p we denote the input partitioning function as v_p and the output signal partitioning function as v'_p.

Formally, v and v' are functions of i, the firing cycle counter, but as we have seen above, v really depends on the process state, w_i, and v' depends on the state and the current input sequence, $(w_i, \dot{a}_i)$.

If the output signal of a process p has the same type as the input signal of another process p', they can be connected together and we define their *match* as the ratio of their partitioning functions:

$$match(p, p') = \frac{v'_p(i)}{v_{p'}(i)}$$

We speak of a *perfect match* if this number is a constant 1. It is a *rational match* if this number is a constant rational or integer number. If this number is not constant, we say it is a *varying match*. In the case of a varying match the composition of two processes becomes a tedious task because we have to realize all the buffering and synchronization internally in the newly composed process, which are otherwise handled at the process boundaries. However, in the case of perfect or rational matches the processes run already synchronized to a large extent, and it is often straightforward to merge them.

Note that in the case of perfect and rational matches, the v factors $v_p(i)$ and $v_{p'}(i)$ need not be constant; only their ratio $v'_p(i)/v_{p'}(i)$ must be.

Example 3.10 **Amplifier** The type of process A_1 is $TYPE(A_1) = \langle TI, TO, NI, NO \rangle$ with $TI = \{Integer, Integer\}, TO = \{Integer\}, NI = \{1, 5\}, NO = \{1\}$.

For process A_2 we have $TI = \{Integer\}, TO = \{Integer\}, NI = \{1\}, NO = \{5\}$. For process A_3 we have $TI = \{Integer\}, TO = \{Integer\}, NI = \{5\}, NO = \{1\}$. For process A_4 we have $TI = \{Integer\}, TO = \{Integer\}, NI = \{1\}, NO = \{1\}$. In all cases, the v functions for input and output are constant.

3.10 Process Up-rating

Before we will discuss the composition of processes, we will introduce how to change the partitioning of input and output signals of a process, which we call *up-rating*. We do this first for map-, scan-, and Mealy-based processes, and then for zip and unzip processes.

Definition 3.5 Let ϱ be a natural number, $\varrho > 0$, and let p be a U-MoC process with one input and one output and the input signal s is partitioned $\pi(v_p, s) = \langle a_i \rangle$. Process p is *up-rated* by a factor ϱ, resulting in another process q, if q is continuous and the input signal partitioning of q is $\pi(v_q, s) = \langle b_j \rangle$ with

$$b_j = \bigoplus_{i=0}^{\varrho-1} a_{j\varrho+i} \qquad \forall j \in \mathbb{N}_0$$

and p behaves identically to q for all increasing prefixes of an input signal defined by b_j:

$$p\left(\bigoplus_{i=0}^{j-1} b_i\right) = q\left(\bigoplus_{i=0}^{j-1} b_i\right) \qquad \forall j \in \mathbb{N}_0$$

We also say that q is a ρ-up-rated variant of p.

This definition defines the conditions when one process is an up-rated version of another process. It indicates neither how to up-rate a process nor even whether it can be done at all.

To derive an up-rating operation, we need two supporting functions. The first extracts subsequences from a given sequence. The function $\mathsf{sseq}(r, b, l)$ denotes a subsequence of sequence r that starts at position b in r and has length l. For example, $\mathsf{sseq}(\langle e_1, e_2, e_3, e_4 \rangle, 0, 2) = \langle e_1, e_2 \rangle$.

The function parm allows us to apply a function repeatedly onto an event sequence, as illustrated in Figure 3-16.

parm applies a function on subsequences of the input to produce subsequences of the output and concatenates the partial results into the result sequence. The function f must accept sequences with l events as input and return sequences of j events.

$$\mathsf{parm}(f, r, l, n) = r' = \bigoplus_{i=0}^{n-1} f(\mathsf{sseq}(r, il, l))$$

FIGURE 3-16

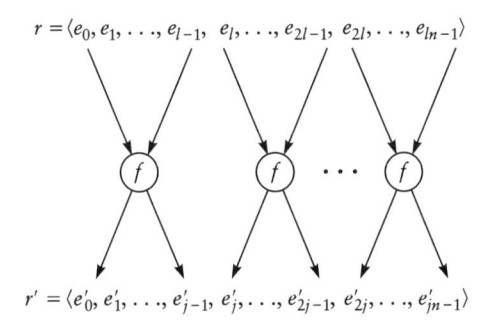

$r = \langle e_0, e_1, \ldots, e_{l-1}, e_l, \ldots, e_{2l-1}, e_{2l}, \ldots, e_{ln-1} \rangle$

$r' = \langle e'_0, e'_1, \ldots, e'_{j-1}, e'_j, \ldots, e'_{2j-1}, e'_{2j}, \ldots, e'_{jn-1} \rangle$

parm *applies function f several times on an event sequence.*

3.10.1 Map-Based Processes

Definition 3.6 Consider a process p defined by $\mathsf{mapU}(c, f) = p$. Up-rating of p by a factor ϱ, denoted as $\mathsf{uprate}(p, \varrho)$, defines another process p' as follows:

$$p' = \mathsf{mapU}(c\varrho, f')$$
$$f'(\dot{a}'_i) = \mathsf{parm}(f, \dot{a}'_i, c, \varrho)$$
$$\dot{a}'_i = \bigoplus_{j=0}^{\varrho-1} \dot{a}_{i\varrho+j}$$

As the following theorem shows, process p' will always generate the same output signal as p when given the same input signal if the input signal is infinite, and the size of the partitions of p' is ϱ times the size of the partitions of p for both the input and output signal. If the input signal is finite, p' will produce a possibly shorter subsignal of the one produced by process p.

Theorem 3.6 For a mapU-based process p and a number $\varrho \in \mathbb{N}$, the process $q = \mathsf{uprate}(p, \varrho)$ is a ϱ-up-rated variant of p.

Proof Let $p = \mathsf{mapU}(c, f)$ and $q = \mathsf{mapU}(c', f')$. From Definition 3.6, we have $c' = c\varrho$. Further, let $\pi(v_p, \dot{s}) = \langle \dot{a}_i \rangle$ and $\pi(v_q, \dot{s}) = \langle \dot{b}_i \rangle$ be the partitionings of the input signal $\dot{s}$. Obviously, we have

$$\dot{b}_i = \bigoplus_{j=0}^{\varrho-1} \dot{a}_{i\varrho+j} \qquad \forall i \in \mathbb{N}_0$$

and we have to show that

$$p\left(\bigoplus_{i=0}^{k} \dot{b}_i\right) = q\left(\bigoplus_{i=0}^{k} \dot{b}_i\right) \qquad \forall k \in \mathbb{N}_0$$

We prove this by induction.
Base case $k = 0$:

$$p(\dot{b}_0) = q(\dot{b}_0)$$

$$p\left(\bigoplus_{j=0}^{\varrho-1} \dot{a}_j\right) = q(\dot{b}_0)$$

$$\bigoplus_{j=0}^{\varrho-1} f(\dot{a}_j) = f'(\dot{b}_0)$$

$$= \mathsf{parm}(f, \dot{b}_0, c, \varrho)$$

$$= \bigoplus_{j=0}^{\varrho-1} f(\mathsf{sseq}(\dot{b}_0, jc, c))$$

$$= \bigoplus_{j=0}^{\varrho-1} f(\dot{a}_j) \quad \text{(since } \mathsf{sseq}(\dot{b}_0, jc, c) = \dot{a}_j \quad \text{for } 0 \leq j \leq \varrho - 1\text{)}$$

Induction step $k > 0$:

$$p\left(\bigoplus_{i=0}^{k} \dot{b}_i\right) = q\left(\bigoplus_{i=0}^{k} \dot{b}_i\right)$$

$$p\left(\bigoplus_{i=0}^{k-1} \dot{b}_i \oplus \dot{b}_k\right) = q\left(\bigoplus_{i=0}^{k-1} \dot{b}_i \oplus \dot{b}_k\right)$$

$$p\left(\bigoplus_{i=0}^{k-1} \dot{b}_i \oplus \left(\bigoplus_{j=0}^{\varrho-1} \dot{a}_{k\varrho+j}\right)\right) = q\left(\bigoplus_{i=0}^{k-1} \dot{b}_i \oplus \dot{b}_k\right)$$

$$p\left(\bigoplus_{i=0}^{k-1} \dot{b}_i\right) \oplus \left(\bigoplus_{j=0}^{\varrho-1} f(\dot{a}_{k\varrho+j})\right) = q\left(\bigoplus_{i=0}^{k-1} \dot{b}_i\right) \oplus (f'(\dot{b}_k))$$

Since from the base case we may assume safely that

$$p\left(\bigoplus_{i=0}^{k-1} \dot{b}_i\right) = q\left(\bigoplus_{i=0}^{k-1} \dot{b}_i\right)$$

we need to show that

$$\bigoplus_{j=0}^{\varrho-1} f(\dot{a}_{k\varrho+j}) = f'(\dot{b}_k)$$

$$= \mathsf{parm}(f, \dot{b}_k, c, \varrho)$$

$$= \bigoplus_{j=0}^{\varrho-1} f(\mathsf{sseq}(\dot{b}_k, jc, c))$$

$$= \bigoplus_{j=0}^{\varrho-1} f(\dot{a}_{k\varrho+j})$$

since $\mathsf{sseq}(\dot{b}_k, jc, c) = \dot{a}_{k\varrho+j}$ for $0 \leq j \leq \varrho - 1$.

Example 3.11 **Amplifier** We up-rate process A_2 by a factor 2 to derive process $A'_2 = \mathsf{uprate}(A_2, 2)$:

$$A'_2 = \mathit{mapU}(2, f')$$
$$f'(\dot{a}'_i) = \mathsf{parm}(f, \dot{a}'_i, 1, 2)$$
$$\dot{a}'_i = \bigoplus_{j=0}^{1} \dot{a}_{2i+j}$$

When we expand $\dot{a}'_i$ and f' we get

$$\dot{a}'_i = \bigoplus_{j=0}^{1} \dot{a}_{2i+j} = \dot{a}_{2i} \oplus \dot{a}_{2i+1}$$

$$f'(\langle \dot{a}'_{2i}, \dot{a}'_{2i+1} \rangle) = \bigoplus_{j=0}^{1} f(\mathsf{sseq}(\dot{a}'_i, j, 1))$$
$$= f(\mathsf{sseq}(\dot{a}'_i, 0, 1)) \oplus f(\mathsf{sseq}(\dot{a}'_i, 1, 1))$$
$$= f(\dot{a}_{2i}) \oplus f(\dot{a}_{2i+1})$$

Assuming $\dot{s} = \langle (\langle 10 \rangle, \langle 1, 2, 3, 4, 5 \rangle), (\langle 10 \rangle, \langle 6, 7, 8, 9, 10 \rangle) \rangle$, we have
$$A_2(\dot{s}) = \langle f((\langle 10 \rangle, \langle 1, 2, 3, 4, 5 \rangle)), f((\langle 10 \rangle, \langle 6, 7, 8, 9, 10 \rangle)) \rangle$$
$$= \langle 10, 20, 30, 40, 50, 60, 70, 80, 90, 100 \rangle$$
$$A'_2(\dot{s}) = \langle f'(\langle (\langle 10 \rangle, \langle 1, 2, 3, 4, 5 \rangle), (\langle 10 \rangle, \langle 6, 7, 8, 9, 10 \rangle) \rangle) \rangle$$
$$= \langle f(\mathsf{sseq}(\langle (\langle 10 \rangle, \langle 1, 2, 3, 4, 5 \rangle), (\langle 10 \rangle, \langle 6, 7, 8, 9, 10 \rangle) \rangle, 0, 1))$$
$$\oplus f(\mathsf{sseq}(\langle (\langle 10 \rangle, \langle 1, 2, 3, 4, 5 \rangle), (\langle 10 \rangle, \langle 6, 7, 8, 9, 10 \rangle) \rangle, 1, 1)) \rangle$$
$$= \langle f((\langle 10 \rangle, \langle 1, 2, 3, 4, 5 \rangle)) \oplus f((\langle 10 \rangle, \langle 6, 7, 8, 9, 10 \rangle)) \rangle$$
$$= \langle 10, 20, 30, 40, 50, 60, 70, 80, 90, 100 \rangle$$

3.10.2 Scan-Based Processes

Up-rating scan-based processes results in Mealy-based processes because the up-rated process maintains as internal state only a part of the state sequence of the original process. However, the entire original state sequence must be emitted at the output. Hence, an output encoding function is required.

Definition 3.7 Let $p = \mathtt{scanU}(\gamma, g, w_0)$ with $\pi(v_p, \dot{s}) = \langle \dot{a}_i \rangle$ and $\pi(v'_p, \dot{s}') = \langle \dot{a}'_i \rangle$ defining the partitionings of the input and output signals $\dot{s}$ and $\dot{s}'$, respectively. Then the process $p_1 = \mathtt{uprate}(p, \varrho)$ is defined as follows:

$$p_1 = \mathtt{mealyU}(\gamma_1, g_1, f_1, v_0)$$

where

$$\pi(v_{p_1}, \dot{s}) = \langle u_i \rangle, u_i = \bigoplus_{j=0}^{\varrho-1} \dot{a}_{i\varrho+j}$$

$$\pi(v'_{p_1}, \dot{s}') = \langle u'_i \rangle, u'_i = \bigoplus_{j=0}^{\varrho-1} \dot{a}'_{i\varrho+j}$$

$$v_{p_1}(i) = \sum_{j=0}^{\varrho-1} v_p(i\varrho + j)$$

$$\gamma_1(v_i) = \sum_{j=0}^{\varrho-1} \gamma(w_{i\varrho+j})$$

$$g_1(v_i, u_i) = w_{(i+1)\varrho} = v_{i+1}$$

$$f_1(v_i, u_i) = \bigoplus_{j=1}^{\varrho} w_{i\varrho+j}$$

with

$$w_{i\varrho} = v_i$$

and

$$w_{i\varrho+j} = g(\dot{a}_{i\varrho+j}, w_{i\varrho+j-1}), \ 1 \leq j \leq \varrho$$

provided there exists a function γ_1 such that

$$\gamma_1(v_i) = \sum_{j=0}^{\varrho-1} \gamma(w_{i\varrho+j}) \qquad (3.16)$$

Because neither the number of events consumed nor the number of events produced in each firing cycle is constant, these numbers cannot be merely multiplied by ϱ to derive the corresponding numbers for the up-rated process. Rather we take ϱ number of input sequences of p to get one input sequence of p_1. Similarly, p_1 produces ϱ number of output sequences of p in each firing cycle.

Note that the new state v_i represents only every ϱth state of the original process p. The intermediate state values are calculated on the fly by functions g_1 and f_1. The alternative of incorporating the entire original state $w_{i\varrho} \ldots w_{i\varrho+(\varrho-1)}$ into the new state vector v_i would give rise to difficulties in determining the initial state vector v_0, which would not be known completely before the computation sets off.

3.10 Process Up-rating

Note finally that condition (3.16) suggests that not every scan-based process can be up-rated. If in a particular state the number of events to be consumed next cannot be known because it depends on the first few events to be read, the process cannot be up-rated and condition (3.16) is not met.

We will not prove that the up-rate operator of Definition 3.7 in fact always results in an up-rated process variant because the proof runs similarly to the map case in the previous section and we also prove the more general case for Mealy-based processes below.

Example 3.12 **Amplifier** We up-rate process A_3 by a factor of 2 to derive process $A_3' = \mathtt{uprate}(A_3, 2)$.

Recall the definition of A_3:

$$A_3 = \mathtt{scanU}(\gamma, g, w_0)$$

where

$\dot{s}' = A_3(\dot{s})$

$w_0 = 10$

$\gamma(w_i) = 5 \; \forall i \in \mathbb{N}_0$

$\pi(5, \dot{s}) = \langle \dot{a}_i \rangle$ and $\pi(1, \dot{s}') = \langle \dot{a}_i' \rangle$

$$g(\langle x_1, x_2, x_3, x_4, x_5 \rangle, w_i) = \begin{cases} w_i - 1 & \text{if } x_1 + x_2 + x_3 + x_4 + x_5 > 500 \\ w_i + 1 & \text{if } x_1 + x_2 + x_3 + x_4 + x_5 < 400 \\ w_i & \text{otherwise} \end{cases}$$

A_3' can then be defined as follows:

$$A_3' = \mathtt{mealyU}(2 \cdot 5, g', f', v_0)$$

where

$v_0 = w_0 = 10$

$\pi(10, \dot{s}) = \langle u_i \rangle, u_i = \dot{a}_{2i} \oplus \dot{a}_{2i+1}$

$\pi(2, \dot{s}') = \langle u_i' \rangle, u_i' = \dot{a}_{2i}' \oplus \dot{a}_{2i+1}'$

$f'(u_i, v_i) = u_i' = g(\dot{a}_{2i}, v_i) \oplus g(\dot{a}_{2i+1}, g(\dot{a}_{2i}, v_i))$

$g'(u_i, v_i) = v_{i+1} = g(\dot{a}_{2i+1}, g(\dot{a}_{2i}, v_i))$

When we feed A_3' with the first 10 inputs, we derive

$A_3'(\langle 10, 20, 30, 40, 50, 60, 70, 80, 90, 100 \rangle)$

$= f'(u_0, v_0) = f'(\langle 10, 20, 30, 40, 50, 60, 70, 80, 90, 100 \rangle, 10)$

$= g(\langle 10, 20, 30, 40, 50 \rangle, 10)$

$\oplus g(\langle 60, 70, 80, 90, 100 \rangle, g(\langle 10, 20, 30, 40, 50 \rangle, 10))$

$= 11 \oplus g(\langle 60, 70, 80, 90, 100 \rangle, 11) = \langle 11, 11 \rangle$

which is what we expect.

3.10.3 Mealy-Based Processes

Mealy-based processes are up-rated in a similar way as scan-based processes, but we also have to deal with the output encoding function.

Definition 3.8 Consider a process $p = \text{mealyU}(\gamma, g, f, w_0)$ with $\pi(v_p, \dot{s}) = \langle \dot{a}_i \rangle$ and $\pi(v'_p, \dot{s}') = \langle \dot{a}'_i \rangle$ defining the partitionings of the input and output signals $\dot{s}$ and $\dot{s}'$, respectively. Then the process $q = \text{uprate}(p, \varrho)$ is defined as follows:

$$q = \text{mealyU}(\gamma_1, g_1, f_1, v_0)$$

$$v_i = w_{i\varrho}$$

$$\pi(v_q, \dot{s}) = \langle \dot{b}_i \rangle, \dot{b}_i = \bigoplus_{j=0}^{\varrho-1} \dot{a}_{i\varrho+j}$$

$$\pi(v'_q, \dot{s}') = \langle \dot{b}'_i \rangle, \dot{b}'_i = \bigoplus_{j=0}^{\varrho-1} \dot{a}'_{i\varrho+j}$$

$$v_q(i) = \gamma_1(v_i)$$

$$f_1(\dot{b}_i, v_i) = \bigoplus_{j=0}^{\varrho-1} f(\dot{a}_{i\varrho+j}, w_{i\varrho+j})$$

$$g_1(\dot{b}_i, v_i) = g(\dot{a}_{i\varrho+\varrho-1}, w_{i\varrho+\varrho-1})$$

with

$$w_{i\varrho} = v_i$$

and

$$w_{i\varrho+j} = g(w_{i\varrho+j-1}, \dot{a}_{i\varrho+j-1}),\ 1 \leq j \leq \varrho - 1$$

provided there exists a function γ_1 such that

$$\gamma_1(v_i) = \sum_{j=0}^{\varrho-1} \gamma(w_{i\varrho+j}) \quad \forall i \geq 0 \qquad (3.17)$$

As was the case with scan-based processes, condition (3.17) requires that it is possible in every evaluation cycle to know how many events to consume before the process is activated. This is not possible if γ depends on the event values of the previous evaluation cycle. Consequently, up-rating is not always possible for arbitrary ϱ. It is possible, however, for a number of interesting processes. Up-rating is inherently related to the parallelization of computation. Typically, the more data that can be consumed in one evaluation cycle, the more that can be done in parallel. If up-rating is not possible, there exists a data dependency that also prohibits further parallelization of the computation.

3.10 Process Up-rating

Theorem 3.7 For a *mealyU*-based process p and a number $\varrho \in \mathbb{N}$, the process $q = $ uprate(p, ϱ), if it exists, is an ϱ-up-rated variant of p.

Proof For $q = $ mealyU$(\gamma_1, g_1, f_1, v_0)$, $\dot{s}' = q(\dot{s})$ with $\pi(v_q, \dot{s}) = \langle \dot{b}_i \rangle$ and $\pi(v'_q, \dot{s}') = \langle \dot{b}' \rangle$, we have to show that

$$p\left(\bigoplus_{i=0}^{j} \dot{b}_j\right) = q\left(\bigoplus_{i=0}^{j} \dot{b}_j\right) \quad \forall j \in \mathbb{N}_0$$

We prove this by induction.

Base case $j = 0$:

$$p(\dot{b}_0) = q(\dot{b}_0)$$

$$p(\dot{b}_0) = p\left(\bigoplus_{k=0}^{\varrho-1} \dot{a}_k\right) = \bigoplus_{k=0}^{\varrho-1} f(w_k, \dot{a}_k)$$

with

$$w_{k+1} = g(w_k, \dot{a}_k)$$

$$q(\dot{b}_0) = f_1(v_0, \dot{b}_0) = \bigoplus_{k=0}^{\varrho-1} f(\dot{a}_k, w_k)$$

with

$$v_0 = w_0$$

and

$$w_{k+1} = g(w_k, \dot{a}_k)$$

As a consequence we also obtain the following relation, which we will need in the induction step:

$$v_1 = g_1(v_0, \dot{b}_0) = g(w_{\varrho-1}, \dot{a}_{\varrho-1}) = w_\varrho$$

Induction step: As a result from the base step, we can also assume that $v_j = w_{j\varrho}$. Hence,

$$v_j = w_{j\varrho} \wedge p\left(\bigoplus_{i=0}^{j} \dot{b}_i\right) = q\left(\bigoplus_{i=0}^{j} \dot{b}_i\right) \Rightarrow p\left(\bigoplus_{i=0}^{j+1} \dot{b}_i\right) = q\left(\bigoplus_{i=0}^{j+1} \dot{b}_i\right)$$

$$p\left(\bigoplus_{i=0}^{j+1} \dot{b}_i\right) = p\left(\bigoplus_{i=0}^{j} \dot{b}_i \oplus \dot{b}_{j+1}\right)$$

$$= p\left(\bigoplus_{i=0}^{j} \dot{b}_i\right) \oplus \bigoplus_{k=0}^{\varrho-1} f(w_{(j+1)\varrho+k}, \dot{a}_{(j+1)\varrho+k})$$

with
$$w_{(j+1)\varrho+k+1} = g(w_{(j+1)\varrho+k}, \dot{a}_{(j+1)\varrho+k})$$

$$q\left(\bigoplus_{i=0}^{j+1} \dot{b}_i\right) = q\left(\bigoplus_{i=0}^{j} \dot{b}_i \oplus \dot{b}_{j+1}\right)$$

$$= q\left(\bigoplus_{i=0}^{j} \dot{b}_i\right) \oplus f_1(v_{j+1}, \dot{b}_{j+1})$$

$$= q\left(\bigoplus_{i=0}^{j} \dot{b}_i\right) \oplus \bigoplus_{k=0}^{\varrho-1} f(w_{(j+1)\varrho+k}, \dot{a}_{(j+1)\varrho+k})$$

with
$$w_{(j+1)\varrho} = v_{j+1}$$

and
$$w_{(j+1)\varrho+k+1} = g(w_{(j+1)\varrho+k}, \dot{a}_{(j+1)\varrho+k})$$

Example 3.13   Consider a process B that emits the sum of a window of five input values. It emits one output for every two inputs.

$$B = \mathtt{mealyU}(\gamma, g, f, w_0)$$

where
$$w_0 = \langle 0, 0, 0, 0, 0 \rangle$$
$$\gamma(w_i) = 2 \ \forall i \in \mathbb{N}_0$$
$$g(\langle v_1, v_2, v_3, v_4, v_5 \rangle, \langle x_1, x_2 \rangle) = \langle v_3, v_4, v_5, x_1, x_2 \rangle$$
$$f(\langle v_1, v_2, v_3, v_4, v_5 \rangle, \langle x_1, x_2 \rangle) = \langle v_1 + v_2 + v_3 + v_4 + v_5 \rangle$$

We up-rate this process by a factor of 2 to derive process $B' = \mathtt{uprate}(B, 2)$.

$$B' = \mathtt{mealyU}(2 \cdot 2, g', f', v_0)$$
$$v_0 = w_0 = \langle 0, 0, 0, 0, 0 \rangle$$
$$\gamma_1 = \gamma + \gamma = 4$$
$$u_i = \dot{a}_{2i} \oplus \dot{a}_{2i+1}$$
$$u'_i = \dot{a}'_{2i} \oplus \dot{a}'_{2i+1}$$
$$f'(u_i, v_i) = f(\dot{a}_{2i}, w_{2i}) \oplus f(\dot{a}_{2i+1}, w_{2i+1})$$
$$g'(u_i, v_i) = g(\dot{a}_{2i+1}, w_{2i+1})$$

with
$$w_{2i} = v_i$$

and

$$w_{2i+1} = g(\dot{a}_{2i}, w_{2i}) \quad \text{for } i \in \mathbb{N}_0$$

Assuming $\dot{s} = \langle 2, 4, 6, 8 \rangle$, we have

$$\begin{aligned}
B(\dot{s}) &= \langle f(\langle 2,4 \rangle, \langle 0,0,0,0,0 \rangle), f(\langle 6,8 \rangle, g(\langle 2,4 \rangle, \langle 0,0,0,0,0 \rangle)) \rangle \\
&= \langle 0, 6 \rangle \\
B'(\dot{s}) &= \langle f'(u_0, v_0) \rangle \\
&= \langle f'(\langle 2,4,6,8 \rangle, \langle 0,0,0,0,0 \rangle) \rangle \\
&= f(\langle 2,4 \rangle, \langle 0,0,0,0,0 \rangle) \oplus f(\langle 6,8 \rangle, g(\langle 2,4 \rangle, \langle 0,0,0,0,0 \rangle)) \\
&= \langle 0, 6 \rangle
\end{aligned}$$

3.10.4 Processes with Multiple Inputs

The generalization of up-rating to processes with multiple inputs is straightforward.

Definition 3.9 Let ϱ be a natural number, $\varrho > 0$, and let p be a U-MoC process with n inputs and m outputs and the input signals s_l, $1 \leq l \leq n$ are partitioned $\pi(v_{p,l}, s_l) = \langle a_{i,l} \rangle$. Process p is *up-rated* by a factor ϱ, resulting in another process q, if q is continuous and the input signal partitioning of q is $\pi(v_{q,l}, s_l) = \langle b_{j,l} \rangle$ with

$$b_{j,l} = \bigoplus_{i=0}^{\varrho-1} a_{j\varrho+i,l} \quad \forall j \in \mathbb{N}_0, 1 \leq l \leq n$$

and p behaves identically to q for all increasing prefixes of input signals defined by $b_{j,l}$:

$$p\left(\left(\bigoplus_{i=0}^{j-1} b_{i,1}\right), \cdots, \left(\bigoplus_{i=0}^{j-1} b_{i,n}\right)\right) = q\left(\left(\bigoplus_{i=0}^{j-1} b_{i,1}\right), \cdots, \left(\bigoplus_{i=0}^{j-1} b_{i,n}\right)\right) \quad \forall j \in \mathbb{N}_0$$

`zipU` processes cannot be up-rated in a simple way because the values on the control input determine the partitionings of the data signals. Consuming ϱ times the control values does not mean that ϱ should be consumed. More fundamentally, this problem arises because `zipU` processes violate our stated ideal of separating the computation of a process from its communication interface. Whenever we do this we pay a price in terms of weaker formal properties.

However, we can safely up-rate the `zipUs` and `unzipU` processes. Consider a process $p = \mathtt{zipUs}(c_1, c_2)$. The process $q = \mathsf{uprate}(p, \varrho)$ is defined by

$$q = q_2 \circ q_1$$
$$q_1 = \mathtt{zipUs}(\varrho c_1, \varrho c_2)$$
$$q_2 = \mathtt{mapU}(1, f)$$

$$f((\dot a_1, \dot a_2)) = \begin{cases} \langle\rangle & \text{if length}(\dot a_1) < c_1 \vee \text{length}(\dot a_2) < c_2 \\ \langle(\mathsf{take}(c_1, \dot a_1), \mathsf{take}(c_2, \dot a_2))\rangle \\ \oplus f(\mathsf{drop}(c_1, \dot a_1), \mathsf{drop}(c_2, \dot a_2)) & \text{otherwise} \end{cases}$$

In each evaluation cycle, q consumes ϱ times as many events on each input as p, and it produces exactly ϱ output values since p produces one event.

Consider a process $p = \mathtt{unzipU}()$. The process $q = \mathsf{uprate}(p, \varrho)$ is defined by

$$q = q_2 \circ q_1$$
$$q_1 = \mathtt{mapU}(\varrho, f)$$
$$f(\langle(\dot a_1, \dot b_1), \cdots, (\dot a_\varrho, \dot b_\varrho)\rangle) = (\dot a_1 \oplus \dot a_2 \oplus \cdots \oplus \dot a_\varrho, \dot b_1 \oplus \dot b_2 \oplus \cdots \oplus \dot b_\varrho)$$
$$q_2 = \mathtt{unzipU}()$$

First, q_1 assembles ϱ events, which are pairs, into an appropriate pair of sequences. Then, q_2 unzips them and emits ϱ times as many output events in each evaluation cycle as p.

3.10.5 Up-rating and Process Composition

Up-rating a process preserves its behavior up to the remainder of the input signal. For a process $p : S \to S$ and its up-rated variant $q = \mathsf{uprate}(p, \varrho)$, we have

$$p(\pi(\nu_q, s)) = q(s) \quad \text{for all } s \in S$$

In general, q may operate only on a prefix of s, that is, on $\pi(\nu_q, s)$. But for this prefix of s both processes compute the same output. This follows directly from the definition of the (uprate(,)) operator, and therefore it complies with Definition 3.5.

Up-rating can be applied to compound processes as well:

$$\mathsf{uprate}((p \circ q), \varrho) = (\mathsf{uprate}(p, \varrho)) \circ (\mathsf{uprate}(q, \varrho))$$
$$\mathsf{uprate}((p \parallel q), \varrho) = (\mathsf{uprate}(p, \varrho)) \parallel (\mathsf{uprate}(q, \varrho))$$

Unfortunately, $\mathsf{uprate}(\mathbf{FB_P}(p), \varrho) = \mathbf{FB_P}(\mathsf{uprate}(p, \varrho))$ does not hold because (uprate(p, ϱ)) awaits more input data than p before computing output, which may never arrive, since, being in a feedback loop, it has to provide this data itself. This has the unfortunate consequence that the techniques for process merging and process migration, to be developed in the next sections, cannot be applied to move processes into and out of feedback loops. More elaborate techniques, which are subject of current research but beyond the scope of this book, must be used.

Note, however, that we can still up-rate process $\mathbf{FB_P}(p)$ by placing wrapper processes on its inputs and outputs, which collect and emit the appropriate number of events. But this is of little value because it still does not allow us to merge a process from outside the feedback loop with a process inside.

3.11 Process Down-rating

Definition 3.10   Let p be a U-MoC process and ϱ be a natural number, $\varrho > 0$. If there exists as process q such that $p = \mathsf{uprate}(q, \varrho)$, we say that p can be *down-rated* by a factor ϱ and we write $\mathsf{downrate}(p, \varrho) = q$.

As for up-rating, down-rating may not always be possible. Clearly, it is not possible if a process consumes one event in each evaluation cycle. It may also not be possible if it consumes more than one event because its characteristic function needs several values to evaluate. To rectify this, individual events could be read in and stored internally until there are sufficient data. When more information about the process and its characteristic functions is available, more sensible down-rating methods can be developed.

The up-rating and down-rating operations together define equivalence classes for processes.

Definition 3.11   Let $\rightsquigarrow$ be the transitive closure of $\mathsf{uprate}(,)$ and $\mathsf{downrate}(,)$, that is, an arbitrary sequence of up-rating and down-rating operations with arbitrary factors greater than zero. If $p \rightsquigarrow q$ for U-MoC processes p and q, we say p and q are *rate equivalent*. The set of processes that is rate equivalent to p is denoted by $\stackrel{p}{\rightsquigarrow} = \{q \mid p \rightsquigarrow q\}$. There is a unique minimal element in every rate equivalence class, which is defined by $\stackrel{p}{\rightsquigarrow}_m = q \in \stackrel{p}{\rightsquigarrow}$ with $\forall q' \in \stackrel{p}{\rightsquigarrow}, \langle r_i \rangle = \pi(v_q, s), \langle r'_i \rangle = \pi(v_{q'}, s) : r_i < r'_i \, \forall i \geq 0, s \in S$.

The minimal element cannot be down-rated, but not every process that cannot be down-rated is a minimal element of its equivalent class.

3.12 Process Merge

Merging and splitting of processes should not change the behavior but can have a significant influence on non-functional properties such as performance, cost, and power consumption. Processes that form a perfect or rational match can be easily merged.

3.12.1 Perfect Match

We start by defining the merge transformation for simple, state-less processes. Then we discuss scan-based processes and, finally, the general case with Mealy machines.

FIGURE 3-17

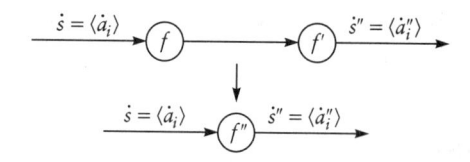

Composition of two map-based processes.

Map-Based Processes

Theorem 3.8 Given are two processes, p with one output signal and p' with one input signal. The two signals have the same type and form a perfect match. They are defined by $p = mapU(c, f)$ and $p' = mapU(c', f')$, respectively.

The composition of the two processes $p' \circ p$ (Figure 3-17) is identical with $p'' = mapU(c'', f'')$, defined as follows:

$$f''(\dot{a}_i) = \dot{a}_i'' = f'(f(\dot{a}_i))$$
$$c'' = c$$
$$v'_{p''} = v'_{p'}$$

Proof Both the input signal and output signal partitionings are identical for both processes p'' and $(p' \circ p)$. Thus, we only have to show that they behave identically in all activation cycles, that is, $f''(\dot{a}_i) = f'(f(\dot{a}_i)), \forall i \in \mathbb{N}_0$, which is obviously true.

Scan-Based Processes

Theorem 3.9 Given are two processes, p with one output signal and p' with one input signal. The two signals have the same type and form a perfect match. They are defined by $p = scanU(\gamma, g, w_0)$ and $p' = scanU(\gamma', g', w'_0)$, respectively.

The composition $p' \circ p$ (Figure 3-18) is identical with $p'' = mealyU(\gamma'', g'', f'', v''_0)$, defined as follows:

$$\gamma''(v''_i) = \gamma(w_i)$$
$$v''_0 = (v_0, v'_0) = (w_0, w'_0)$$
$$g''((v_i, v'_i), \dot{a}_i) = (v_{i+1}, v'_{i+1}) = (g(v_i, \dot{a}_i), g'(v'_i, v_i))$$
$$f''((v_i, v'_i), \dot{a}_i) = v'_i$$

Proof Again, since the input signal and output signal partitionings are identical for the two processes p'' and $(p' \circ p)$, we only have to show that they behave identically in all activation cycles. We do this by induction on i.

FIGURE 3-18

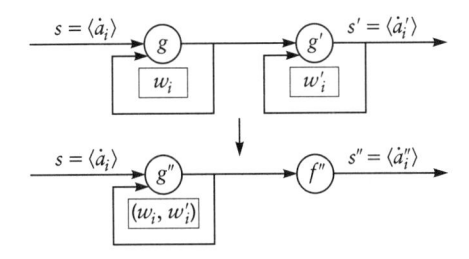

Composition of two scan-based processes.

Let $s' = \langle \dot{a}'_i \rangle = (p' \circ p)(s)$ and $s'' = \langle \dot{a}''_i \rangle = p''(s)$.

We show that $\dot{a}'_i = \dot{a}''_i, \forall i \in \mathbb{N}_0$.

$i = 0:$ $(w_0, w'_0) = (v_0, v'_0)$
$\dot{a}'_0 = w'_0$
$\dot{a}''_0 = f((v_0, v'_0), \dot{a}_0) = v'_0 = w'_0$

$i > 0:$ $\dot{a}'_i = g'(w'_{i-1}, w_{i-1})$
$\dot{a}''_i = f''((v_i, v'_i), \dot{a}_i) = v'_i = g'(v'_{i-1}, v_{i-1}) = g'(w'_{i-1}, w_{i-1})$

Not unexpectedly, the new internal state is (w_i, w'_i), which is the composition of the states of the two processes p and p'.

Mealy-Based Processes

Merging map-based processes yields map-based processes, and merging scan-based processes yields again scan-based processes. But merging a scan-based process with a map-based process results in a Moore or Mealy process. However, we don't yet know how to merge these kinds of processes. We must also cover these cases if we want to merge arbitrary processes. It suffices to discuss the merging of Mealy-type processes.

Theorem 3.10 Given are two processes, p with one output signal and p' with one input signal. The two signals have the same type and form a perfect match. They are defined by $p = \texttt{mealyU}(\gamma, g, f, w_0)$ and $p' = \texttt{mealyU}(\gamma', g', f', w'_0)$, respectively.

The composition $p' \circ p$ (Figure 3-19) is identical with $p'' = \texttt{mealyU}(\gamma'', g'', f'', v''_0)$, defined as follows:

$$v''_0 = (v_0, v'_0) = (w_0, w'_0)$$
$$g''((v_i, v'_i), \dot{a}_i) = (v_{i+1}, v'_{i+1}) = (g(v_i, \dot{a}_i), g'(v'_i, f(v_i, \dot{a}_i)))$$
$$f''((v_i, v'_i), \dot{a}_i) = \dot{a}''_i = f'(v'_i, f(v_i, \dot{a}_i))$$
$$\gamma''(v''_i) = \gamma(w_i)$$

FIGURE 3-19

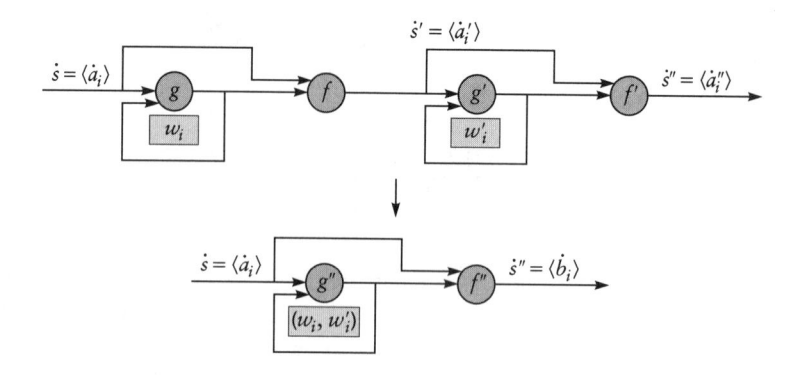

Composition of two Mealy-based processes into a new Mealy-based process.

Proof Again, since the input signal and output signal partitioning are identical for the two processes p'' and $(p' \circ p)$, we only have to show that they behave identically in all activation cycles.

We do this by induction on i. Let $\dot{s}'' = \langle \dot{a}_i'' \rangle = (p' \circ p)(\dot{s})$ and $\dot{r} = \langle \dot{b}_i \rangle = p''(\dot{s})$. We show that $\dot{a}_i'' = \dot{b}_i, \forall i \in \mathbb{N}_0$.

$i = 0$: $(v_0, v_0') = (w_0, w_0')$

$\dot{a}_0'' = f'(w_0', \dot{a}_0') = f'(w_0', f(w_0, \dot{a}_0))$

$\dot{b}_0 = f''(v_0'', \dot{a}_0) = f''((v_0, v_0'), \dot{a}_0) = f'((f(w_0, \dot{a}_0), w_0'))$

$i > 0$: $(v_i, v_i') = g''((v_{i-1}, v_{i-1}'), \dot{a}_{i-1})$

$= (g(v_{i-1}, \dot{a}_{i-1}), g'(v_{i-1}', f(v_{i-1}, \dot{a}_{i-1})))$

$= (w_i, w_i')$

$\dot{a}_i'' = f'(w_i', \dot{a}_i') = f'(w_i', f(w_i, \dot{a}_i))$

$\dot{b}_i = f''((v_i, v_i'), \dot{a}_i) = f'(v_i', f(v_i, \dot{a}_i))$

Again, as in the case of scan-based processes, the new internal state is (w_i, w_i'), the composition of the states of the two processes p and p'.

Example 3.14 **Amplifier** To be able to merge two processes in our amplifier example, we first copy process A_2 as shown in Figure 3-20 (a).

Recall the definitions of processes A_2 and A_3:

$$A_2 = mapU(c, f)$$

where

$$c = 1$$

$$f((\langle x \rangle, \langle y_1, y_2, y_3, y_4, y_5 \rangle)) = \langle xy_1, xy_2, xy_3, xy_4, xy_5 \rangle$$

3.12 Process Merge

FIGURE 3-20

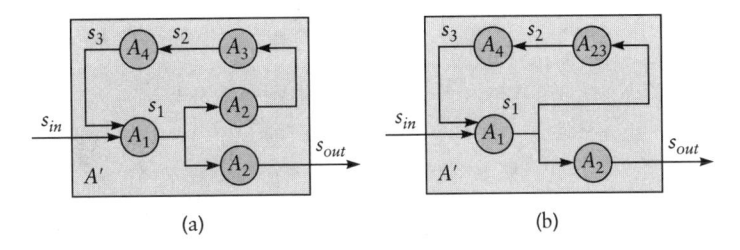

(a) We copy the process A_2 to be able to merge it with process A_3, (b) which yields process A_{23}.

and
$$A_3 = \text{scanU}(\gamma, g, w_0)$$

where
$$w_0 = 10$$
$$\gamma(w_i) = 5 \quad \forall i \in \mathbb{N}_0$$
$$g(w_i, \langle x_1, x_2, x_3, x_4, x_5 \rangle) = \begin{cases} w_i - 1 & \text{if } x_1 + x_2 + x_3 + x_4 + x_5 > 500 \\ w_i + 1 & \text{if } x_1 + x_2 + x_3 + x_4 + x_5 < 400 \\ w_i & \text{otherwise} \end{cases}$$

First, we model both processes as Mealy machines:
$$A'_2 = \text{mealyU}(\gamma_2, g_2, f_2, 0)$$

where
$$\gamma_2(0) = 1$$
$$f_2((\langle x \rangle, \langle y_1, y_2, y_3, y_4, y_5 \rangle), 0) = \langle xy_1, xy_2, xy_3, xy_4, xy_5 \rangle$$
$$g_2(x, 0) = 0$$

and
$$A'_3 = \text{mealyU}(\gamma_3, g_3, f_3, w_0)$$

where
$$w_0 = 10$$
$$\gamma_3(w_i) = 5 \quad \forall i \in \mathbb{N}_0$$
$$f_3(x, w_i) = w_i$$

154 chapter three *The Untimed Model of Computation*

$$g_3(\langle x_1, x_2, x_3, x_4, x_5 \rangle, w_i) = \begin{cases} w_i - 1 & \text{if } x_1 + x_2 + x_3 + x_4 + x_5 > 500 \\ w_i + 1 & \text{if } x_1 + x_2 + x_3 + x_4 + x_5 < 400 \\ w_i & \text{otherwise} \end{cases}$$

Next, we merge them to obtain process A_{23}:

$$A_{23} = \texttt{mealyU}(\gamma, f, g, (0, w_0))$$

where

$$\gamma((0, w_i)) = 1 \quad \forall i \in \mathbb{N}_0$$
$$g(\dot{a}_i, (0, w_i)) = (0, g_3(f_2(\dot{a}_i, 0), w_i))$$
$$f(\dot{a}_i, (0, w_i)) = f_3(f_2(\dot{a}_i, 0), w_i)$$

This process can be significantly optimized by avoiding the multiplication of each input value:

$$A'_{23} = \texttt{mealyU}(\gamma', f', g', w_0)$$

where

$$\gamma(w_i) = 1 \; \forall i \in \mathbb{N}_0$$

$$g'(((x), \langle y_1, y_2, y_3, y_4, y_5 \rangle), 0) = \begin{cases} w_i - 1 & \text{if } y_1 + y_2 + y_3 + y_4 + y_5 \\ & > 500/x \\ w_i + 1 & \text{if } y_1 + y_2 + y_3 + y_4 + y_5 \\ & > 500/x \\ w_i & \text{otherwise} \end{cases}$$

$$f'(x, w_i) = w_i$$

However, we have replaced five multiplications by one division. Hence, we have only gained something if this replacement is a true benefit. Moreover, A'_{23} can be even further simplified into a scan-based process.

3.12.2 Rational Match

Consider again two processes p_1 and p_2 that are connected by one signal (Figure 3-21). The two processes may be arbitrary; that is, they may be map-based, scan-based, or compound processes. We only require that their match is rational; thus

$$\text{match}(p_1, p_2) = \frac{v'_{p_1}(i)}{v_{p_2}(i)}$$

is a constant rational number.

When we merge the two processes, the compound process p_3 will require a different partitioning of the input and output signals.

FIGURE 3-21

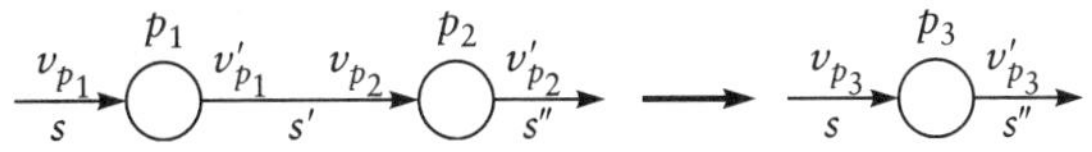

Composition of two processes with a rational match of $(v'_{p_1}(i))/(v_{p_2}(i))$. The signal names are shown below the signal lines and the signal partitioning numbers are above.

We can up-rate the two processes such that the resulting processes form a perfect match. Let

$$\text{match}(p_1, p_2) = \frac{v'_{p_1}(i)}{v_{p_2}(i)} = \frac{\varrho_2}{\varrho_1}$$

with two constant integers ϱ_1 and ϱ_2 that have no common integer divisor. Note that this is the case even if $v'_{p_1}(i)$ and $v_{p_2}(i)$ are not constant. Then the processes uprate(p_1, ϱ_1) and uprate(p_2, ϱ_2) form a perfect match and can be directly merged into $p_3 = (\text{uprate}(p_2, \varrho_2)) \circ (\text{uprate}(p_1, \varrho_1))$ with

$$v_{p_3}(i) = \sum_{j=0}^{\varrho_1 - 1} v_{p_1}(i\varrho_1 + j)$$

$$v'_{p_3}(i) = \sum_{j=0}^{\varrho_2 - 1} v'_{p_2}(i\varrho_2 + j)$$

3.13 Rugby Coordinates

In the terminology of the Rugby metamodel, untimed models are denoted as ⟨[Alg-LB], [IPC-Top], [Sym-LV], Caus⟩. Thus, the timing abstraction is the characteristic feature of untimed models. In the computation domain the algorithm or state machine is the most common means to express the behavior of individual processes. Logic blocks and logic equations are also reasonable for this purpose and have been used. Since in the untimed MoC the behavior is already broken down into processes, the system functions are not directly visible anymore. Hence, its Rugby coordinate does not include SystemFunctions although the individual processes may be modeled as functions.

In the other two domains, the untimed MoC is also positioned in the middle. Neither the lower levels (Layout, ContValue) nor the highest levels (InterfaceConstraints, DataTypeConstraints) are covered.

3.14 The Untimed Computational Model and Petri Nets

Recall that the signature of a process is expressed as a pair of sets (NI, NO), where NI contains the partitioning functions for the inputs and NO the partitioning functions

FIGURE 3-22

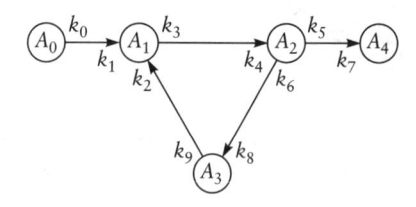

A process network with constant partitioning functions k_0 through k_9.

for the outputs. For a process network, where all processes have constant signatures, we have a useful relationship to Petri nets. We can map such a process network to a Petri net by representing each process by a transition and each signal by a place. The mapping results in an abstraction of the original process network because the data is abstracted into indistinguishable tokens.

Consider the process network in Figure 3-22 with signatures (NI_i, NO_i) for process A_i, $0 \leq i \leq 4$, and let

$$NI_0 = \{\} \qquad NO_0 = \{k_0\}$$
$$NI_1 = \{k_1, k_2\} \qquad NO_1 = \{k_3\}$$
$$NI_2 = \{k_4\} \qquad NO_2 = \{k_5, k_6\}$$
$$NI_3 = \{k_8\} \qquad NO_3 = \{k_9\}$$
$$NI_4 = \{k_7\} \qquad NO_4 = \{\}$$

with all k_j being constant natural numbers. We construct a Petri net where each process is represented by a transition and each signal by a place. Each arc in the Petri net represents an input or output of a process, and the weight of an arc is the corresponding constant partitioning function. Transition t_0 is an input and corresponds to the source process A_0; transition t_4 is an output corresponding to the sink process A_4.

The Petri net in Figure 3-23 represents the process network of Figure 3-22. The transition t_i represents process A_i, and the arcs are annotated by their weight factors, which represent the partitioning of the corresponding input or output signal.

This correspondence has the delightful consequence that we can apply all Petri-net-based analysis techniques, for example, those we introduced in Section 2.3. Unfortunately, the restriction on the process signatures is rather severe. We can relax it a bit by observing that we only need rational matches between all processes. We can then use these ratios to determine the weights on the Petri net arcs.

For instance, let A_1 and A_2 be two processes with $NO_1 = \{v_1\}$ and $NI_2 = \{v_2\}$, and let them be connected as shown in Figure 3-24(a). If the processes form a rational match with $v_1(i)/v_2(i) = k_1/k_2$ for all i, and k_1, k_2 are two nonnegative integer constants, we can represent them by the Petri net of Figure 3-24(b).

Of course, it would be desirable to map arbitrary process networks onto Petri nets. However, this is only possible if the functional behavior of the processes is taken into account because the firing of transitions depends on the number of tokens on its

FIGURE 3-23

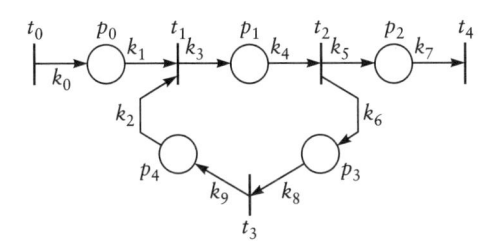

A Petri net that represents the process network of Figure 3-22.

FIGURE 3-24

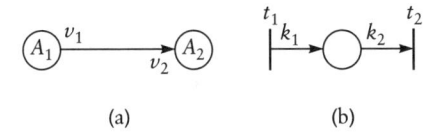

(a)   (b)

A process network with only rational matches can be represented by a Petri net. In this case the process network in (a) can be represented by the Petri net in (b) if $v_1(i)/v_2(i) = k_1/k_2$ for all i.

input places, which in turn depends on the behavior of the corresponding process. We can do this with any accuracy we need. To illustrate this, let's consider the situation where a process has n different internal states; i_q is the number of tokens consumed in state q, and o_q is the number of tokens produced in state q, $1 \leq q \leq n$. We avoid complete accuracy by modeling the state transition function nondeterministically. We assume that in a given state q the number of input tokens is always the same and designate it as k_q; similarly, the number of output tokens is l_q in state q. This is an additional restriction for general processes because for the processes based on constructor `mealyU` (Equation (3.3)) the number of output tokens depends both on the current state and on the current input. But under this additional assumption we can use a Petri net that, for each state, consumes and produces the correct number of tokens while it switches between the states nondeterministically.

The main idea is illustrated by the Petri net in Figure 3-25, which models a process with two states. Place p_1 is the input of the process, and p_7 is the output. When a token is in p_4, the net is in state 1; when a token is in place p_5, the process is in state 2. In the beginning there is a token in place p_6 and either transition t_5 or t_6 can fire. The choice is nondeterministic. If t_5 fires, the process enters state 1 and fills p_4 with k_1 tokens; if t_6 fires, state 2 is entered and p_5 obtains k_2 tokens. Let's consider the consumption and production of tokens in state 1. We have k_1 tokens in place p_4. Hence, when tokens appear at input place p_1, transition t_1 fires and fills place p_2. It can fire at most k_1 times; then p_4 becomes empty and p_2 contains k_1 tokens. At this moment transition t_3 can

FIGURE 3-25

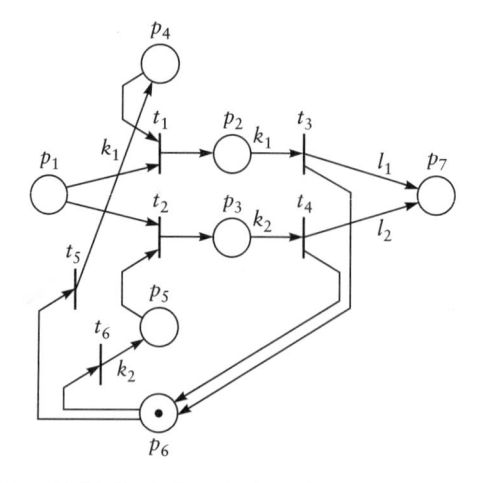

This Petri net represents a process with two states, that are randomly selected. p_1 is the input place, and p_7 is the output place. In state 1, k_1 tokens are consumed from p_1 and l_1 are produced in p_7; in state 2, k_2 tokens are consumed and l_2 are produced.

fire the first time; it consumes all k_1 tokens from p_2 and adds l_1 tokens to the output place p_7. It also sends a token to place p_6, which corresponds to the Petri net's original state, but it has consumed k_1 tokens from p_1 and it has added l_1 tokens to place p_7. This is exactly what our process is supposed to do in state 1. Now the Petri net can again nondeterministically select to fire transition t_5 or t_6, corresponding to the activation of state 1 or 2, respectively.

The usefulness of such a model depends of course on our purposes and objectives. The more our purposes concern the detailed internal behavior of the process, the more accurate it has to be modeled by a Petri net. There are many possible techniques that cannot be described here. But we hope the example has illustrated how a more abstract Petri net model can be obtained from a more detailed model of process.

3.15 Synchronous Dataflow

Synchronous dataflow (SDF)[2] is a special case of the untimed model of computation. It significantly restricts its expressiveness, but in return we obtain some nice properties that allow efficient solutions to important problems. In particular we will investigate the problem of finding static schedules for single- and multi-processor implementations and the problem of buffer dimensioning.

---

[2]Here the term "synchronous" means "regular" or "static" and should not be confused with another common meaning of the the word denoting "at the same time." The expression "static dataflow" has been proposed, but that name is reserved for dataflow networks with a static number of processes and connections, in contrast to *dynamic dataflow*, where connections and processes can be created and terminated dynamically.

FIGURE 3-26

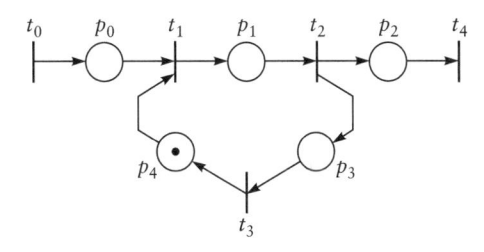

A Petri net that represents the process network of Figure 3-22 with all constants $k_i = 1$.

Definition 3.12 *Synchronous dataflow* is an untimed MoC where all processes define only *constant partitionings* for all their input and output signals; that is, all partitioning functions are constant, and all process signatures are constant.

A more formal definition along the lines of Section 3.7 can easily be given (see Exercise 3.4).

Recall that we have defined a mapping to Petri nets for process networks with this property (Section 3.14). We use this mapping to derive the incidence matrix (Section 2.3) for SDF models, which we can use to compute schedules and maximum buffer sizes.

Example 3.15 Consider again the process network in Figure 3-22 and let all $k_i = 1$, $0 \le i \le 9$. Its Petri net representation is shown in Figure 3-26 with an initial marking of one token in place p_4.

Its incidence matrix is

$$\mathcal{A} = \begin{bmatrix} k_0 & 0 & 0 & 0 & 0 \\ -k_1 & k_3 & 0 & 0 & -k_2 \\ 0 & -k_4 & k_5 & k_6 & 0 \\ 0 & 0 & 0 & -k_8 & k_9 \\ 0 & 0 & -k_7 & 0 & 0 \end{bmatrix} = \begin{bmatrix} 1 & 0 & 0 & 0 & 0 \\ -1 & 1 & 0 & 0 & -1 \\ 0 & -1 & 1 & 1 & 0 \\ 0 & 0 & 0 & -1 & 1 \\ 0 & 0 & -1 & 0 & 0 \end{bmatrix}$$

where the ith row represents transition t_i (process A_j), $0 \le i \le 4$, and the jth column represents place p_j, $0 \le j \le 4$.

The initial state of the Petri net is $\vec{x}_0 = [0, 0, 0, 0, 1]$, and the firing vector $\vec{u}_0 = [1, 0, 0, 0, 0]$ designates the firing of transition t_0. More generally, let $\vec{u}_i$ be the firing vector that designates the firing of transition t_i. Then the sequence $\vec{u}_0, \vec{u}_1, \vec{u}_2, \vec{u}_3, \vec{u}_4$ represents the sequential execution of the processes A_0, A_1, A_2, A_3, A_4. If we follow the evolution of the Petri net by evaluating the equation

$$\vec{x} = \vec{x}_0 + (\vec{u}_0 + \vec{u}_1 + \vec{u}_2 + \vec{u}_3 + \vec{u}_4) \mathcal{A}$$

$$= [0,0,0,0,1] + [1,1,1,1,1] \begin{bmatrix} 1 & 0 & 0 & 0 & 0 \\ -1 & 1 & 0 & 0 & -1 \\ 0 & -1 & 1 & 1 & 0 \\ 0 & 0 & 0 & -1 & 1 \\ 0 & 0 & -1 & 0 & 0 \end{bmatrix}$$

$$= [0,0,0,0,1] + [0,0,0,0,0]$$

$$= [0,0,0,0,1]$$

we observe that the Petri net again reaches its original state $\vec{x}_0$.

This example illustrates that we can use the Petri net to evaluate a synchronous dataflow model without the need to evaluate the individual nodes.

Even though synchronous dataflow is a restricted model, there are important applications that can be expressed in terms of SDF. Image-processing, audio-processing, and signal-processing applications for wireless communication devices are examples where very regular streams of data are processed at a constant rate. The amount of data received, processed, and emitted per time unit is constant without fluctuations in the data rate. Because of this feature, SDF is very appropriate for modeling these applications.

An SDF process network can be implemented by mapping each process on its own exclusive resource, which can be a custom hardware block or a processor. But more commonly, more than one process will share a single resource. If the performance requirements are moderate, all processes can be mapped on a single processor, which is a very cost-effective solution. In this case the question is how to schedule the processes. There are two principal ways to schedule multiple processes on shared resources. *Static scheduling* computes a schedule once and applies it periodically during run time. *Dynamic scheduling* computes a schedule dynamically during run time based on the status of the processes and the availability of resources. Dynamic scheduling is more flexible and can adapt to varying computation loads and resource configurations. However, it implies a sometimes significant run-time overhead since scheduling decisions have to be computed frequently. Static scheduling does not incur any run-time overhead, but it cannot always be applied. If applicable, static scheduling is clearly preferred due to its cost effectiveness and simpler design. Whenever we have regular, predictable performance requirements, a static schedule may be an option and should be investigated. Obviously, SDF models fit very well with these requirements, and it turns out that we can always find a static schedule for an SDF process network if any schedule, dynamic or static, exists.

Since we assume very long, possibly infinite streams of input data, any static schedule has to be applied periodically. In Example 3.15 we saw that firing the sequence t_0, t_1, t_2, t_3, t_4 returns the Petri net to its original state. This implies we can fire this sequence again and again ad infinitum. The corresponding schedule, that is, the sequence of processes to be activated, A_0, A_1, A_2, A_3, A_4, therefore constitutes a periodic schedule that can be applied as long as there is input data to be processed.

In the following sections we will

- give necessary and sufficient conditions for the existence of periodic, static schedules

- give a practical algorithm to construct a periodic, sequential schedule
- give a practical algorithm to construct a periodic, parallel schedule

3.15.1 Single-Processor Schedule

In order to execute or schedule a particular process, there must be sufficient data on its input signals. To capture this requirement we somehow intuitively define the status of the buffers between processes.

Definition 3.13 Let s be a signal between two processes A and B, where s is an output of A and an input of B. The *initial buffer condition of s* is the number of events in s before A and B are executed the first time.

The number of events buffered in s is the number of events initially in s or produced by executions of A but not yet consumed by B.

Definition 3.14 An *admissible, sequential schedule* ϕ is a nonempty sequence of processes such that if the processes are executed in the sequence given by ϕ, the number of events buffered in any signal will remain nonnegative and bounded.

A *periodic, admissible, sequential schedule* (PASS) is a periodic and infinite admissible sequential schedule. It is specified by a list ϕ that is the list of processes executed in one period.

In the terminology of Petri nets, a PASS is a periodic, infinite sequence of transitions where the number of tokens in all places remain nonnegative and bounded. For instance, $\phi = \langle A_0, A_1, A_2, A_3, A_4, A_5 \rangle$ is a PASS for the process network in Example 3.15 but neither $\langle A_0, A_2, A_1, A_3, A_4 \rangle$ nor $\langle A_0, A_1, A_1, A_2, A_3, A_4 \rangle$ is.

The Rank Test

In Example 3.15 we saw that when we multiplied the sum of the invocation vectors of a schedule with the incidence matrix, we ended up in the original state of the Petri net. We can generalize this observation. Let ϕ be a schedule, that is, a sequence of transitions, $\phi = \langle t_i, t_k, \ldots, \rangle$. A transition of a Petri net may occur zero, one, or several times. Each transition t_i is represented by a firing vector $\vec{u}_i$ as defined in Definition 2.16. Let $\vec{q}_\phi$ be the firing vector representing the schedule ϕ, that is,

$$\vec{q}_\phi = \sum_{t_i \in \phi} \vec{u}_i$$

Apparently, ϕ can only be a periodic schedule with all places being bounded, if

$$\vec{q}_\phi \mathcal{A} = \vec{0} \qquad (3.18)$$

where $\vec{0}$ is the vector with all zeros.

162 chapter three *The Untimed Model of Computation*

FIGURE 3-27

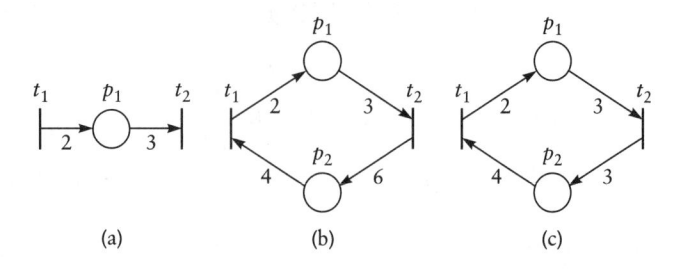

(a) (b) (c)

For both Petri nets in (a) and (b), the relative transition frequency of the transitions t_1 and t_2 must be 2/3 for any admissible periodic schedule. For the net in (c) there exists no periodic admissible schedule.

We can view this as a set of equations where $\vec{q}_\phi$ is the vector of variables we want to solve the equations for. Thus, we have N_p equations and N_t variables, where N_p is the number of places and N_t the number of transitions. In general we need at most n independent equations to solve them for n variables. However, in our case not all the equations can be mutually independent. Each equation couples exactly two variables because each place connects exactly two transitions.

Example 3.16 Consider the Petri net in Figure 3-27(a). The incidence matrix is

$$\mathcal{A}_a = \begin{bmatrix} 2 \\ -3 \end{bmatrix}$$

and the equation we can set up according to equation (3.18) is

$$2q_1 - 3q_2 = 0$$

where $\vec{q}_\phi = [q_1, q_2]$. This is not sufficient for unambiguous solutions for q_1 and q_2. However, remember that we are interested in a periodic schedule, which means we are in fact interested in an infinite number of solutions, each representing a certain number of times the schedule is executed. We represent these solutions with the smallest, strictly positive integer solutions for q_1 and q_2, which is

$$q_1 = 3$$
$$q_2 = 2$$

Thus, this tells us how often a transition has to fire in a single period of the schedule. Possible schedules are then $\phi = [t_1, t_1, t_1, t_2, t_2]$ or $\phi = [t_1, t_1, t_2, t_1, t_2]$.

Further, consider the net in Figure 3-27(b), which has an additional place. Its incidence matrix is

$$\mathcal{A}_b = \begin{bmatrix} 2 & -4 \\ -3 & 6 \end{bmatrix}$$

and the equations we can set up are

$$2q_1 - 3q_2 = 0$$
$$-4q_1 + 6q_2 = 0$$

Fortunately, these two equations are not independent, but both define the same set of solutions as for the net in Figure 3-27(a).

In contrast, consider the net in Figure 3-27(c) with the incidence matrix

$$\mathcal{A}_c = \begin{bmatrix} 2 & -4 \\ -3 & 3 \end{bmatrix}$$

This net does not allow for any periodic, admissible schedule because the equations

$$2q_1 - 3q_2 = 0$$
$$-4q_1 + 3q_2 = 0$$

have only one solution: $q_1 = q_2 = 0$.

This example shows that in the equation system set up by Equation (3.18) we should have less than N_t independent equations. In fact the following theorem gives us the precise number of independent equations that we need for a PASS to exist.

Theorem 3.11 For a connected SDF process network with N_t processes and its corresponding incidence matrix $\mathcal{A}$, rank($\mathcal{A}^T$) = $N_t - 1$ is a necessary condition for a PASS to exist.

The rank of a matrix is the number of independent equations it defines. This theorem states that the number of independent equations for the transposed incidence matrix must be exactly $N_t - 1$ for a PASS to exist. (A proof for it can be found in Lee and Messerschmitt 1987a.) Intuitively, this can be understood from the observation that each equation relates exactly two processes or transitions. Adding a new process, which connects to one or more existing processes, requires exactly one more independent equation to determine the firing frequency of that process.

If the rank of the transposed incidence matrix is not $N_t - 1$, any schedule applied periodically would accumulate tokens in one or more places. Hence, this is essentially a test that, if passed, guarantees that there exists a schedule that can be implemented with a fixed, finite buffer size between processes. However, an appropriate rank of the transposed incidence matrix does not guarantee the existence of a schedule that does not temporarily require a negative number of tokens in one or more places.

Initial Buffer Conditions

Consider again the Petri net in Figure 3-27(b) that has passed the rank test. However, no transition can ever fire because there is not a sufficient number of tokens in the places. Even with an initial marking of $\vec{x}_0 = [3, 0]$, we would not get very far because after firing $\langle t_2, t_1 \rangle$ the net would be in a deadlock. Only from an initial state $\vec{x}_0 = [6, 0]$ could we apply a periodic schedule $\phi = [t_2, t_2, t_1, t_1, t_1, t_1]$. Apparently, the problem is to find the initial buffer conditions for all signals.

To find appropriate initial buffer conditions, we select an arbitrary but positive, nonzero integer vector $\vec{q}$ such that $\vec{q}\mathcal{A} = \vec{0}$.

To find $\vec{q}$ we first note that, under the assumption $\text{rank}(\mathcal{A}) = N_t - 1$, there exists a vector $\vec{v} = [v_1, \ldots, v_{N_t}]$ with $\vec{v}\mathcal{A} = \vec{0}$, $v_1 = 1$, all $v_i \geq 0, 1 \leq i \leq N_t$, and all v_i are rational numbers such that their numerator and denominator are mutually prime. It exists because the rank of $\mathcal{A}$ is $N_t - 1$, and therefore we can set v_1 to 1 and as a consequence derive solutions for all the other variables v_i. These solutions are positive rational numbers. To see why, recall that each equation, representing an arc in the original process network, is of the form $kv_i - lv_j = 0$, which can be written as $kv_i = lv_j$, with l and k being strictly positive integers.

Let η be the least common multiple of all denominators in all v_i. The multiplication of all v_i with η results in a vector of strictly positive integers $\vec{q} = \eta \vec{v}$.

The vector $\vec{q}$ reveals how often a transition has to fire in each cycle of the schedule. Next we select an arbitrary schedule ϕ, where each transition is fired as many times as defined by $\vec{q}$. To obtain the initial buffer conditions that allow the execution of a full first cycle, we select $\vec{x} = [0, \ldots, 0]$ as the initial marking. Then we execute one full cycle of ϕ while we modify the execution semantics of Petri nets in the following way. If a place does not have enough tokens for a transition to fire, we still fire the transition and note the missing number of tokens as a negative marking of that place. We also memorize for each place the most negative marking that occurs. After the execution of the first cycle of ϕ the minimum occurred markings, multiplied by -1, designate the initial buffer conditions for that place.

Example 3.17 To illustrate the procedure, let's consider again the Petri net in Figure 3-27(b). Recall that the incidence matrix is

$$\mathcal{A}_b = \begin{bmatrix} 2 & -4 \\ -3 & 6 \end{bmatrix}$$

which defines the equations

$$2v_1 - 3v_2 = 0$$
$$-4v_1 + 6v_2 = 0$$

We set $v_1 = 1$ and derive $v_2 = 2/3$. The least common multiple of all denominators of v_i is $\eta = 3$. Thus we get

$$\eta \vec{v} = 3[1, 2/3] = [3, 2] = \vec{q}$$

Next we select an arbitrary schedule based on $\vec{q}$: $\phi = \langle t_1, t_1, t_1, t_2, t_2 \rangle$, which we apply to the Petri net with an initial state $\vec{x}_0 = [0, 0]$:

$$\vec{x}_1 = \vec{x}_0 + \vec{u}_1 \, \mathcal{A} = [2, -4]$$
$$\vec{x}_2 = \vec{x}_1 + \vec{u}_1 \, \mathcal{A} = [4, -8]$$
$$\vec{x}_3 = \vec{x}_2 + \vec{u}_1 \, \mathcal{A} = [6, -12]$$
$$\vec{x}_4 = \vec{x}_3 + \vec{u}_2 \, \mathcal{A} = [3, -6]$$
$$\vec{x}_5 = \vec{x}_4 + \vec{u}_2 \, \mathcal{A} = [0, 0]$$

As expected we arrive again at state $\vec{x}_0$, and we observe that place p_2 requires 12 tokens as an initial buffer condition for ϕ to be a PASS.

Note that we have made one arbitrary choice when we selected one particular schedule. This would not have an impact on the overall run time because the number of times each process is executed would be the same in all schedules that we can choose, since all of them have to comply with the vector $\vec{q}$ and in a single-processor implementation this processor would be utilized to 100% all of the time in all schedules. However, it may have an impact on the initial buffer conditions and on the maximum buffer sizes.

Example 3.18 For instance, assume we had selected the schedule $\phi = \langle t_2, t_1, t_2, t_1, t_1 \rangle$ in Example 3.17. The sequence of states during one cycle of the schedule would be

$$\vec{x}_1 = \vec{x}_0 + \vec{u}_2 \, \mathcal{A} = [-3, 6]$$
$$\vec{x}_2 = \vec{x}_1 + \vec{u}_1 \, \mathcal{A} = [-1, 2]$$
$$\vec{x}_3 = \vec{x}_2 + \vec{u}_2 \, \mathcal{A} = [-4, 8]$$
$$\vec{x}_4 = \vec{x}_3 + \vec{u}_1 \, \mathcal{A} = [-2, 4]$$
$$\vec{x}_5 = \vec{x}_4 + \vec{u}_1 \, \mathcal{A} = [0, 0]$$

with the initial buffer condition of 4 tokens in place p_1 and maximum buffer requirements of 4 tokens in place p_1 and 8 tokens in place p_2. However, keep in mind that these figures must be related to the size of each token, which may be different on each signal. Hence, an optimality criteria can only be formulated with respect to the token sizes.

Minimum Buffer Scheduling

To find a schedule with the minimum combined buffer requirements on all edges is a NP-complete problem (Bhattacharyya et al. 1996, Chapter 3). However, there exist simple heuristics that work well in most practical cases. In the following we present a variant of the algorithm described by Bhattacharyya et al. (1996, page 54). The main idea is to always select a transition that does not accumulate data on its outputs unnecessarily. We call a transition *deferrable* if any of its output places has at least as many tokens as is required for a successor transition to fire. In Figure 3-28 transition t_0 is deferrable

FIGURE 3-28

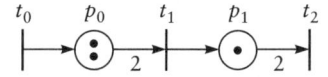

Transition t_0 is deferrable, but t_1 is not.

because place p_0 contains enough tokens for t_1 to fire. However, t_1 is not deferrable because, unless it fires, t_2 cannot be triggered.

Algorithm 3.1 (*Minimum buffer scheduling*)

$F := \{\text{firable transitions}\}$ // set of firable transitions
$D := \{\text{deferrable}(F)\}$ // set of deferrable, firable transitions
while $(F \neq \{\})$ // if there are nondeferrable transitions
    if $(F \setminus D \neq \{\})$ // fire a nondeferrable transition
    then fire a transition from $F \setminus D$
    else fire a transition that increases total number of tokens the least
    end if
end while

3.15.2 Multiprocessor Schedule

The problem of finding a schedule for a parallel architecture is complicated significantly by a number of factors. The processors may not be of the same kind, and not every processor may be able to execute all processes. The run time of a given process may be different on different processors. The communication time to send data between processors may be significant. There are many different approaches to this problem with a wide variety of different assumptions. We will illustrate the problem by discussing only one specific approach based on a specific set of assumptions. For a very complete account of scheduling and synchronization problems for multiprocessor systems, see Sriram and Bhattacharrya (2000). Our assumptions are as follows:

- There is a homogeneous set of resources; all resources are able to execute all processes in the same time.
- Communication time is negligible; the resources are tightly coupled such that the communication of data between them takes a constant and relatively small amount of time.
- There is a statically fixed, constant number of resources.

Based on these assumptions, we define a periodic schedule for a parallel architecture as follows:

Definition 3.15 Given is an SDF process network and n resources. A *periodic, admissible, parallel schedule* (PAPS) is a periodic and infinite admissible sequential schedule for each of the resources available such that the data dependencies between the processes are respected. It is specified by a list $\{\psi_1, \ldots, \psi_n\}$, where ψ_i is a sequential schedule for processor i.

The approach we describe here contains the following steps:

1. Compute a PASS schedule.
2. Determine a PASS "unroll factor" J, that is, how many PASS cycles form a single PAPS cycle.
3. Construct a precedence graph.
4. Compute the PAPS based on the Hu-level algorithm.

Compute a PASS

We have shown in the previous section how to compute a PASS. Now we use a particular PASS as a starting point and determine how often to unroll it and how to distribute the processes in the unrolled PASS over the available resources to balance the load and optimize the performance.

Determine the Unroll Factor J

A trivial solution is not to unroll the PASS, that is, $J = 1$. However, better solutions can be found if the PASS is unrolled a few times.

Consider the SDF process network represented as a Petri net in Figure 3-29. Assume we have two processors and the runtime, measured in arbitrary time units, of the three transitions t_1, t_2, and t_3 is $T_1 = 1$, $T_2 = 2$, and $T_3 = 3$, respectively.

For $J = 1$ the best schedule we can achieve has length 4, as shown in Figure 3-30(a). But if we unroll it once, we obtain a schedule of length 7 for two elementary PASS

FIGURE 3-29

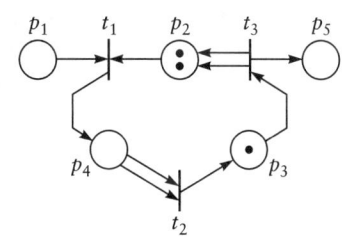

An SDF network represented as a Petri net with its initial buffer conditions. p_1 and p_5 are input and output places, respectively. A possible PASS is $\phi = \langle t_1, t_1, t_2, t_3 \rangle$.

FIGURE 3-30

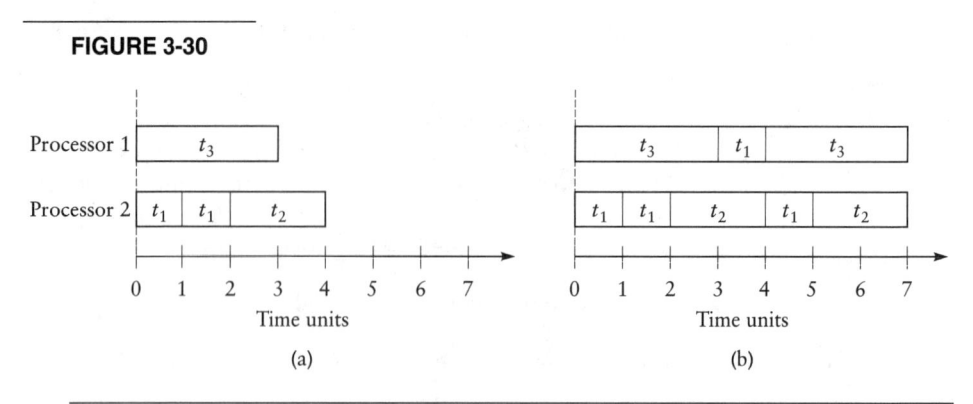

Two possible schedules for two processors with (a) $J = 1$ and (b) $J = 2$.

cycles, as shown in Figure 3-30(b). By intertwining two or several elementary cycles we can achieve schedules with a shorter average run time. But we have to be careful because it is not obvious how early we can schedule a transition without violating data dependences. For example, can we safely schedule the third occurrence of transition t_1 before t_2 has completed as we have done in Figure 3-30(b)? To determine this question we construct a precedence graph, which makes these dependences explicit.

Our problem now is that we have to consult the precedence graph to determine a good value for J, but we have to know J to know which precedence graph to construct. Based on the observation that J is usually a relatively small number, we propose the following heuristics: start with $J = 1$, construct the precedence graph and the PAPS, and increment J until no further improvement on the run time of the PAPS can be obtained.

Construct the Precedence Graph

The precedence graph is specific for each value of J. It shows all the data dependences of processes in the PASS that has been unrolled J times.

Example 3.19 Figure 3-31(a) shows the precedence graph for $J = 1$. Because of the initial buffer conditions, transitions t_1 and t_3 can fire without depending on any other transition. t_1 can even fire twice. Hence, the precedence graph shows two instances of t_1 and one instance of t_3 at the very left without incoming arcs. The superscript k in t_i^k is used to denote the kth firing of transition t_i. In contrast, t_2 requires transition t_1 to fire twice before it becomes enabled. Hence, it has an incoming arc from the second firing of t_1. There is a weak precedence between t_1^1 and t_1^2, indicated by a dotted arc. None is really dependent on the other, but all transitions depending on t_1^2 also depend on t_1^1. Hence, t_2^1 really depends on both invocations of t_1, but we only have one direct dependence on t_1^2.

The precedence graph for $J = 2$ has to contain four instances of t_1, two of t_2, and two of t_3 because unrolling the basic PASS once results in these numbers of occurrences. Figure 3-31(b) shows the correct dependences

FIGURE 3-31

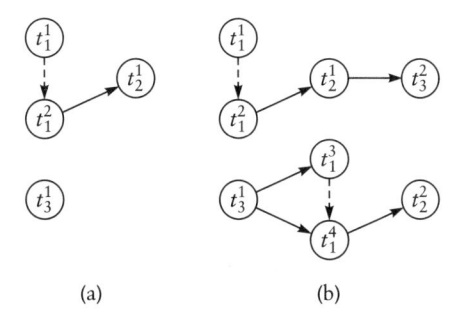

The precedence graphs of the example in Figure 3-29 for (a) $J = 1$ and (b) $J = 2$.

for each occurrence of all transitions. For instance, the third and forth firing of transition t_1 depends on t_3 having fired once, and the second firing of t_3 depends on two firings of t_1 and one firing of t_2. In this way each occurrence assumes its own unique identity in the precedence graph indicated by the superscript.

To make this idea precise, we will define exactly what a "precedence graph" is. For that some notation will be convenient. Let $\vec{p}_i$, $1 \le i \le m$, be the column vector with a 1 on position i and 0s elsewhere, that is, $\vec{p}_i = [0, \cdots, 0, 1, 0, \cdots, 0]^T$. It will be used to extract the number of tokens in a particular place. Further, recall that $\vec{u}_i$ is the row vector representing a firing of transition i. We designate k firings of transition t_i by the row vector $\vec{u}_i^k$, with $\vec{u}_i^0$ denoting the zero-vector.

Definition 3.16 Given is an SDF process network and its corresponding Petri net representation with n transitions $t_1, \ldots, t_n$ and m places $p_1, \ldots, p_m$. Further given is a PASS for this Petri net and an unroll factor $J > 0$. Let $K(t_i)$ be the number of occurrences of transition t_i in the PASS. Hence, $JK(t_i)$ is the number of occurrences of transition t_i in the J-unrolled pass. We denote the kth occurrence of transition t_i by t_i^k, with $0 < k \le JK(t_i)$.

A precedence graph is a graph where the nodes represent the occurrences of transitions in the J-unrolled PASS. Hence, the number of nodes is $\sum_{i=1}^{n} JK(t_i)$, and the nodes are labeled t_i^k. There is a *strong precedence arc* or just *precedence arc* from node t_i^k to node t_j^l if and only if

(i) there is a place p_r in the Petri net with an arc from t_i to p_r and an arc from p_r to t_j

(ii) and if the following conditions hold:

$$(\vec{u}_i^{k-1} \mathcal{A} + \vec{x}_0) \vec{p}_r < -(\vec{u}_j^l \mathcal{A} \vec{p}_r) \quad (3.19)$$

$$(\vec{u}_i^k \mathcal{A} + \vec{x}_0) \vec{p}_r \ge -(\vec{u}_j^l \mathcal{A} \vec{p}_r) \quad (3.20)$$

These conditions capture that t_j^l should not be able to fire before t_i^k (3.19), and that it should be able to fire after t_i^k (3.20).

In addition to strong precedence arcs, there is a *weak precedence arc* from node t_i^k to node t_i^{k+1} if and only if both nodes share all incoming strong precedence arcs, that is, if for all arcs (t_j^l, t_i^k), there exists an arc $(t_j^l, t_i^{+1}k)$, and vice versa.

Example 3.20 The intuition of this definition can best be acquired by going through an example. Consider again the SDF network represented by the Petri net of Figure 3-29, a PASS $\phi = \langle t_1, t_1, t_2, t_3 \rangle$, and an unroll factor $J = 1$. Condition (i) above tells us we have to investigate if precedence arcs should be placed between the pairs (t_1^1, t_2^1), (t_1^2, t_2^1), (t_2^1, t_3^1), (t_3^1, t_1^1), and (t_3^1, t_1^2).

Should there be an arc between t_1^1 and t_2^1? To answer this we set up the equations.

(i)
$$(\vec{u}_1^0 \, \mathcal{A} + \vec{x}_0) \vec{p}_4 < -(\vec{u}_2^1 \mathcal{A} \vec{p}_4)$$

$$(\vec{u}_1^0 \, \mathcal{A} + \vec{x}_0) \vec{p}_4 = \left([0,0,0] \begin{bmatrix} -1 & -1 & 0 & 1 & 0 \\ 0 & 0 & 1 & -2 & 0 \\ 0 & 2 & -1 & 0 & 1 \end{bmatrix} \right.$$

$$\left. + [0, 2, 1, 0, 0] \right) \begin{bmatrix} 0 \\ 0 \\ 0 \\ 1 \\ 0 \end{bmatrix}$$

$$= ([0,0,0,0,0] + [0,2,1,0,0]) \begin{bmatrix} 0 \\ 0 \\ 0 \\ 1 \\ 0 \end{bmatrix}$$

$$= 0$$

$$-(\vec{u}_2^1 \mathcal{A} \vec{p}_4) = -\left([0,1,0] \begin{bmatrix} -1 & -1 & 0 & 1 & 0 \\ 0 & 0 & 1 & -2 & 0 \\ 0 & 2 & -1 & 0 & 1 \end{bmatrix} \begin{bmatrix} 0 \\ 0 \\ 0 \\ 1 \\ 0 \end{bmatrix} \right)$$

3.15 Synchronous Dataflow

$$= -\left([0,0,1,-2,0] \begin{bmatrix} 0 \\ 0 \\ 0 \\ 1 \\ 0 \end{bmatrix} \right)$$

$$= 2$$

Hence, condition (i) is fulfilled. Let's check condition (ii).

(ii)
$$(\vec{u}_1^1 \mathcal{A} + \vec{x}_0) \vec{p}_4 \geq -(\vec{u}_2^1 \mathcal{A} \vec{p}_4)$$
$$([-1,-1,0,1,0] + [0,2,1,0,0]) \vec{p}_4 \geq -([0,0,1,-2,0] \vec{p}_4)$$
$$[-1,1,1,1,0] \vec{p}_4 \geq -([0,0,1,-2,0] \vec{p}_4)$$
$$1 \geq 2 \,!!$$

This condition is not fulfilled, which means firing t_1 once is not sufficient to enable t_2^1.

As a second example we also answer the following question: *Should there be an arc between t_1^2 and t_2^1?*

(i)
$$(\vec{u}_1^1 \mathcal{A} + \vec{x}_0) \vec{p}_4 < -(\vec{u}_2^1 \mathcal{A} \vec{p}_4)$$
$$([1,0,0] \mathcal{A} + [0,2,1,0,0]) \vec{p}_4 < -([0,1,0] \mathcal{A} \vec{p}_4)$$
$$([-1,-1,0,1,0] + [0,2,1,0,0]) \vec{p}_4 < -([0,0,1,-2,0] \vec{p}_4)$$
$$1 < 2$$

(ii)
$$(\vec{u}_1^2 \mathcal{A} + \vec{x}_0) \vec{p}_4 \geq -(\vec{u}_2^1 \mathcal{A} \vec{p}_4)$$
$$([-2,-2,0,2,0] + [0,2,1,0,0]) \vec{p}_4 \geq -([0,0,1,-2,0] \vec{p}_4)$$
$$[-2,0,1,2,0] \vec{p}_4 \geq -([0,0,1,-2,0] \vec{p}_4)$$
$$2 \geq 2$$

This condition is also fulfilled, which gives us the only strong precedence arc in the precedence graph, as shown in Figure 3-31(a).

From these definitions and examples we see that we can construct the precedence graph by testing conditions (i) and (ii) for each arc in the SDF process network and then adding the weak precedence arcs. A detailed and efficient algorithm is provided in Lee and Messerschmitt (1987a).

Note that even though we assume a specific PASS, the precedence graph is the same for all PASSs because we do not use the particular order of transitions in the PASS, only the total number of each transition.

FIGURE 3-32

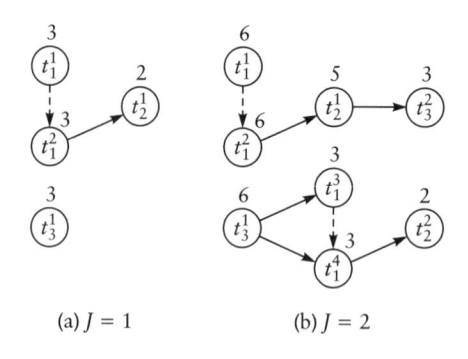

(a) $J = 1$ (b) $J = 2$

The precedence graphs of Figure 3-31 annotated with the Hu-levels.

Compute a PAPS

Now we have everything in place to construct a PAPS. We do this based on a Hu-level scheduling algorithm (Hu 1961), which is an effective heuristic for most practical problems.

The nodes in the precedence graph are annotated with a number that designates the maximum run time of this node and all its successors in the graph, as illustrated in Figure 3-32. This annotation is called *Hu-level* or *Hu-level annotation*. As above we assume that we have two processors and that the run time, measured in arbitrary time units, of the three transitions t_1, t_2, and t_3 is $T_1 = 1$, $T_2 = 2$, and $T_3 = 3$, respectively. Then we annotate all nodes that have no successor node with their run time. In Figure 3-32(b) the nodes t_3^2 and t_2^2 have no successor nodes. Hence they are annotated by their run time, which is 3 and 2, respectively. Then we work toward the root of the graph by adding the Hu-level annotation of a successor node to the run time of a node to compute the Hu-level annotation of a node. For instance, the run time of node t_2^1 is 2, and the maximum Hu-level annotation of all its successors is 3; hence, the Hu-level annotation for t_2^1 becomes 5. If a node has only a weak precedence arc to a successor node, it assumes the same Hu-level annotation as its successor, motivated by the fact that they can run in parallel but all successor nodes depend on both these nodes.

The PAPS is constructed by going through the precedence graph left to right and gradually scheduling nodes for which all predecessor nodes have been scheduled and that have the highest Hu-levels of all unscheduled nodes.

For example, in Figure 3-32(b) we would start with the nodes t_1^1, t_1^2, and t_1^3 because they all have no unscheduled predecessor and they have equal Hu-levels. Applying this procedure to the precedence graphs in Figure 3-32 will lead us to the schedules shown in Figure 3-30(a) and (b), respectively.

Example 3.21 To illustrate this procedure in more detail, we derive a three-processor schedule for the precedence graph of Figure 3-32(b). The four steps correspond to the four distinct Hu-levels as shown in Figure 3-33.

3.15 Synchronous Dataflow

FIGURE 3-33

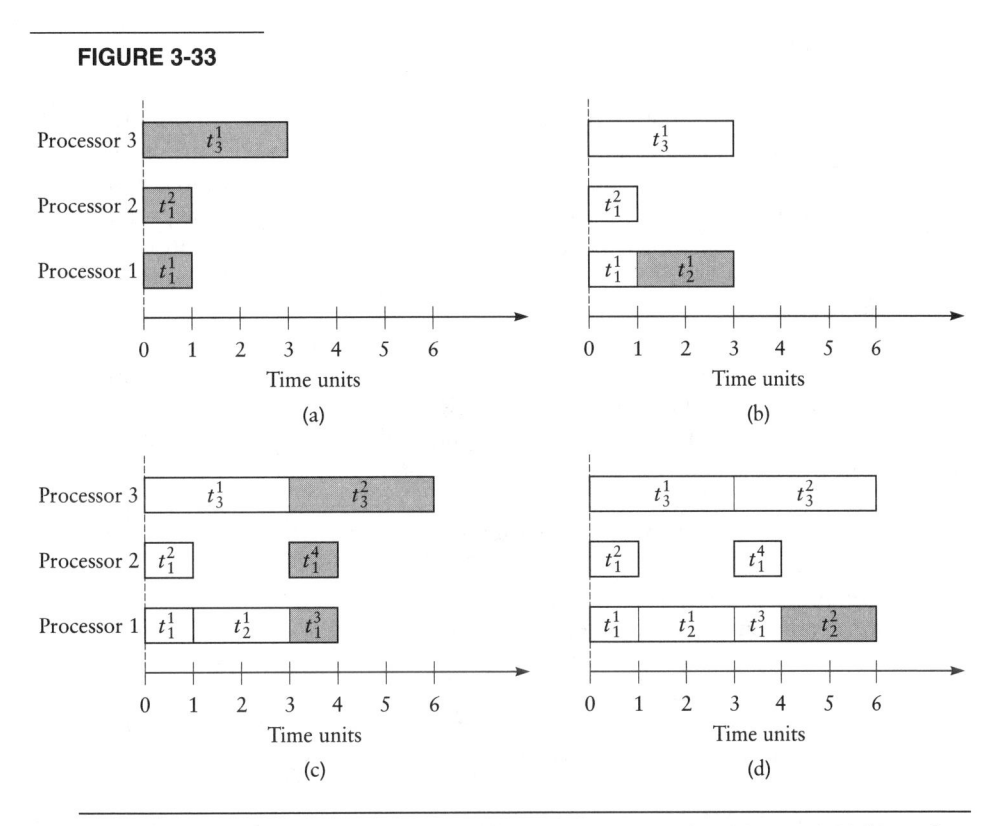

Deriving the three-processor schedule for precedence graph in Figure 3-32(b) in four steps: (a) step 1, (b) step 2, (c) step 3, and (d) step 4.

In step 1 we schedule the nodes with Hu-level 6 by assigning each of them to a different processor. In step 2 the node with Hu-level 5 is assigned to processor 1. In step 3 we have three nodes with Hu-level 3. Since we have dependences on earlier nodes, we cannot schedule them as soon as a processor is idle. For example, nodes t_1^3 and t_1^4 cannot be scheduled before node t_3^1 has completed. Complying with these dependency constraints we assign the three level 3 nodes to the three processors (Figure 3-33(c)). Finally, in step 4 we schedule node t_2^2 at time 4 on processor 1.

Note that the Hu-level annotated precedence graph gives the length of the shortest possible schedule that equals the maximum Hu-level in the graph. Figure 3-32(b) has three nodes with Hu-level 6. Thus, the minimum length of any schedule with any number of processors is 6. In this sense the schedule derived in Example 3.21 is optimal even though it does not fully utilize the processors. It also means that more processors would not improve on the performance. On the other hand, the two-processor schedule in Figure 3-30(b) is optimal in the sense that all processors are 100% utilized and it is the shortest possible schedule for two processors.

3.16 Variants of the Untimed MoC

The synchronous dataflow model is a prominent example of an untimed MoC where restrictions lead to powerful analysis and synthesis techniques. Through the years many related models have been proposed that have attempted to equip the SDF with some control features while still maintaining most of its nice theoretical properties.

We will informally and very briefly discuss a small selection of other untimed dataflow models. More formal definitions can easily be developed (see Exercises 3.5 and 3.6).

The *boolean dataflow* (BDF) model, proposed by Lee (1991) and further developed by Buck (1993), extends the SDF model with two special process types called SWITCH and SELECT(see Figure 3-34). The SWITCH process reads its data input and, depending on the boolean value of the control input, it emits the data either to the T or to the F output. The SELECT process reads either the T or the F input, depending on the control input, and emits the consumed data to its output. They can be used to build arbitrary data-dependent control structures such as if-then-else and loops (Figure 3-35). In fact Buck (1993) has shown that BDF is as expressive as Turing machines. As a consequence many scheduling and buffer optimization problems become NP-complete so that efficient algorithms for their optimal solution cannot be found. Hence, Buck develops a number of heuristics or makes constraining assumptions to address these problems in BDF graphs.

In the *cyclo-static dataflow* (CSDF) model, proposed by Bilsen (1995), the number of tokens consumed and produced by a process can vary periodically. Thus, the number of tokens can be different for different evaluations but has to go through a statically known cycle. CSDF allows us to conveniently express fluctuating consumption and production rates of processes. In Figure 3-36 this is denoted by a vector at each process input and output. Process D always consumes one token from its input, but it produces alternately one or zero tokens at its two outputs. Process C is its complement. The bullet on the arc from C to D indicates an initial buffer condition of one token.

Figure 3-37 shows a corresponding SDF graph, where for each process the cycle of activations is compressed into one activation. For instance process D' consumes two tokens and produces one token on each output during one activation. Process D performs the same in two consecutive activations. The problem with the SDF graph is

FIGURE 3-34

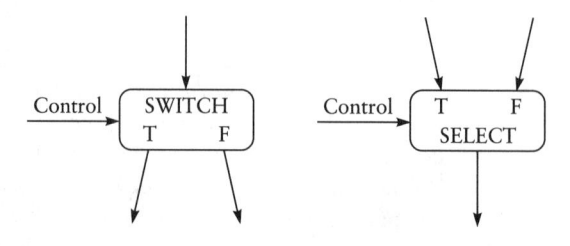

The BDF processes SWITCH *and* SELECT *can represent data-dependent control.*

FIGURE 3-35

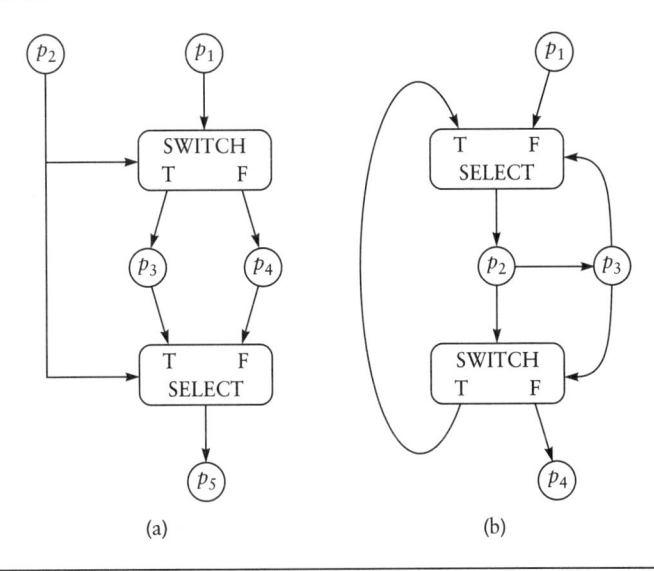

The SWITCH and SELECT processes can be used to build (a) branching and (b) loop constructs.

FIGURE 3-36

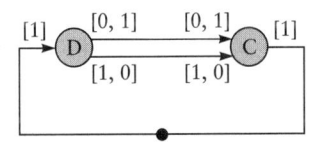

In the cyclo-static dataflow the processes consume and produce a cyclically varying number of tokens, which is denoted as a vector at the process inputs and outputs (Parks et al. 1995).

that it deadlocks because the initial buffer condition of one token on the arc from C to D is not sufficient to get the process network started. Thus, although CSDF and SDF are in principle of equal expressive power, their usage results in important practical differences such as different buffer requirements. Other differences, as elaborated by Parks et al. (1995), concern scheduling, dead code elimination, and the exploitation of parallelism.

The *synchronous piggy-backed dataflow* (SPDF) model, proposed by Park et al. (2002), allows control information to flow over the data arcs. Figure 3-38 illustrates a communication between processes A and D, which are not directly connected. They can still communicate because there exists a dataflow path from A to D. In this

FIGURE 3-37

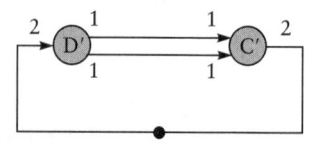

In the SDF graph that corresponds to the cyclo-static dataflow graph of Figure 3-36, the entire cycle of activations of each process is collapsed into one activation. Hence, process D' accomplishes in one activation what D performs in two consecutive activations. However, with the same initial buffer conditions the SDF graph deadlocks.

FIGURE 3-38

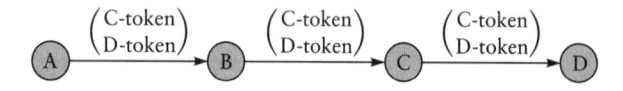

In the synchronous piggy-backed dataflow model, control tokens (C-tokens) can be sent as an attachment to the data tokens (D-tokens) to allow processes A and D to exchange control information.

model every data token comes with a control token even though control information is exchanged irregularly and at a rate orders of magnitude below the data rate. To rectify this efficiency problem Park et al. propose an implementation technique based on shared memory. Only processes that generate control information actually write to the shared memory. Similarly, only processes that use control information read from the shared memory. Thus, processes B and C would not be bothered with the control tokens if they are not using them. Also, control information is accessed in the shared memory only at the required rate, not at the data rate. Although the implementation is based on a shared, global memory, its usage is restricted such that processes do not cause side effects and are thus faithful to the principles of the untimed MoC.

In summary, the SPDF MoC does not extend the expression power of SDF but provides a means to formulate control communication, without adding new arcs and communication links.

3.17 Further Reading

The untimed model of computation, as it is presented here, is derived from the work on dataflow process networks originally proposed in the 1970s by Dennis (1974), Kahn (1974), and others. This theory has further been developed and successfully applied to signal-processing applications in the 1980s and 1990s (Lee and Messerschmitt 1987a, 1987b). A good discussion of this area can be found in Lee and Parks (1995). A comprehensive account of scheduling and software generation from SDF graphs is given by Bhattacharyya et al. (1996), and fairly complete coverage of the scheduling and synchronization for multiprocessor systems is provided by Sriram and Bhattacharyya (2000).

Frequently, untimed MoCs have been the starting point for modeling and designing control-dominated and reactive applications. However, this always leads to the introduction of time either at the clocked-time or at the physical-time level (e.g., Lee 2000; Jantsch and Bjuréus 2000)—issues we take up in the next chapters.

Several attempts to unify different computational models into a common framework have been described (Lee and Sangiovanni-Vincentelli 1998; Girault et al. 1999).

3.18 Exercises

3.1 A process p translates characters of the English alphabet and digits into their binary ASCII representation. Hence, the character A is translated to "01100101", while a is translated to "10010111". Process q performs the reverse operation. For instance,

$$p(\langle A3b1 \rangle) = \langle 0, 1, 1, 0, 0, 1, 0, 1, 1, 0, 1, 0, 0, 0, 1, 1, 0, 0, 1, 1,$$
$$0, 0, 0, 0, 1, 0, 0, 1, 0, 0, 1 \rangle$$

and

$$qp(\langle A3b1 \rangle) = \langle A3b1 \rangle$$

a. Model p and q by means of untimed process constructors.

b. Give the signature of both processes.

c. Up-rate both processes by a factor of 3.

3.2 Consider a process p' that counts the number of 0s and 1s in sequences of consecutive 0s and 1s. The input signal consists only of 0s and 1s. When the input changes from 1 to 0 or from 0 to 1, the process outputs the number of consecutive 1s or 0s, respectively, just received. For example, $p'(\langle 0, 0, 0, 1, 1, 0, 0 \rangle) = \langle 3, 2, 2 \rangle$.

a. Model p' by means of an untimed process constructor.

b. Up-rate p' by a factor of 2.

c. Give the signature of uprate(p', 2).

d. Merge p' with p from Exercise 3.1 into p_1 such that $p_1 = p' \circ p$.

e. Merge processes (uprate(p, 2)) and (uprate(p', 2)) to get $p_2 =$ (uprate(p', 2)) $\circ$ (uprate(p, 2)).

3.3 Model a function that calculates the checksum of the Internet Control Message Protocol (ICMP), which is part of the TCP/IP protocol suite.

An ICMP packet consists of the following fields:

- Source address: 128 bits
- Destination address: 128 bits
- Payload length: 32 bits
- Next header byte which has a fixed value of 58: 32 bits
- ICMP type: 8 bits
- ICMP code: 8 bits
- Checksum: 16 bits
- Message body: variable size between 128 and 5056 bits but always an integer number of bytes

For calculation of the checksum, the checksum field is set to 0.

If the size of the message in bytes is not even, an extra byte is appended to the message with a value of 0. This is necessary to get an integer number of 16-bit words.

To calculate the checksum, ICMP performs a 16-bit ones complement addition of this data. It ignores any carries and places the 16-bit sum in the checksum field of the ICMP header.

For more details of the ICMP protocol refer to *community.roxen.com/developers/idocs/rfc/rfc792.html*.

a. Model the ICMP checksum generation for a sender node.

b. For a receiver, model a function that checks the ICMP checksum and discards the packet if the checksum is incorrect.

3.4 Define formally the synchronous dataflow MoC, similarly to the definition of the untimed MoC (Definition 3.4). Do this by deriving appropriate process constructors from the untimed process constructors.

3.5 Define formally the boolean dataflow MoC, similarly to the definition of the untimed MoC (Definition 3.4).

3.6 Define formally the cyclo-static dataflow MoC, similarly to the definition of the untimed MoC (Definition 3.4).

3.7 Consider the synchronous dataflow graph in Figure 3-39. Determine the following:

a. The incidence matrix Γ and its rank.

b. The smallest $\mathbf{q}$ such that $\Gamma \mathbf{q} = \mathbf{0}$.

3.8 For the SDF in Figure 3-39 determine a periodic admissible sequential schedule and the initial buffer conditions.

FIGURE 3-39

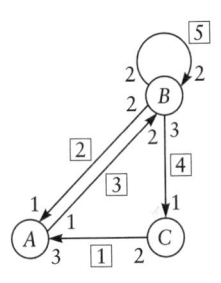

A synchronous dataflow graph with one self-loop.

FIGURE 3-40

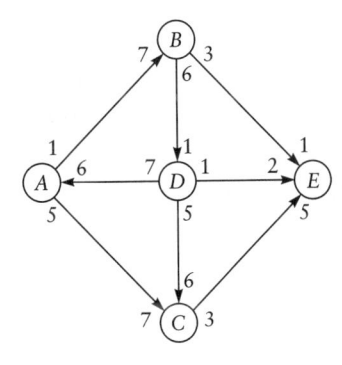

A synchronous dataflow graph with complex interaction between nodes.

3.9 For the SDF in Figure 3-39 determine a periodic admissible parallel schedule for two processors. The execution time for task A is 10 ms, for task B 20 ms, and for task C 10 ms.

3.10 Consider the synchronous dataflow graph in Figure 3-40.
Determine the following:

a. The incidence matrix.

b. The vector **q**.

3.11 For the synchronous dataflow graph in Figure 3-40 find a periodic admissible sequential schedule and the initial buffer conditions.

3.12 For the synchronous dataflow graph in Figure 3-40 and for the following task execution times, $t(A) = 10, t(B) = 20, t(C) = 40, t(D) = 20, t(E) = 30$, find initial

FIGURE 3-41

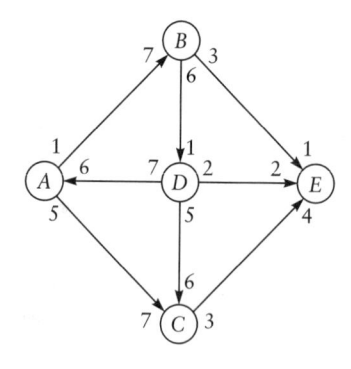

A synchronous dataflow graph similar to but distinct from the one in Figure 3-40.

buffer conditions and an optimal periodic admissible parallel schedule for $J = 1$ and three processors.

3.13 Consider the synchronous dataflow graph in Figure 3-41.

Either determine the vector **q** and a PASS or prove that there can be no admissible sequential schedule.

3.14 Model a switch with two input ports and two output ports. The router has a routing table that defines to which output port a packet with a particular address should go. Special packets are used to set up the routing table. Thus each packet consist of three parts: the packet type (routing table definition packet (RTD) or user packet (UP)), the target address, and the payload. The router itself has an address too, which is 250. It interprets RTD packets only if they are addressed to 250. If there is no entry in the routing table, the default is output port C. RTD packets have a payload with two parts: the target address and the output port.

chapter four
The Synchronous Model of Computation

Synchronous models of computation divide the time axis into slots. *Everything inside a slot occurs at the same time, but the slots are totally ordered along the time axis and are often enumerated by the natural numbers. In synchronous MoCs the evaluation cycle of processes lasts exactly one time slot. We distinguish between two important synchronous MoCs.*

In the perfectly synchronous MoC *the output events of a process occur in the same time slot as the corresponding input events. Moreover, they are instantaneously distributed in the entire system and are available to all other processes in the same slot. Receiving processes in turn consume the events and emit output events again in the same time slot. This leads to interesting situations when an output event of a process becomes also an input event of the same process in the same time slot and in this way contributes to its own creation. This situation corresponds to a set of recursive equations that constrain the values of all the involved events. It is resolved by adopting a unique solution that meets all the constraints, which is the* least fixed point *of the recursive equations. Recall that in the previous chapter we also adopted a least fixed-point semantics for determining the signals in feedback loops in the untimed MoC. The difference is that in the untimed MoC we based the semantics on the prefix order of signals and the resulting fixed-point solution defined the entire signals. In contrast, for the perfectly synchronous MoC we define the fixed-point semantics for individual events in each evaluation cycle based on the Scott order of event values. We give an example to illustrate this important difference.*

The second MoC presented in this chapter is the clocked synchronous MoC. *It differs from the perfectly synchronous MoC in that every process incurs a delay from an input to an output event. The delay is equivalent to the duration of an evaluation cycle. Consequently, feedback loops lose their difficulty because there is always a delay of at least one evaluation cycle in each loop and events cannot affect their own generation. The analysis of a process network is complicated, however, because delays and event timing have to be taken into account explicitly.*

As an application example, a monitor-and-assertion-based validation technique is reviewed. It has the appealing feature that it can be used with advantage in simulation-based validation as well as in formal verification.

4.1 Perfect Synchrony

Abstraction is a mechanism for selecting which information should be taken into consideration and which shouldn't. Discrete event models provide a very general timing model that allows us to represent precisely the delays observed on real entities. However, for many design tasks this is unnecessary information. Indeed, for many tasks all the detailed timing information would cloud the relevant issues and make precise answers to important questions infeasible. It would be like applying the laws of quantum mechanics to predict planetary motion. Clearly, writing down all the required equations would be impossible, let alone solving them. If we want to predict the motion of planets for the next several days, years, or millennia, we have to make some assumptions and find another, more suitable abstraction level. Newtonian physics, for example, is very suitable for this. Even though Newtonian physics is less accurate and even makes incorrect predictions in certain cases, it can be applied to a great variety of problems that cannot be tackled by quantum mechanics due to the sheer amount of details involved. Newtonian physics is successful for two important reasons. First, it ignores a huge amount of information that is just not relevant for the question at hand. Ignoring this information gives us the same answer, within the desired accuracy, as if we had taken it fully into account. Second, we know exactly when the laws of Newtonian physics can be applied and when they fail. We know that for very tiny distances and for very high velocities the laws of Newtonian physics are not reliable.

Similarly, the perfectly synchronous model of concurrency ignores all details concerning timing behavior. It makes the following assumption, which is called the *perfect synchrony hypothesis:*

Perfect synchrony hypothesis: *Neither computation nor communication takes time.*

This assumption has several far-reaching implications. But before we explore them and identify applications of this model, we first understand why we can build useful models based on this assumption. If nothing takes time, is time totally excluded? No, because a certain time structure is maintained by the sequence of input events. In fact, time is represented to a higher degree than in the untimed model because the untimed

FIGURE 4-1

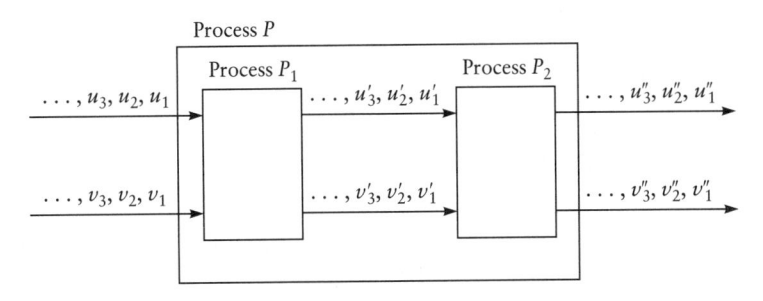

A process consisting of two subprocesses in the synchronous model.

model employs a much weaker assumption, essentially saying that both computation and communication take an arbitrary and unknown amount of time.

In a perfectly synchronous model, inputs arrive in a particular order, and even though we cannot say anything about the concrete time instances when this happens, we can observe that event u_1 arrives at the system before u_2, as shown in Figure 4-1. We also know that u_1 and v_1 appear at the same time as inputs to process P. Hence, the time structure is given by the input events. P reacts to the events u_1 and v_1 immediately by computing outputs u_1'' and v_1'' and then it waits for the next inputs. So we observe a sequence of (read inputs, compute outputs) cycles. Since neither computation nor communication takes time, the outputs occur at exactly the same time as the inputs; thus u_1, v_1, u_1'', v_1'' and all corresponding intermediate events u_1', v_1' occur simultaneously. Thus, the time instances of the occurrence of the output events are fully determined by the time instances of the occurrence of the input events.

There is in fact a large class of practical systems that conduct an endless sequence of (read inputs, compute outputs) cycles:

- *Reactive systems* receive inputs, react to them by computing outputs, and wait for the next inputs to arrive.
- Many embedded control systems, which are connected to sensors and actuators, continuously read their inputs with a predefined frequency and compute new data for their actuators before they read the next input sample.
- Wireless communication devices, which are connected to an A/D converter, receive new samples with a fixed, predefined frequency, apply a signal-processing algorithm, and output a result before they read the next sample.
- A switch in a telecommunication backbone network reads all the data packets at its inputs, determines at which output ports the packets should be emitted, and then accepts the next input packets.

Apparently, a real physical system cannot comply with the perfect synchrony assumption. But simulated models can, and a model based on the perfect synchrony assumption would behave exactly the same as the real implementation, as long as the

FIGURE 4-2

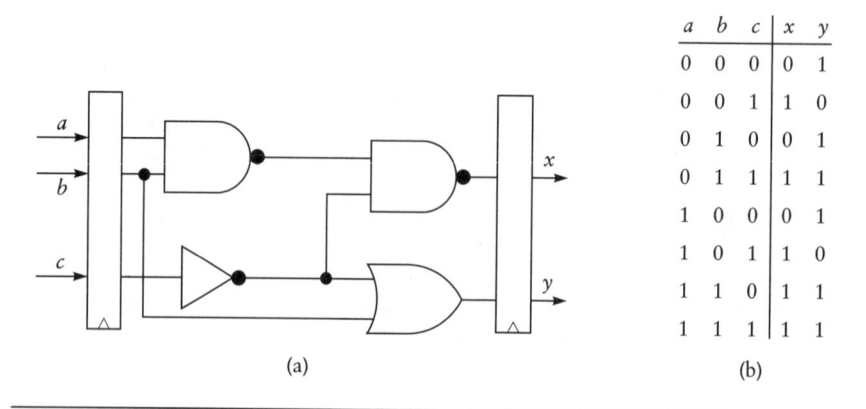

a	b	c	x	y
0	0	0	0	1
0	0	1	1	0
0	1	0	0	1
0	1	1	1	1
1	0	0	0	1
1	0	1	1	0
1	1	0	1	1
1	1	1	1	1

(a)   (b)

(a) A network of gates with registers on the boundaries. (b) Its input-output functionality is described as a truth table.

latter reacts "fast enough." The perfect synchrony hypothesis of the models translates into the real-world assumption that the system is fast enough to compute outputs before the next inputs arrive. Essentially we have separated functional behavior from timing behavior. Therefore we can concentrate on these two issues one by one, and both issues become easier and more manageable. First, we concentrate on getting the functionality right by working on a perfectly synchronous model. Then, during the implementation, we have to validate the assumption that the system is fast enough compared to its environment. This can be done statically without involving simulation. Consider the netlist of gates in Figure 4-2. Its *static* input-output behavior is described as a truth table. Dynamically, different output patterns could be observed if the propagation delay on different paths from inputs to outputs varies. For instance, it is likely that we observe the pattern $x = 0, y = 0$ on the output during a short transition time, even though it does not appear in the truth table. Such undefined patterns may trigger unwanted reactions in other parts of the system and the environment. We can avoid this by placing registers on the inputs and outputs as shown in Figure 4.2. If the registers are controlled by a common clock signal and the clock period is longer than the propagation delay of the longest path in the network, we will always observe the correct output patterns matching the static truth table. Consequently, we can in the first step assume that our circuit is "fast enough" and concentrate solely on the functionality as defined by the static truth table. This means for a given input pattern the functionality is defined by the corresponding entry in the truth table.

Apparently, to get the truth table right is much easier than to design the circuit such that it behaves properly for all possible gate delays that an implementation in a given manufacturing process may exhibit. In the second step, when we are satisfied with the functionality as defined by the truth tables, we can validate our timing assumption by means of static timing analysis. Static timing analysis takes the worst-case delays on the gates and sums them up to identify the longest path from an input to an output. If the worst-case gate delays are defined by Table 4-1, the longest path through the circuit would be from input *a* or *b* to output *x*, and it would take 3.6 ns. We could therefore be

TABLE 4-1: *Delays of gates of Figure 4-2.*

Gate	Delay (ns)
Inverter	1.5
NAND gate	1.8
OR gate	2.1

assured that our circuit implementation behaves exactly according to its specification by the truth table if the clock period is greater than 3.6 ns.

This methodology of separating functional design from timing analysis has been so successful in circuit design that today virtually all ASICs are designed in this so-called *synchronous design style*. Just like Newtonian physics the synchronous assumption has been successful because it can be used for a large class of important problems, that is, the design of many electronic systems, and we can determine exactly when the assumptions are safe and when they are not. If the clock frequency is above 277 MHz in the previous example, the behavior of the implemented circuit may deviate from the simulated model.

However, as we will see later, the assumptions made in the synchronous design style in circuit design are not identical to the perfect synchrony hypothesis. The difference and its implications will be elaborated on in Section 4.6. For now we will continue to investigate the perfect synchrony assumption.

4.2 Process Constructors

Formally, we develop synchronous processes as a special case of untimed processes. This will allow us later to easily connect different domains and migrate processes from one domain to another.

Synchronous processes have two specific characteristics. First, all synchronous processes consume and produce exactly one event on each input or output in each evaluation cycle; that is, the signature is always $\langle \{1,\ldots\}, \{1,\ldots\} \rangle$. Second, in addition to the value set V, events can carry the special value $\sqcup$, which denotes the *absence* of an event; this is the way we defined synchronous events $\bar{E}$ and signals $\bar{S}$ in Section 3.2. Both the processes and their contained functions must be able to deal with these events.

All synchronous process constructors and processes operate exclusively on synchronous signals.

By analogy with the *mapU* constructor, the *mapS* process constructor creates a process that takes one input signal, generates one output signal, and has no internal state:

$$mapS(f) = mapU(1, f) \qquad (4.1)$$

with

$$\exists \bar{e}' \in \bar{E} : f(\sqcup) = \bar{e}'$$
$$\forall \bar{e} \in \bar{E} : \text{length}(f(\bar{e})) = 1$$

Thus, `mapS` processes are like `mapU` processes, with the difference that f must be defined on $\sqcup$ and it always takes a single event as argument and produces a single event.

Note, that we could define synchronous processes in a different way, such that f is in fact not defined on synchronous events but only on untimed events:

$$\mathtt{mapSstrict}(f) = \mathtt{mapU}(1, f') \qquad (4.2)$$

with

$$\forall \dot{e} \in \dot{E} : \mathsf{length}(f(\dot{e})) = 1$$

$$f'(\hat{e}) = \begin{cases} \sqcup & \text{if } \hat{e} = \sqcup \\ f(\hat{e}) & \text{otherwise} \end{cases}$$

We call such processes *strict* because they always map an absent event to an absent event, $p(\sqcup) = \sqcup$. This approach would be consistent with our outspoken objective to separate process interfaces from their internal behavior. It would prove very convenient when we want to migrate a process from one timing domain to another. However, it also limits expressiveness and hampers one main objective of synchronous models—to have some control over timing issues, in particular to detect the absence of events. In the case of `mapS(f)` the function f can do this, but not in the case of `mapSstrict(f)`. Consequently we follow approach (4.1), but in practice a viable approach would be to provide both alternatives to the designer. This would also have the benefit that there is an explicit representation in the model of when a process interface exclusively handles timing and synchronization issues and when not.

The definitions of `scanS`, `scandS`, `mooreS`, and `mealyS` follow the idea of (4.1) and are based on the untimed processes defined in Section 3.4.

$$\mathtt{scanS}(g, w_0) = \mathtt{scanU}(1, g, w_0) \qquad (4.3)$$

with

$$\forall w \in V, \exists w' \in V : g(w, \sqcup) = w'$$

$$\mathtt{scandS}(g, w_0) = \mathtt{scandU}(1, g, w_0) \qquad (4.4)$$

with

$$\forall w \in V, \exists w' \in V : g(w, \sqcup) = w'$$

$$\mathtt{mooreS}(g, f, w_0) = \mathtt{mooreU}(1, g, f, w_0) \qquad (4.5)$$

with

$$\forall w \in V, \exists w' \in V : g(w, \sqcup) = w'$$
$$\forall w \in V, \exists \bar{e}' \in \bar{E} : f(w, \sqcup) = \bar{e}'$$
$$\forall w \in V, \bar{e} \in \bar{E} : \mathsf{length}(f(w, \bar{e})) = 1$$

$$\mathtt{mealyS}(g, f, w_0) = \mathtt{mealyU}(1, g, f, w_0) \qquad (4.6)$$

with

$$\forall w \in V, \exists w' \in V : g(w, \sqcup) = w'$$
$$\forall w \in V, \exists \bar{e}' \in \bar{E} : f(w, \sqcup) = \bar{e}'$$
$$\forall w \in V, \bar{e} \in \bar{E} : \text{length}(f(w, \bar{e})) = 1$$

We only require that g and f are defined for absent input events and that the output signal partitioning is a constant 1.

When we merge two signals into one, we have to decide how to represent the absence of an event in one input signal in the compound signal. We choose to use the $\sqcup$ symbol for this purpose also, which has the consequence that $\sqcup$ also appears in tuples together with normal values.

$$\text{zipS}() = p \tag{4.7}$$

with

$$p(\bar{s}_a, \bar{s}_b) = \bar{s}_c$$

$$\langle \bar{c}_i \rangle = \begin{cases} \sqcup & \text{if } \bar{a}_i = \sqcup \text{ and } \bar{b}_i = \sqcup \\ (\bar{a}_i, \bar{b}_i) & \text{otherwise} \end{cases}$$

$$\pi(v_a, \bar{s}_a) = \langle \bar{a}_i \rangle, \ v_a(i) = 1$$
$$\pi(v_b, \bar{s}_b) = \langle \bar{b}_i \rangle, \ v_b(i) = 1$$
$$\pi(v_c, \bar{s}_c) = \langle \bar{c}_i \rangle, \ v'(i) = 1$$

and

$$\text{unzipS}() = p \tag{4.8}$$

where

$$p(\bar{s}) = \langle \bar{s}', \bar{s}'' \rangle$$

$$\bar{a}_i = \begin{cases} \sqcup & \text{if } \bar{c}_i = \sqcup \text{ or } \bar{c}_i = (\sqcup, v_b) \\ v_a & \text{otherwise, where } \bar{c}_i = (v_a, v_b) \end{cases}$$

$$\bar{b}_i = \begin{cases} \sqcup & \text{if } \bar{c}_i = \sqcup \text{ or } \bar{c}_i = (v_a, \sqcup) \\ v_b & \text{otherwise, where } \bar{c}_i = (v_a, v_b) \end{cases}$$

$$\pi(v, \bar{s}) = \langle \bar{c}_i \rangle, \ v(i) = 1$$
$$\pi(v', \bar{s}') = \langle \bar{a}_i \rangle, \ v'(i) = 1$$
$$\pi(v'', \bar{s}'') = \langle \bar{b}_i \rangle, \ v''(i) = 1$$

In addition we also define a zipWith constructor that is based solely on its untimed equivalent:

$$\text{zipWithS}(f) = \text{zipWithU}(1, 1, f) \tag{4.9}$$

As in the untimed model we need sources and sinks and the possibility of initializing a signal before any other processing commences.

$$\text{sourceS}(g, w_0) = p \tag{4.10}$$

where

$$p() = \bar{s}'$$
$$w_i = \bar{e}'_i$$
$$g(w_i) = w_{i+1}$$
$$\pi(v', \bar{s}') = \langle\langle \bar{e}'_i \rangle\rangle, \ v'(i) = 1$$

$$\text{sinkS}(g, w_0) = p \tag{4.11}$$

where

$$p(\bar{s}) = \langle\rangle$$
$$g(w_i) = w_{i+1}$$
$$\pi(v, \bar{s}) = \langle \bar{a}_i \rangle, \ v(i) = 1$$

$$\text{initS}(\bar{r}) = p \tag{4.12}$$

where

$$p(\bar{s}) = \bar{r} \oplus \bar{s}$$
$$v = v' = 1$$
$$\bar{r}, \bar{s} \in \bar{S}$$

4.3 Feedback Loops

The untimed MoC deals with time at the causality level, which essentially means that the output events of a process emerge "after" the input events that triggered the computation. It is not specific about the length of this "after," but it is in fact strict in that it really is an "after" and it cannot be an "at the same time." In contrast, the perfect synchrony assumption explicitly defines the time instants of output events and causing input events to be identical. This difference has no practical implications as long as there are no feedback loops involved. To model feedback loops in the untimed MoC, we introduced the feedback operator **FB**$_\textbf{P}$ in Section 3.6.30. Recall Example 3.9, where the input of the stateless process $p_1 = \text{zipWithU}(1, 1, g)$ was directly dependent on its own output. **FB**$_\textbf{P}$ was defined over the prefix order of signals such that we started the computation with an empty signal on the feedback link to find the behavior of **FB**$_\textbf{P}$(p_1). Since zipWithU processes block when one of their inputs is empty, process p_1 has never been evaluated and the output of **FB**$_\textbf{P}$(p_1) was the empty signal independent of its input. In contrast to this approach, the perfectly synchronous MoC, and with it all the synchronous languages such as Esterel, Signal, Lustre, and so on, define the behavior of feedback loops such that all signals carry consistent values *in each evaluation cycle*. As a consequence the distinction between cause and effect is blurred, which is consistent with the hypothesis that input and output events occur at the very same time instant and, in a feedback loop, the input is as much a cause for the output as vice versa.

To elaborate this concept we define a feedback operator that is not based on the prefix order of signals, but on the Scott order of the event values in each evaluation cycle. This essentially amounts to finding the least fixed point for each individual event of the signal separately.

Recall that we have defined the set of synchronous events as $\bar{E} = V \cup \{\sqcup\}$. We further extend this domain with another special element, $\bot$ (pronounced "bottom"), to capture the situation when we do not know if the event occurred or which value it has: $\bar{E}_\bot = \bar{E} \cup \{\bot\}$.

The order relation $\preceq$ (pronounced "weaker") on $\bar{E}_\bot$ is intended to correspond to the notion "less defined than or equal to" and is defined as follows:

$$\bot \preceq \bar{e} \text{ and } \bar{e} \preceq \bar{e} \quad \text{for all } \bar{e} \in \bar{E}_\bot$$

Distinct elements in $\bar{E}$ are unrelated by $\preceq$; that is, for $\bar{e}_1, \bar{e}_2 \in \bar{E}$ we have $\bar{e}_1 \neq \bar{e}_2 \implies \bar{e}_1 \not\preceq \bar{e}_2$ and $\bar{e}_2 \not\preceq \bar{e}_1$. Figure 4-3 illustrates the structure of this order relation.

We extend this relation to infinite signals by applying $\preceq$ to the events in the same position. Let $\bar{S}_\bot$ be the set of signals consisting of elements from $\bar{E}_\bot$, and let $\bar{S}_\bot^\infty \subseteq \bar{S}_\bot$ be the set of infinite signals. For signals of length n, we define

$$\langle \bar{e}_1, \ldots, \bar{e}_n \rangle \preceq \langle \bar{e}'_1, \ldots, \bar{e}'_n \rangle \text{ iff } \bar{e}_i \preceq \bar{e}'_i \quad \text{for all } i, 1 \leq i \leq n$$

Example 4.1

$$\langle \bot, \bot \rangle \preceq \langle \bot, \bot \rangle \preceq \langle \bar{e}_1, \bot \rangle \preceq \langle \bar{e}_1, \bar{e}_2 \rangle$$

$$\langle \bot, \bot \rangle \preceq \langle \bot, \bar{e}_2 \rangle \preceq \langle \bar{e}_1, \bar{e}_2 \rangle$$

$$\langle \bot, \bar{e} \rangle \not\preceq \langle \bar{e}, \bot \rangle$$

$$\langle \bot, \bar{e}_1 \rangle \not\preceq \langle \bot, \bar{e}_2 \rangle \quad \text{unless } \bar{e}_1 = \bar{e}_2$$

$$\langle \bot, \bot, \bot \rangle \preceq \langle \bar{e}_1, \bot, \bot \rangle \preceq \langle \bar{e}_1, \bar{e}_2, \bot \rangle \preceq \langle \bar{e}_1, \bar{e}_2, \bar{e}_3 \rangle$$

FIGURE 4-3

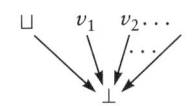

The partial order relation $\preceq$ in $\bar{E}_\bot$, where the arrows point to the weaker element in the relation.

In the following we will only deal with infinite signals. Finite signals can be made infinite by appending either infinitely many absent events if we know that no event will occur, or infinitely many bottom events if we do not know. Let $\langle \bot \rangle^\infty$ be the infinite signal consisting only of $\bot$, that is, being undefined everywhere, and we have $\langle \bot \rangle^\infty \preceq \bar{s}$ for all $\bar{s} \in \bar{S}_\bot^\infty$.

Let $(\bar{E}_\bot)^n$ be the set of n-tuples over $\bar{E}_\bot$. An $f : (\bar{E}_\bot)^n \to (\bar{E}_\bot)^m$ is monotonic if

$$(\bar{e}_1, \ldots, \bar{e}_n) \preceq (\bar{e}'_1 \ldots \bar{e}'_n) \implies f(\bar{e}_1, \ldots, \bar{e}_n) \preceq f(\bar{e}'_1 \ldots \bar{e}'_n)$$

$$\text{for all } (\bar{e}_1, \ldots, \bar{e}_n), (\bar{e}'_1 \ldots \bar{e}'_n) \in (\bar{E}_\bot)^n$$

Intuitively, this means that if the second of two inputs is "more defined" than the first, then the result of f should be "more defined" for the second input than for the first.

Example 4.2 Let $V = \mathbb{Z} \cup \{\bot\}$. Consider functions $f_1, f_2 : V^2 \to V$, which both extend the addition operation on integers:

$$f_1(x, y) = \begin{cases} \bot & \text{if } x = \bot \text{ or } y = \bot \\ x + y & \text{otherwise} \end{cases}$$

f_1 is monotonic in its second argument because for all $x, y \in \mathbb{Z}$ we have

$$f_1(x, \bot) = \bot$$
$$f_1(x, y) = x + y$$
$$(x, \bot) \preceq (x, y)$$
$$\bot \preceq x + y$$

Similarly we could show that it is also monotonic in its first argument. In contrast, f_2 is not monotonic:

$$f_2(x, y) = \begin{cases} x & \text{if } y = \bot \\ y & \text{if } x = \bot \\ x + y & \text{otherwise} \end{cases}$$

The following example illustrates why not:

$$f_2(1, \bot) = 1$$
$$f_2(1, 1) = 2$$
$$(1, \bot) \preceq (1, 1)$$
$$1 \not\preceq 2$$

A set of event tuples is a *chain* $C = \langle c_0, c_1, \ldots, c_n \rangle$ if every event tuple in the chain is more well defined than or equally well defined as its predecessor; thus $c_i \preceq c_{i+1}$ for $0 \leq i < n$. Since our relation $\preceq$ defines a *complete partial order* (Stoy 1989; Davey and Priestley 1997), we know that every chain has a *least upper bound* (LUB), which is the weakest event tuple c with $c_i \preceq c$ for all $c_i \in C$. We write $\text{LUB}(C) = c$.

4.3 Feedback Loops

A monotonic function $f : (\bar{E}_\perp)^n \to (\bar{E}_\perp)^m$ is said to be *continuous* if for every chain $C = \langle c_i \rangle, c_i \in (\bar{E}_\perp)^n$,

$$f(\text{LUB}(C)) = \text{LUB}(f(c_i))$$

These definitions can easily be extended to signals and processes. A process $p : (\bar{S}_\perp^\infty)^n \to (\bar{S}_\perp^\infty)^m$ is monotonic if

$$(\bar{s}_1, \ldots, \bar{s}_n) \preceq (\bar{s}_1', \ldots \bar{s}_n') \implies p(\bar{s}_1, \ldots, \bar{s}_n) \preceq p(\bar{s}_1' \ldots \bar{s}_n')$$

$$\text{for all } (\bar{s}_1, \ldots, \bar{s}_n), (\bar{s}_1' \ldots \bar{s}_n') \in (\bar{S}_\perp^\infty)^n$$

It is also continuous if for every chain $C = \langle c_i \rangle, c_i \in (\bar{S}_\perp^\infty)^n$

$$p(\text{LUB}(C)) = \text{LUB}(p(c_i))$$

Now we are in the position to define the feedback operator for the synchronous MoC (Figure 4.4). Given a process $p : (\bar{S}_\perp^\infty \times \bar{S}_\perp^\infty) \to (\bar{S}_\perp^\infty \times \bar{S}_\perp^\infty)$ with two input signals and two output signals, we define the process $\mathbf{FB_S}(p) : \bar{S}_\perp^\infty \to \bar{S}_\perp^\infty$ by

$$\mathbf{FB_S}(p)(\bar{s}_1) = \bar{s}_2 \quad \text{where } p(\bar{s}_1, \bar{s}_3) = (\bar{s}_2, \bar{s}_3) \tag{4.13}$$

The events in $\bar{s}_3$ in each evaluation cycle are determined by the least fixed point, which is the weakest of all signals that satisfy the constraint of the feedback loop; that is, the second output of process p is also its second input.

Again, we will not develop the formal theory, but we will use an example to illustrate the essential aspects. Consider the process network in Figure 4-5, where the processes are defined as follows:

$$p_1 = zipS()$$
$$p_2 = mapS(f)$$
$$p_3 = unzip()$$
$$p_4 = p_3 \circ p_2 \circ p_1$$
$$q = \mathbf{FB_S}(p_4)$$

FIGURE 4-4

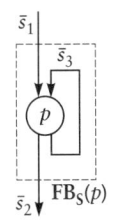

Feedback operator based on the Scott order over event values.

FIGURE 4-5

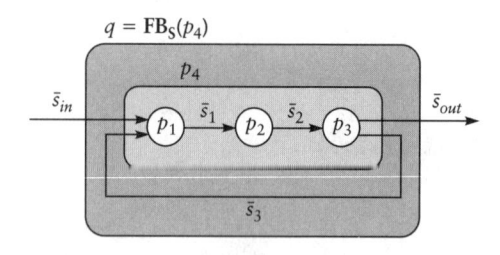

A synchronous model with a feedback loop.

If we had used the **FB_P** operator as in Section 3.6, we would compute the solution by starting with the empty sequence on signal $\bar{s}_3$. The $\mathtt{zipS}$ process p_1 would block, and the output of q would be the empty sequence whatever the value of the input signal $\bar{s}_{in}$. In contrast, the operator **FB_S** requires to find the least fixed-point solution for each evaluation cycle separately.

In order to compute the signals $\bar{s}_{out}$ and $\bar{s}_3$, we find the fixed points for every event on these signals. Since p_1 and p_3 only group and ungroup the signals but do not contribute to the actual computation, we can concentrate on function f. To find the event values on $\bar{s}_{out}$ and $\bar{s}_3$ for a given event x on $\bar{s}_{in}$ we go through the following sequence, where $y_0 = \bot$:

$$f(x, y_0) = (z_1, y_1)$$
$$f(x, y_1) = (z_2, y_2)$$
$$\vdots$$
$$f(x, y_{n-1}) = (z_n, y_n)$$
$$f(x, y_n) = (z_n, y_n)$$

The y_i and z_i correspond to the events on $\bar{s}_3$ and $\bar{s}_{out}$, respectively. In fact, they are chains of events, and their LUB corresponds to the desired least fixed point that constitutes the solution of the feedback loop.

Let's investigate several cases depending on the definition of function f:

1. Let $f(x, y) = f_1(x, y) = (2x, 2x)$, where the multiplication is extended such that $\sqcup$ is interpreted as a 0, and a $\bot$ argument always evaluates to $\bot$.

$$f_1(x, \bot) = (2x, 2x)$$
$$f_1(x, 2x) = (2x, 2x)$$

The y-chain is $\langle \bot, 2x, \ldots \rangle$, and the z-chain is $\langle 2x, 2x, \ldots \rangle$. Thus, for the input $\bar{s}_{in} = \langle 1, 2, 3 \rangle \oplus \langle \sqcup \rangle^\infty$ we get

$$q(\langle 1, 2, 3 \rangle \oplus \langle \bot \rangle^\infty) = \langle 2, 4, 6, 0, 0, \ldots \rangle$$

Alternatively, we could operate directly on the signals and we obtain

$$p_1(\langle(1,\bot),(2,\bot),(3,\bot),(\sqcup,\bot),\ldots\rangle) = \langle(2,2),(4,4),(6,6),(0,0),\ldots\rangle$$
$$p_1(\langle(1,2),(2,4),(3,6),(\sqcup,0),\ldots\rangle) = \langle(2,2),(4,4),(6,6),(0,0),\ldots\rangle$$

2. $f(x,y) = f_2(x,y) = (2x, 2y)$:

$$f_2(x,\bot) = (2x,\bot)$$
$$f_2(x,\bot) = (2x,\bot)$$

and we get again

$$q(\langle 1,2,3 \rangle \oplus \langle \sqcup \rangle^\infty) = \langle 2,4,6,0,0,\ldots \rangle$$

However, the feedback signal is all undefined: $\bar{s}_3 = \langle \bot \rangle^\infty$, although the output of process q is well defined everywhere.

3. $f(x,y) = f_3(x,y) = (y,y)$:

$$f_3(x,\bot) = (\bot,\bot)$$
$$f_3(x,\bot) = (\bot,\bot)$$

and we get

$$q(\langle 1,2,3 \rangle \oplus \langle \sqcup \rangle^\infty) = \langle \bot \rangle^\infty$$

which can be considered as a questionable model since nothing can be known about the output, whatever the input may be. But note that f_3 has infinitely many fixed points. Consider, for instance,

$$f_3(x,3) = (3,3)$$
$$f_3(x,3) = (3,3)$$

Any other value for $\bar{s}_3$ would also be a solution for the equations and consistent with all constraints, but the least fixed-point semantics requires the unambiguous solution $\bar{s}_3 = \langle \bot \rangle^\infty$.

4. $f(x,y) = f_4(x,y) = (x+y, y+1)$: Again, the addition is extended such that $\sqcup$ is interpreted as a 0, and a $\bot$ argument always evaluates to $\bot$.

$$f_4(x,\bot) = (\bot,\bot)$$
$$f_4(x,\bot) = (\bot,\bot)$$

and we get

$$q(\langle 1,2,3 \rangle \oplus \langle \sqcup \rangle^\infty) = \langle \bot \rangle^\infty$$

The result is the same as in the previous case. But in contrast to f_3, f_4 has no fixed point. As an example, consider

$$f_4(x,0) = (x,1)$$

$$f_4(x, 1) = (x+1, 2)$$
$$f_4(x, 2) = (x+2, 3)$$
$$\vdots$$

5. $f(x,y) = f_5(x,y) = (2y, 2x)$:

$$f_5(x, \bot) = (\bot, 2x)$$
$$f_5(x, 2x) = (4x, 2x)$$
$$f_5(x, 2x) = (4x, 2x)$$

and we get

$$q(\langle 1, 2, 3 \rangle \oplus \langle \sqcup \rangle^\infty) = \langle 4, 8, 12, 0, \ldots \rangle$$

In this case also, the feedback signal is well defined with $\bar{s}_3 = \langle 2, 4, 6, 0, \ldots \rangle$.

It is important that f is monotonic. Consider the following comparison operation:

$$x \equiv y = \begin{cases} \text{True} & \text{if both arguments are } \bot \text{ or both arguments are not } \bot \text{ but equal} \\ \text{False} & \text{if exactly one argument is } \bot \text{ or none is } \bot \text{ and they are not equal} \end{cases}$$

It is not monotonic because

$$1 \equiv \bot = \text{False}$$
$$1 \equiv 1 = \text{True}$$
$$(1, \bot) \preceq (1, 1)$$

but

$$\text{False} \not\preceq \text{True}$$

6. If it is used in a feedback loop, it may lead to an unspecified behavior. If in our example above (Figure 4-5) we define

$$f(x,y) = f_6(x,y) = \text{if } y \equiv 1 \text{ then } (2x, 2) \text{ else } (4x, 1)$$

Trying to find the least fixed point yields

$$f_6(x, \bot) = \text{if } \bot \equiv 1 \text{ then } (2x, 2) \text{ else } (4x, 1)$$
$$= (4x, 1)$$
$$f_6(x, 1) = \text{if } 1 \equiv 1 \text{ then } (2x, 2) \text{ else } (4x, 1)$$
$$= (2x, 2)$$

$$f_6(x, 2) = \text{if } 2 \doteq 1 \text{ then } (2x, 2) \text{ else } (4x, 1)$$
$$= (4x, 1)$$
$$\vdots$$

The process oscillates with no conclusive result.

In contrast, a monotonic comparison can be defined as follows:

$$x \doteq y = \begin{cases} \bot & \text{if either argument is } \bot \\ \text{True} & \text{if both arguments are not } \bot \text{ and both arguments are equal} \\ \text{False} & \text{if both arguments are not } \bot \text{ and they are not equal} \end{cases}$$

To be able to use it, we have also to extend the if-then-else operation:

$$\text{if } p(x) \text{ then } g(x) \text{ else } h(x) = \begin{cases} \bot & \text{if } p(x) \text{ is } \bot \\ g(x) & \text{if } p(x) \text{ is True} \\ h(x) & \text{if } p(x) \text{ is False} \end{cases} \quad (4.14)$$

7. Using this operation, or in fact an extension of it for tuples, in our feedback loop, we obtain

$$f(x, y) = f_7(x, y) = \text{if } y \doteq 1 \text{ then } (2x, 2) \text{ else}(4x, 1)$$

Trying to find the least fixed point yields

$$f_7(x, \bot) = \text{if } \bot \doteq 1 \text{ then } (2x, 2) \text{ else } (4x, 1)$$
$$= (\bot, \bot)$$

The process quickly converges and tells us that both the output signal and the feedback signal are not defined anywhere.

An interesting monotonic variant of the if-then-else operation is used in Esterel. Let $V = \{\sqcup, 1\}$ be the restricted set of event values, and let $\bar{e}_1, \bar{e}_2 \in V$, $p \in (V \to \{\text{True}, \text{False}\}$.

$$\text{if } p(x) \text{ then } \bar{e}_1 \text{ else } \bar{e}_2 = \begin{cases} \bar{e}_1 & \text{if } p(x) \text{ is True} \\ \bar{e}_2 & \text{if } p(x) \text{ is False} \\ \bar{e}_1 & \text{if } p(x) \text{ is } \bot \text{ and } \bar{e}_1 = \bar{e}_2 \\ \bot & \text{otherwise} \end{cases} \quad (4.15)$$

Essentially, this means that if we cannot determine the value of $p(x)$, we have to check if $\bar{e}_1 = \bar{e}_2$. If they are equal, the result will be $\bar{e}_1$, independent of the value of $p(x)$. This definition allows to express functions as the following:

8. $f(x, y) = f_8(x, y) = \text{if } y \doteq 1 \text{ then } (1, 1) \text{ else } (1, 1)$

The least fixed point is calculated as follows:

$$f_8(x, \bot) = \text{if } \bot \doteq 1 \text{ then } (1, 1) \text{ else } (1, 1)$$

$$= \text{if } \perp \text{ then } (1,1) \text{ else } (1,1)$$
$$= (1,1)$$
$$f_8(x,1) = \text{if } 1 \doteq 1 \text{ then } (1,1) \text{ else } (1,1)$$
$$= (1,1)$$

The Esterel compiler conducts a sophisticated case analysis to identify the different possibilities in case the condition $p(x)$ cannot be calculated immediately. If it finds that the result would be identical in both branches, it adopts this result. Most other languages adopt (4.14) and thus cannot emulate the behavior of function f_8.

Because it is important to use monotonic functions, it is useful to know that the composition of monotonic functions results in a monotonic function. Since our functions must also be defined on the element $\perp$, we usually have to extend them. A standard procedure to extend a function is called *natural extension*. A naturally extended function maps a $\perp$ argument always to $\perp$. Thus, multiplication, addition, and the $\doteq$ comparison operator we used above are naturally extended. All naturally extended functions are monotonic.

This account of this theory is only an intuitive introduction and leaves many interesting questions open, such as why unique least fixed points must exist for continuous functions and how they can be computed. Manna (1974, Chapter 5) gives an excellent, basic introduction to the theory, which can be absorbed in its full beauty from Stoy (1989) and Davey and Priestley (1997). The semantics of Esterel is described in great detail by Berry (1999).

4.4 Perfectly Synchronous MoC

Finally, we can now make precise what we mean by the synchronous model of computation:

Definition 4.1 The *synchronous model of computation* (synchronous MoC) is defined as synchronous MoC $= (C, O)$, where

$$C = \{\texttt{mapS}, \texttt{mapSstrict}, \texttt{scanS}, \texttt{scandS}, \texttt{mooreS}, \texttt{mealyS},$$
$$\texttt{zipS}, \texttt{unzipS}, \texttt{zipWithS}, \texttt{sourceS}, \texttt{sinkS}, \texttt{initS}\}$$

$$O = \{\parallel, \circ, \mathbf{FB_S}\}$$

In other words, a process or a process network belongs to the *synchronous MoC domain* iff all its processes and process compositions are constructed either by one of the named process constructors or by one of the composition operators. We call such processes *S-MoC processes*.

4.5 Process Merge

The merging of processes is simplest for synchronous processes, because all processes always form perfect matches. No up-rating is required or even possible. Because it is simple in the synchronous context, we will elaborate on the merging of processes and on feedback loops in more detail here.

We begin by giving a list of equivalences for later use. Note that we always use the *scandS* version of scan processes because it matches the way Moore and Mealy processes are defined. The proofs are left as an exercise.

$$mapS(f_1) \circ mapS(f_2) = mapS(f_1 \circ f_2) \tag{4.16}$$

$$mapS(f_1) \circ scandS(g_2, w_0) = mooreS(g_2, f_1, w_0) \tag{4.17}$$

$$mapS(f_1) \circ mooreS(g_2, f_2, w_0) = mooreS(g_2, f_1 \circ f_2, w_0) \tag{4.18}$$

$$mapS(f_1) \circ mealyS(g_2, f_2, w_0) = mealyS(g_2, f_1 \circ f_2, w_0) \tag{4.19}$$

$$scandS(g_1, v_0) \circ mapS(f_2) = scandS(g, v_0) \tag{4.20}$$

where

$$g(v, \bar{e}) = g_1(v, f_2(\bar{e}))$$

$$scandS(g_1, v_0) \circ scandS(g_2, w_0) = mooreS(g, f, (v_0, w_0)) \tag{4.21}$$

where

$$g((v, w), \bar{e}) = (g_1(v, w), g_2(w, \bar{e}))$$
$$f((v, w)) = v$$

$$scandS(g_1, v_0) \circ mooreS(g_2, f_2, w_0) = mooreS(g, f, (v_0, w_0)) \tag{4.22}$$

where

$$g((v, w), \bar{e}) = (g_1(v, (f_2(w))), g_2(w, \bar{e}))$$
$$f((v, w)) = v$$

$$scandS(g_1, v_0) \circ mealyS(g_2, f_2, w_0) = mooreS(g, f, (v_0, w_0)) \tag{4.23}$$

where

$$g((v, w), \bar{e}) = (g_1(v, (f_2(w, \bar{e}))), g_2(w, \bar{e}))$$
$$f((v, w)) = v$$

$$mooreS(g_1,f_1,v_0) \circ mapS(f_2) = mooreS(g,f_1,v_0) \quad (4.24)$$

where

$$g(v,\bar{e}) = g_1(v,f_2(\bar{e}))$$

$$mooreS(g_1,f_1,v_0) \circ scandS(g_2,w_0) = mooreS(g,f,(v_0,w_0)) \quad (4.25)$$

where

$$g((v,w),\bar{e}) = (g_1(v,w),g_2(w,\bar{e}))$$
$$f((v,w)) = f_1(v)$$

$$mooreS(g_1,f_1,v_0) \circ mooreS(g_2,f_2,w_0) = mooreS(g,f,(v_0,w_0)) \quad (4.26)$$

where

$$g((v,w),\bar{e}) = (g_1(v,(f_2(w))),g_2(w,\bar{e}))$$
$$f((v,w)) = f_1(v)$$

$$mooreS(g_1,f_1,v_0) \circ mealyS(g_2,f_2,w_0) = mooreS(g,f,(v_0,w_0)) \quad (4.27)$$

where

$$g((v,w),\bar{e}) = (g_1(v,(f_2(w,\bar{e}))),g_2(w,\bar{e}))$$
$$f((v,w)) = f_1(v)$$

$$mealyS(g_1,f_1,v_0) \circ mapS(f_2) = mealyS(g,f,v_0) \quad (4.28)$$

where

$$g(v,\bar{e}) = g_1(v,f_2(\bar{e}))$$
$$f(v,\bar{e}) = f_1(v,f_2(\bar{e}))$$

$$mealyS(g_1,f_1,v_0) \circ scandS(g_2,w_0) = mooreS(g,f,(v_0,w_0)) \quad (4.29)$$

where

$$g((v,w),\bar{e}) = (g_1(v,w),g_2(w,\bar{e}))$$
$$f((v,w)) = f_1(v,w)$$

$$mealyS(g_1,f_1,v_0) \circ mooreS(g_2,f_2,w_0) = mooreS(g,f,(v_0,w_0)) \quad (4.30)$$

where

$$g((v,w),\bar{e}) = (g_1(v,f_2(w)),g_2(w,\bar{e}))$$
$$f((v,w)) = f_1(v,f_2(w))$$

$$mealyS(g_1,f_1,v_0) \circ mealyS(g_2,f_2,w_0) = mealyS(g,f,(v_0,w_0)) \quad (4.31)$$

where

$$g((v,w),\bar{e}) = (g_1(v,f_2(w,\bar{e})), g_2(w,\bar{e}))$$
$$f((v,w),\bar{e}) = f_1(v,f_2(w,\bar{e}))$$

Example 4.3 Consider the example in Figure 4-1. Let $P_1 = mapS(f_1)$ and $P_2 = mapS(f_2)$, with

$$f_1((x,y)) = (x+y, x-y)$$
$$f_2((x,y)) = (x-y, x+y)$$

Process P is then defined by $mapS(f_P)$ with

$$f_P((x,y)) = f_2(f_1((x,y))) = f_2((x+y, x-y))$$
$$= (x+y-(x-y), x+y+x-y) = (2y, 2x)$$

4.6 Clocked Synchronous Models

A desirable property for every model of concurrency is to easily understand a compound behavior from the behaviors of the constituent parts. We have seen that we could derive the behaviors of processes P in Figure 4-1 and q in Figure 4-5 from the characteristic functions of the subprocesses, by composing the functions and solving the resulting equations. While this may not always be trivial, it is still simpler than a situation where the compound behavior also depends on the delays of the components.

As mentioned earlier, the assumption made for the design of synchronous circuits is weaker than the perfect synchrony hypothesis. The *clocked synchronous models* are based on the following assumption:

Clocked synchronous hypothesis: *There is a global clock signal controlling the start of each computation in the system. Communication takes no time, and computation takes one clock cycle.*

To describe clocked synchronous processes we introduce a delay function Δ, which delays each input by one cycle.

$$\Delta = scandS(g, \sqcup) \quad (4.32)$$

where

$$g(w, \bar{e}) = \bar{e}$$

FIGURE 4-6

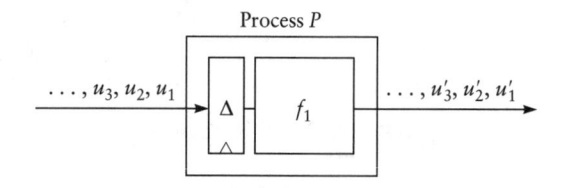

A simple clocked synchronous process P.

Based on this delay process, we define the constructors for the clocked synchronous model.

$$mapCS(f) = mapS(f) \circ \Delta$$
$$scanCS(g, w_0) = scanS(g, w_0) \circ \Delta$$
$$mooreCS(g, f, w_0) = mooreS(g, f, w_0) \circ \Delta$$
$$mealyCS(g, f, w_0) = mealyS(g, f, w_0) \circ \Delta$$
$$zipCS()(\bar{s}_1, \bar{s}_2) = zipS()(\Delta(\bar{s}_1), \Delta(\bar{s}_2)) \qquad (4.33)$$
$$unzipCS() = unzipS() \circ \Delta$$
$$sourceCS = sourceS$$
$$sinkCS = sinkS$$
$$initCS = initS$$

Thus, elementary processes are composed of a combinatorial function and a delay function that represents a latch at the inputs. Process P in Figure 4-6 would be defined as

$$P = mapCS(f_1)$$

Again, we can now precisely define the clocked synchronous MoC.

Definition 4.2 The *clocked synchronous model of computation* is defined as clocked synchronous MoC = (C, O), where

$$C = \{\Delta, mapCS, scanCS, mooreCS, mealyCS, sipCS, unzipCS,$$
$$sourceCS, sinkCS, initCS\}$$
$$O = \{\|, \circ, \mathbf{FB_P}\}$$

In other words, a process or a process network belongs to the *clocked synchronous MoC domain* iff all its processes and process compositions are constructed either by one of the named process constructors or by one of the composition operators. We call such processes *CS-MoC processes*.

Note that we have used the prefix-order-based feedback operator because it seems to be more in line with the intention of this computational model, which essentially tries to avoid instantaneous feedback loops. However, as long as there is a delay operator in every feedback loop, there is no practical difference between **FB_P** and **FB_S**.

4.7 Extended Characteristic Function

For the CS-MoC it is convenient to relate "neighboring" events of the input and output signals via the characteristic function of a process. To this end we index the events in the signals. Let $p = \mathit{mapS}(f)$, and let $\bar{s}' = p(\bar{s})$, $\bar{s} = \langle \bar{e}_i \rangle$, and $\bar{s}' = \langle \bar{e}'_i \rangle$. Then we have

$$\bar{e}'_i = f(\bar{e}_i) \quad \forall i \in \mathbb{N}_0 \tag{4.34}$$

Similarly, let $\bar{s}' = \Delta(\bar{s})$, $\bar{s} = \langle \bar{e}_i \rangle$, and $\bar{s}' = \langle \bar{e}'_i \rangle$. Then we have

$$\begin{aligned} \bar{e}'_i &= \bar{e}_{i-1} \quad \forall i \in \mathbb{N} \\ \bar{e}'_0 &= \sqcup \end{aligned} \tag{4.35}$$

We stretch the concept of a characteristic function (Section 3.8) to express the dependence of an output event on a few past input events. This is convenient for handling processes with a constrained internal state.

Definition 4.3 Let p be a CS-MoC process.

$$f_{p,\bar{s}=\langle \bar{e}_i \rangle}(\bar{e}_j) = \bar{e}'_j$$

for

$$\begin{aligned} p(\bar{s}) &= \bar{s}' \\ \bar{s}, \bar{s}' &\in \bar{S} \\ \langle \bar{e}_i \rangle &= \bar{s} \\ \langle \bar{e}'_i \rangle &= \bar{s}' \end{aligned}$$

is an *extended characteristic function* of process p if it is defined in terms of an arbitrary functional expression of $\bar{e}_i, i \leq j$.

If there is no danger of confusion we may just write $f_{p,s}$ or even f_p as an abbreviation. Likewise, we may write "characteristic function" when we mean "extended characteristic function."

Note that $f_{p,sys=\langle \bar{e}_i \rangle}(\bar{e}_j)$ is specific for a signal $\bar{s}$ and may use any $\bar{e}_i$ as long as $i \leq j$, that is, any event consumed by the process in the past. So in principle we can use this notation to express any process with state, but in practice we use it only for processes where the current output depends on a few past inputs.

For instance, the extended characteristic function of the Δ process is

$$\begin{aligned} f_{\Delta,s}(\bar{e}_i) &= \bar{e}_{i-1}, \quad i > 0 \\ f_{\Delta,s}(\bar{e}_0) &= \sqcup \end{aligned}$$

FIGURE 4-7

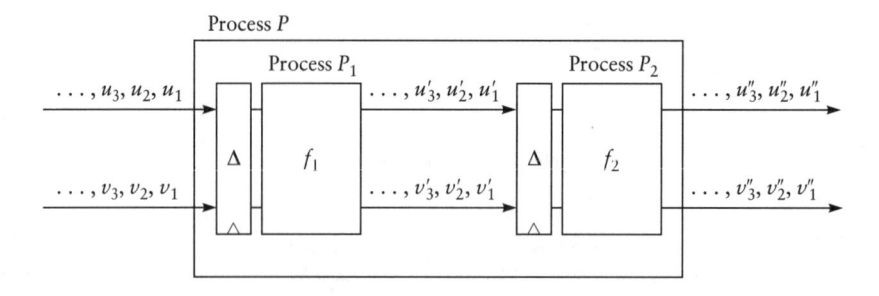

Process P composed of subprocesses P_1 and P_2 in the clocked synchronous model.

Example 4.4 We model a similar process as in Example 4.3. Consider process P in Figure 4-7 with the two constituent processes $P_1 = map2CS(f_1)$ and $P_2 = map2CS(f_2)$ where we define the constructor $map2CS$:

$$map2CS(f) = mapS(f) \circ zipS() \circ \Delta$$

Let all u and v be integers and

$$f_1((\sqcup, \sqcup)) = (0, 0)$$
$$f_1((x, \sqcup)) = (x, x)$$
$$f_1((\sqcup, y)) = (y, -y)$$
$$f_1((x, y)) = (x + y, x - y)$$
$$f_1((\sqcup, \sqcup)) = (0, 0)$$
$$f_2((x, \sqcup)) = (x, x)$$
$$f_2((\sqcup, y)) = (-y, y)$$
$$f_2((x, y)) = (x - y, x + y)$$

Consequently, the characteristic function f_P of P is

$$\begin{aligned}
f_P((x_i, y_i)) &= f_2(f_\Delta(f_1(f_\Delta((x_i, y_i))))) \\
&= f_2(f_\Delta(f_1((x_{i-1}, y_{i-1})))) \\
&= f_2(f_\Delta((x_{i-1} + y_{i-1}, x_{i-1} - y_{i-1}))) \\
&= f_2((x_{i-2} + y_{i-2}, x_{i-2} - y_{i-2})) \\
&= (x_{i-2} + y_{i-2} - x_{i-2} + y_{i-2}, x_{i-2} + y_{i-2} + x_{i-2} - y_{i-2}) \\
&= (2y_{i-2}, 2x_{i-2}) \quad \text{for } i > 1
\end{aligned}$$

$$f_P((x_i, y_i)) = 0 \quad \text{for } 0 \le i \le 1$$

4.7 Extended Characteristic Function

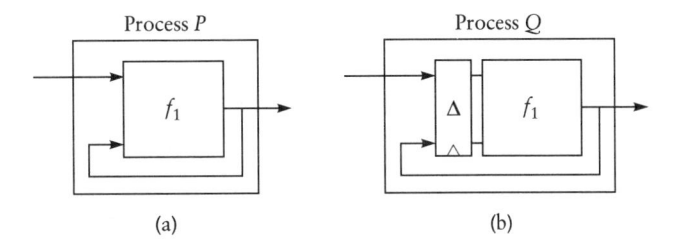

FIGURE 4-8

(a)   (b)

(a) A perfectly synchronous and (b) a clocked synchronous feedback loop.

We get the same result as for the perfectly synchronous case in Example 4.3 but with a delay of two cycles, and the initial two output values depend on the initial values of the latches and how f_1 and f_2 interpret them.

These equations are more complicated than the respective equations for the perfectly synchronous model because we have to deal with the extra dimension of time. We have to distinguish between the values of a variable at different time instances.

Both models give the same results, at least after an initialization phase for the clocked synchronous model. However, this is not the case in general. Consider a simple feedback process in the perfectly synchronous and in the clocked synchronous model as shown in Figure 4-8. The processes are defined by $P = \mathtt{mapS}(f_1) \circ \mathtt{zip}()$ and $Q = \mathtt{map2CS}(f_1)$. The combinatorial function f_1 is the same in both cases:

$$f_1(x, y) = 2y - 2x$$

The characteristic function f_P for process P, the perfectly synchronous case, is

$$f_P(x) = z$$

where

$$z = f_1(x, z)$$
$$= 2z - 2x$$
$$-z = -2x$$
$$z = 2x$$

For an input sequence $[0, 1, 2, 3]$ the process P would generate $P([0, 1, 2, 3]) = [0, 2, 4, 6]$.

On the other hand, the characteristic function f_Q for process Q is as follows:

$$f_Q(x_i) = z_i \tag{4.36}$$

where

$$z_i = f_1(f_\Delta(x_i, z_i))$$

$$= f_1(x_{i-1}, z_{i-1})$$
$$= 2z_{i-1} - 2x_{i-1} \quad \text{for } i > 0$$
$$z_0 = f_1(0,0) = 0$$

For the same input sequence, process Q would generate $Q([0,1,2,3]) = [0,0,-2, -8,-22]$. The reason for the discrepancy is that the result of f_1 depends on how fast the result travels to the input of f_1.

Because delays are relevant for the functional behavior in clocked synchronous models, we always have to take timing into account when we compose processes out of simple components. If, for instance, we have two processes that implement the same functional behavior, we cannot automatically replace one by the other in a larger process. Whether we can safely do it or not depends on the timing behavior. On the other hand we can always replace one process by another, functionally equivalent process in the perfectly synchronous case.

Example 4.5 To illustrate this further we describe two different processes (Figure 4-9), which both model the function $f_1(x,y) = 2y - 2x$. One process R_1 uses function f_1 directly. The other process R_2 uses two other simple functions, $f_2(x) = 2x$ and $f_3(x,y) = y - x$. Again, we assume that ⊔ is treated as 0 without mentioning it explicitly.

Both processes have the same functionality, but they experience different internal delays. Hence, if we put them into a larger process, the result may be quite different. The larger process may be very simple, for example, adding one feedback connection as illustrated in Figure 4-10.

As we saw in (4.36), the characteristic function for Q_1 is

$$f_{Q_1}(u_i) = z_i$$

where

$$z_i = f_{R_1}(x_i, z_i)$$

FIGURE 4-9

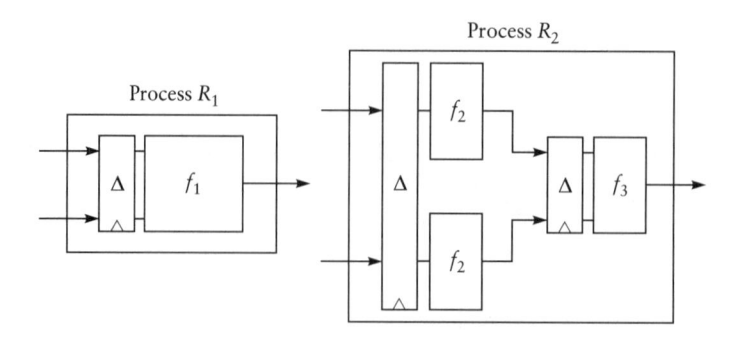

Two processes with different models for the same function f_1.

FIGURE 4-10

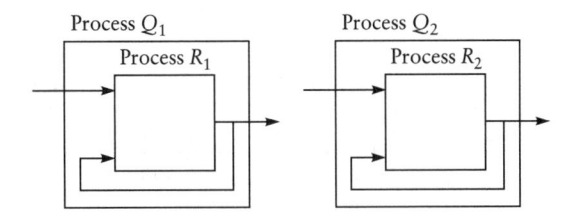

The functionality of a compound process depends both on the functionality and the timing of its constituents.

$$= f_1(f_\Delta(x_i, z_i))$$
$$= f_1(x_{i-1}, z_{i-1})$$
$$= 2z_{i-1} - 2x_{i-1} \quad \text{for } i \le 1$$
$$z_0 = 0$$

and it would generate $Q_1([0, 1, 2, 3]) = [0, 0, -2, -8, -22]$.

But for Q_2 the characteristic function f_{Q_2} is

$$f_{Q_2}(u_0) = 0$$
$$f_{Q_2}(u_1) = 0$$
$$f_{Q_2}(u_i) = z_i$$

where

$$z_i = f_3(f_\Delta((f_2(f_\Delta(u_i)), f_2(f_\Delta(v_i)))))$$
$$= f_3(f_\Delta((f_2(u_{i-1}), f_2(v_{i-1}))))$$
$$= f_3(f_\Delta((2u_{i-1}, 2v_{i-1})))$$
$$= f_3((2u_{i-2}, 2v_{i-2}))$$
$$= 2v_{i-2} - 2u_{i-2}$$
$$z_i = v_i$$
$$z_i = 2v_{i-2} - 2u_{i-2} \quad \text{for } i \ge 2$$

Consequently, Q_2 would also produce a different response:

$$Q_2([0, 1, 2, 3]) = [0, 0, 0, -2, -4, -10]$$

In the clocked synchronous computational model we must be very careful when composing blocks into systems because delays must be taken into account, even when we reason about the functionality. The clocked synchronous model is situated between discrete event models, which we discuss in the next chapter, and perfectly synchronous models. It suppresses part of the timing information but not all. The perfectly synchronous models ignore all timing information but constrain the implementations to be "fast enough." The term "fast enough" is a means to relate the system's timing behavior to the environment's timing. In contrast, the untimed computational model also suppresses all timing information, but does not constrain the performance of an implementation. Its approach is to model the system in a way such that it will always work correctly, independent of the timing behavior of the specific implementation.

4.8 Example: Traffic Light Controller

As an example we will use the familiar traffic light controller. There are two traffic lights: one governing the traffic from the north and south, the other governing the traffic from the east and west (Figure 4-11). The controller has a clock signal as its only input, which transmits one event per second. The outputs control the colors of the two traffic lights. The periods for the different lights are 60 seconds for green, 68 seconds for red, 3 seconds for yellow, and 1 second for both lights being red simultaneously.

Thus the controller state needs two components, one to count the seconds and the other to store the current control signals to the two lights. We denote the first components as `cnt`, the second as `colstate`, and the possible values of `colstate` with two letters, the first denoting the color to the north-south light and the second the color for the east-west light (e.g., `ry1` or `gr`). We distinguish two different states when both lights are red: `rr1`, when the east-west light turns yellow next, and `rr2`,

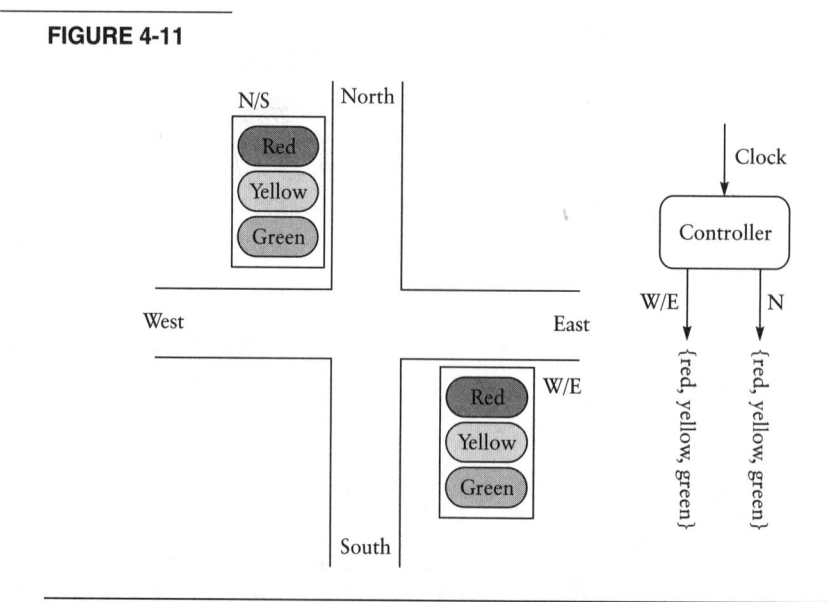

FIGURE 4-11

A traffic light and its controller.

when the north–south light turns yellow next. For similar reasons we have states ry1, ry2, yr1, and yr2.

The controller can be described as a Moore state machine, and we use the *mooreS* to instantiate it:

$$mooreS(nsf, outf, (rr1, 0)) = tlctrl$$

where

$$tlctrl(s_c) = s_l$$

$$s_l = \langle (n_i, e_i) \rangle$$

nsf is the next-state function, *outf* is the output encoding function, s_c is the clock signal, and the output signal consists of two components: the n_i are the control events to the north–south light, and the e_i events control the east–west light.

Now we define the next-state function and the output encoder. In the following definitions the lines with :: are type definitions of the functions; for example, the function *nsf* takes two parameters, the clock signal and the current state, and returns the next-state. The parts between the vertical bar | and the equal sign are additional conditions on the function parameters.

```
nsf :: Clock → State            → State
nsf  c     (rr1, cnt) | cnt < 1  = (rr1, cnt+1)
                      | otherwise = (ry1, 0)
nsf  c     (ry1, cnt) | cnt < 3  = (ry1, cnt+1)
                      | otherwise = (rg, 0)
nsf  c     (rg, cnt)  | cnt < 60 = (rg, cnt+1)
                      | otherwise = (ry2, 0)
nsf  c     (ry2, cnt) | cnt < 3  = (ry2, cnt+1)
                      | otherwise = (rr2, 0)
nsf  c     (rr2, cnt) | cnt < 1  = (rr2, cnt+1)
                      | otherwise = (yr1, 0)
nsf  c     (yr1, cnt) | cnt < 3  = (yr1, cnt+1)
                      | otherwise = (gr, 0)
nsf  c     (gr, cnt)  | cnt < 60 = (gr, cnt+1)
                      | otherwise = (yr2, 0)
nsf  c     (yr2, cnt) | cnt < 3  = (yr2, cnt+1)
                      | otherwise = (rr1, 0)
```

It is easy to see how *nsf* cycles through the different states, based on the number of clock signals received. The state sequence is also illustrated in Figure 4-12.

The function *outf* takes the current state as input and returns a pair of control signals for the traffic lights, denoted as (*Color, Color*). These pairs make up the events in the output signal (n_i, e_i).

```
outf :: State      → (Color, Color)
outf  (rr1, cnt) = (red, red)
outf  (rr2, cnt) = (red, red)
```

FIGURE 4-12

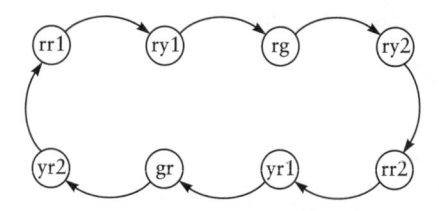

A traffic light controller state sequence.

```
outf  (ry1,cnt) = (red,yellow)
outf  (ry2,cnt) = (red,yellow)
outf  (rg,cnt)  = (red,green)
outf  (yr1,cnt) = (yellow,red)
outf  (yr2,cnt) = (yellow,red)
outf  (gr,cnt)  = (green,red)
```

This example shows that the synchronous model dictates when processes have to consume and produce events. The traffic light controller process has to generate a pair of control events for each received clock event. It does not have the choice of not outputting an event when it receives a clock event. If the two traffic lights would expect only control events when the lights would be switched, we would model this by emitting absent events, denoted as ⊔. In that case we had to use *outf2* as the output encoder.

```
outf2 :: State                    → (Color, Color)
outf2 (rr1,cnt) | cnt == 0   = (red,⊔)
                | otherwise  = (⊔,⊔)
outf2 (rr2,cnt) | cnt == 0   = (⊔,red)
                | otherwise  = (⊔, ⊔)
outf2 (ry1,cnt) | cnt == 0   = (⊔,yellow)
                | otherwise  = (⊔, ⊔)
outf2 (ry2,cnt) | cnt == 0   = (⊔,yellow)
                | otherwise  = (⊔, ⊔)
outf2 (rg,cnt)  | cnt == 0   = (⊔,green)
                | otherwise  = (⊔, ⊔)
outf2 (yr1,cnt) | cnt == 0   = (yellow,⊔)
                | otherwise  = (⊔, ⊔)
outf2 (yr2,cnt) | cnt == 0   = (yellow,⊔)
                | otherwise  = (⊔, ⊔)
outf2 (gr,cnt)  | cnt == 0   = (green,⊔)
                | otherwise  = (⊔, ⊔)
```

Thus, in the synchronous model the number of events produced and consumed is exactly defined because the timing behavior and the synchronization are embedded in the signal structure. This is in contrast to both the timed and the untimed model: neither of them requires expressing the absence of events explicitly.

4.9 Rugby Coordinates

In the Rugby terminology synchronous models are denoted as ⟨[SF-LB], [IPC-Top], [Sym-LV], ClockedTime⟩. Thus, the timing abstraction is the characteristic feature of both discussed synchronous models. In the computation domain, abstractions from the Algorithm level to the LogicBlock level are quite common. Even the SystemFunction level is used together with synchronous models to express abstract system behavior.

In the other two domains, the synchronous models are positioned in the middle. Neither the lower levels (Layout, ContValue) nor the highest levels (InterfaceConstraints, DataTypeConstraints) are covered.

4.10 Validation

As an interesting application example of the perfectly synchronous model, we discuss an approach to validation of both functionality and timing properties based on monitors. Timing properties can be expressed in terms of the evaluation cycle and, consequently, in terms of any time unit that is set in relation to the evaluation cycle. The approach can be used when simulating the system. Moreover, and perhaps even more appropriately, it can be used in conjunction with formal verification tools such as model checkers. However, we will not go into details of formal verification techniques.

4.10.1 A U-Turn Section Controller

We introduce these concepts by way of an example that illustrates the validation of a real-time safety-critical system. We closely follow an article by Halbwachs et al. (1992) where these concepts are introduced in the context of the synchronous language Lustre.

Consider a U-turn section at the end of a subway track, as illustrated in Figure 4-13. Trains arrive in section A, stop in section B, and depart via section C. In S there

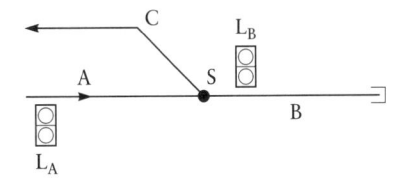

FIGURE 4-13

A subway U-turn section (Halbwachs et al. 1992).

FIGURE 4-14

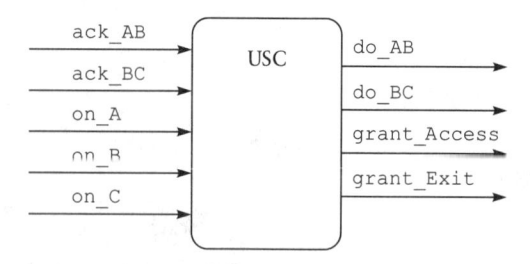

The controller of the subway U-turn section with its inputs and outputs.

is a switch that either connects A to B or B to C. In order to avoid accidents the switch has to be steered properly and trains have to be admitted to the sections in a controlled manner. In the following we model an automatic U-turn section controller (USC) and express the safety conditions that must be fulfilled in all possible conditions.

Figure 4-14 shows the inputs and outputs of the controller:

- ackAB and ackBC indicate the status of the switch. True means that the corresponding sections are connected. Apparently, both signals cannot be True simultaneously. If both are False, trains must not pass the switch.

- onA, onB, and onC are signals from sensors from the three sections. They are True if a train is in the corresponding section.

- doAB and doBC are requests to the switch to connect the corresponding sections.

- grantAccess and grantExit are control signals for traffic lights. grantAccess controls traffic light L_A and thus access of trains to section A. grantExit controls traffic light L_B and thus the exit of trains from section B.

We assume initially that there is no train in any of the three sections.

The switch should connect A and B if there is no train in the sections. B and C should be connected when there is a train in section B. The controller sends the request to the switch only if the switch is not in the right position. This can be expressed by

$$doAB = \neg ackAB \wedge emptySection$$
$$doBC = \neg ackBC \wedge onlyOnB$$
$$emptySection = \neg(onA \vee onB \vee onC)$$
$$onlyOnB = onB \wedge \neg(onA \vee onC)$$

Next, we define how the traffic lights are controlled, that is, when access and exit are granted. Access is granted if all three sections are empty and if the switch connects

FIGURE 4-15

$$\text{USC} = \mathit{mapS}(f)$$

where $f(\texttt{onA}, \texttt{onB}, \texttt{onC}, \texttt{ackAB}, \texttt{ackBC})$

$$= (\texttt{grantAccess}, \texttt{grantExit}, \texttt{doAB}, \texttt{doBC})$$

$$\texttt{grantAccess} = \texttt{emptySection} \wedge \texttt{ackAB}$$
$$\texttt{grantExit} = \texttt{onlyOnB} \wedge \texttt{ackBC}$$
$$\texttt{doAB} = \neg\texttt{ackAB} \wedge \texttt{emptySection}$$
$$\texttt{doBC} = \neg\texttt{ackBC} \wedge \texttt{onlyOnB}$$
$$\texttt{emptySection} = \neg(\texttt{onA} \vee \texttt{onB} \vee \texttt{onC})$$
$$\texttt{onlyOnB} = \texttt{onb} \wedge \neg(\texttt{onA} \vee \texttt{onC})$$

The complete definition of the U-turn section controller.

A to B. Trains are allowed to exit when the switch connects B to C and when there is only a train in section B.

$$\texttt{grantAccess} = \texttt{emptySection} \wedge \texttt{ackAB}$$
$$\texttt{grantExit} = \texttt{onlyOnB} \wedge \texttt{ackBC}$$

This is all we need for defining the controller. Since the functionality is purely combinatorial, a *mapS* process is all we need. The complete controller definition is given in Figure 4-15.

4.10.2 Monitors

Since any failure of the U-turn system including the controller may have severe consequences and even cause loss of life, we would take great comfort if we could prove that our controller cannot enter harmful states or emit sequences of events leading to dangerous situations. In particular we would like to prove the following properties.

- *No collision:* A train is never granted access when another train is in one of the sections A, B, or C.

- *No conflict:* No conflicting control events are ever sent to the switch, that is, say that it should connect both A to B and B to C.

- *No derail:* The switch will never be requested to change its state when a train is driving on the switch region.

There may be other desirable properties of the USC, but these examples suffice to illustrate the principal method.

First, we introduce some monitor processes that are generally useful to express this kind of properties. A monitor process observes signals and outputs boolean events of value True as long as a particular property holds, and False otherwise.

$$mon1S(f_1) = mapS \tag{4.37}$$

where

$$f_1 : Boolean \to Boolean$$

$$mon2S(f_2) = zipWithS \tag{4.38}$$

where

$$f_2 : (Boolean, Boolean) \to Boolean$$

$$mon1sS(g_3, f_3, w_0) = mealyS(g_3, f_3, w_0) \tag{4.39}$$

where

$$g_3, f_3 : (V, Boolean) \to Boolean$$

$$mon2sS(g_4, f_4, w_0) = mealyS(g_3, f_3, w_0) \circ zipS \tag{4.40}$$

where

$$g_4, f_4 : (V, Boolean, Boolean) \to Boolean$$

The monitors with suffix sS maintain an internal state. The $mon1S$ and $mon1sS$ processes operate on one input signal, while the $mon2S$ and $mon2sS$ processes expect two input signals. Hence, monitors are just like other processes, but they exclusively operate on boolean events.

Now it is straightforward to express a few useful monitors:

$$or = mon2S(f)$$

where

$$f(b_1, b_2) = b_1 \vee b_2$$

$$implies = mon2S(f)$$

where

$$f(b_1, b_2) = \neg b_1 \vee b_2$$

$$after = mon1sS(g, f, \text{False})$$

where

$$g(w, b) = w \vee b$$
$$f(w, b) = w \vee b$$

$$alwaysSince = mon2sS(g, f, \text{False})$$

where

$$g(w, b_1, b_2) = (w \wedge b_1) \vee (b_1 \wedge b_2)$$
$$f(w, b_1, b_2) = (w \wedge b_1) \vee (b_1 \wedge b_2)$$

$$onceSince = \text{mon2sS}(g, f, 0)$$

where

$$g(w, b_1, b_2) = \begin{cases} 0 & \text{if } w = 0 \wedge \neg b_2 \\ 1 & \text{if } (w = 0 \wedge \neg b_1 \wedge b_2) \vee (w = 1 \wedge \neg b_1) \vee (w = 2 \wedge \neg b_1 \wedge b_2) \\ 2 & \text{if } (w = 0 \wedge b_1 \wedge b_2) \vee (w = 1 \wedge b_1) \vee (w = 2 \wedge ((b_1 \wedge b_2) \vee b_2)) \end{cases}$$

$$f(w, b_1, b_2) = (w = 2)$$

The *or* process simply implements the logic OR operation.

The *implies* process implements ordinary logic implication; that is, if the first input is True, the second input must also be True.

The *after* process is initially False until its input is True at least once. Then *after* emits True forever.

The *alwaysSince* process relates two inputs. It emits True if and only if its first input (b_2) has been continuously True since the last time its second input (b_1) was True.

The process *onceSince* emits True if and only if its first input has been True at least once since the last time its second input was True. It can be modeled as a three-state state machine as shown in Figure 4-16. State 0 is the initial state. State 1 is the middle state, where the second input b_2 has been True but the first input b_1 has not. State 2 is the final state, where the first input b_1 has been True once since the second input b_2 was True.

The machine will remain in state 0 until b_2 becomes True. If both b_1 and b_2 become True simultaneously, the condition *onceSince* is satisfied and the machine jumps directly to state 2. If in state 0 b_1 is False and b_2 is True, the machine goes to state 1, waiting for b_1 to become True.

In state 1, once b_1 has become True, the machine goes to state 2 independent of the value of b_2. If b_1 is False, it will remain in state 1 and wait for b_1 to become True.

FIGURE 4-16

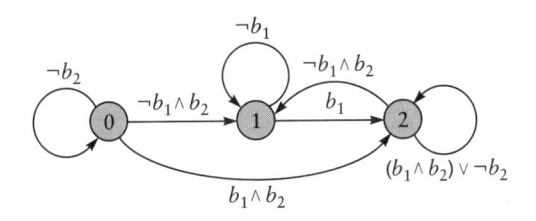

The state transition diagram of the onceSince monitor.

When the machine is in state 2, it will remain there as long as b_2 is False. If both b_1 and b_2 become True simultaneously, the *onceSince* condition is again fulfilled and the machine remains in state 2. Otherwise, if b_2 is True but b_1 is False, the machine jumps back to state 1 to wait until b_1 becomes True again.

Based on these general-purpose monitor processes, we express the desirable safety properties for the USC. First we specify three helper monitors. M_{eS} and M_{oB} resemble the `emptySection` and `onlyOnB` operations we used inside the USC. M_d embodies the derail condition, saying that once a grant event (s_2) has been emitted, either the switch must be in the right position (s_1) ever since the last grant event or the train has passed the switch (s_3).

$$M_{eS}(s_1, s_2, s_3) = mon2S(f_1)(zipS()(s_1, s_2), s_3)$$

where

$$f_1((b_1, b_2), b_3) = \neg(b_1 \vee b_2 \vee b_3)$$

$$M_{oB}(s_1, s_2, s_3) = mon2S(f_2)(zipS()(s_1, s_2), s_3)$$

where

$$f_2((b_1, b_2), b_3) = b_2 \wedge \neg(b_1 \vee b_3)$$

$$M_d(s_1, s_2, s_3) = implies(after(s_2), or(alwaysSince(s_1, s_2), onceSince(s_3, s_2)))$$

- *No collision:* The monitor process M_1 generates the signal `noCollision` and is defined by

$$M_1(\text{grantAccess}, \text{onA}, \text{onB}, \text{onC})$$
$$= implies(\text{grantAccess}, M_{eS}(\text{onA}, \text{onB}, \text{onC}))$$

  `noCollision` is always True provided that access is only granted to a train when the U-turn area is empty.

- *No conflict:* Process M_2 generates the `noConflict` signal, which tells if conflicting requests are sent to the switch:

$$M_2(\text{doAB}, \text{doBC}) = mon2S(f)(\text{doAB}, \text{doBC})$$

where

$$f(b_1, b_2) = \neg(b_1 \wedge b_2)$$

- *No derail AB:* We distinguish between derail conditions when the switch connects A to B and when it connects B to C. M_3 generates the signal `noDerailAB`, which checks the derail condition when the switch connects A to B:

$$M_3(\text{ackAB}, \text{onA}, \text{onB}, \text{onC}, \text{grantAccess})$$
$$= M_d(\text{ackAB}, \text{grantAccess}, M_{oB}(\text{onA}, \text{onB}, \text{onC}))$$

FIGURE 4-17

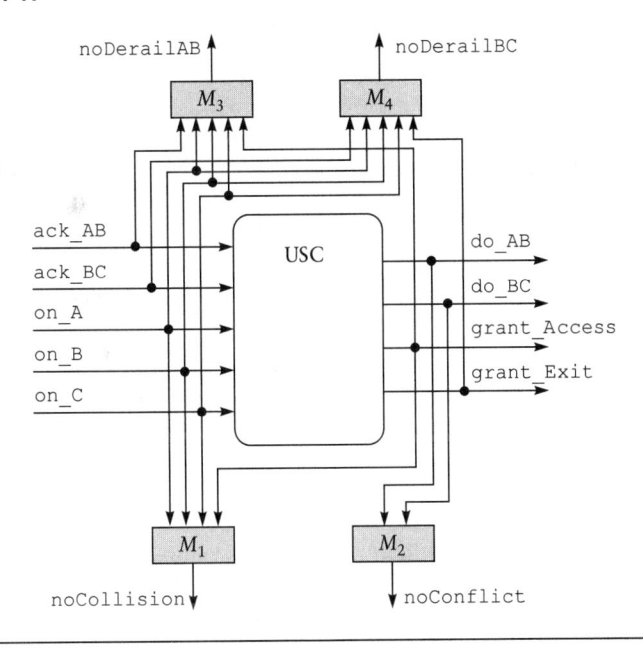

The USC process together with monitor processes.

- *No derail BC:* M_4 generates signal noDerailBC, which checks the derail condition when the switch connects B and C:

$$M_4(\text{ackBC}, \text{onA}, \text{onB}, \text{onC}, \text{grantExit})$$
$$= M_d(\text{ackBC}, \text{grantExit}, M_{oB}(\text{onA}, \text{onB}, \text{onC}))$$

Figure 4-17 depicts the controller with all monitors properly attached to inputs and outputs.

4.10.3 Validation Strategies

Monitors are useful in conjunction with simulation. Whenever a monitor emits a False event, the corresponding property has been violated. This simple relation allows for efficient automatic checks of simulation trace files. This is essentially the only way to approve the results of regression simulations. In regression simulations all already-approved test cases are simulated again and again for every change in the system model to ensure that new additions or modifications to the system model do not introduce errors in unexpected parts of the system.

Monitors can also be used together with random input vectors. The objective of random input vectors is to simulate the system with inputs that are not expected by the validation engineer. Random input vectors can also be generated much faster than carefully crafted test cases. However, one problem with random input vectors is that it is difficult for engineers to predict the "correct" behavior of the system because there is usually no "meaning" associated with randomly generated input vectors. In this case, monitors can check if the specified conditions are fulfilled for any of the input vectors without needing to understand their meaning.

Simulation can prove that a particular property, expressed as a monitor, holds only if all possible input vectors in all possible sequences are applied. This is called *exhaustive simulation*. For any but the tiniest systems, exhaustive simulation cannot be used due to the sheer number of possibilities. Formal verification techniques have been developed to prove given properties independently of particular input vectors. Consequently, such proofs hold for any possible input vector. We do not elaborate these techniques here; we only point out that monitors can be used in conjunction with formal property-checking techniques and tools. A property checker has to prove that the output of a monitor is always `True`. Thus, the same formulation of a desirable property in terms of a monitor can be used in both simulation and formal verification techniques.

However, for formal property checking we typically have to provide further information about the environment. For instance, based on our understanding of a switch on railway tracks, we know that the switch cannot connect A to B and B to C simultaneously. This would be an obvious and necessary piece of information that we have to provide to a formal property-checking tool. Other assumptions about the behavior of the environment for our example are the following:

- The switch remains in a given position until the controller requests a change.
- Initially there is no train in any of the sections A, B, or C.
- Trains obey traffic lights.
- When a train leaves A, it is on B, when a train leaves B, it is either on A or on C.

This rather incomplete list underscores that there are a number of assumptions we can or should make when formally verifying a system model. The explicit formulation of these assumptions usually has the additional benefit of it making the allocation of obligations and responsibilities clearer. Some of the assumptions are simply a consequence of the mechanics and physics involved, but others point to important obligations that may require more attention. For instance, the assumption that "trains obey traffic lights" is not self-evident, but it is a precondition for guaranteeing the safety of the U-turn area. Thus, our assumption for the USC may trigger an obligation for another system, the train controller, to guarantee a particular property.

4.11 Further Reading

The perfectly synchronous models were independently developed several times in the 1980s. That work has been manifested in the synchronous languages StateCharts (Harel 1987), Esterel (Berry et al. 1988), Signal (Le Guernic et al. 1991), Argos, and Lustre (Halbwachs et al. 1991). These languages basically fall into two groups: the control-oriented languages (StateCharts, Esterel, and Argos), and the data-flow-oriented languages (Signal and Lustre). They have been integrated to some extent at the level of the internal representation documented by Poigné et al. (1998).

A number of interesting design and verification methods have been developed. Among these are formal verification (Bouali et al. 1996; Bouali 1997; de Simone and Ressouche 1994; Halbwachs et al. 1992) and synthesis for hardware (Berry 1991) and software (Halbwachs et al. 1991; Berry et al. 1988) implementations and for distributed systems (Caspi et al. 1999).

Good summaries of several of these languages and techniques can be found in a book by Halbwachs (1993) and a special issue of the *Proceedings of the IEEE* (Benveniste and Berry 1991a). A well-written introduction to Esterel is given by Berry (1998). A recent summary of the state of the art is provided by Benveniste et al. 2003.

There are numerous excellent textbooks on principles and practical techniques of clocked synchronous hardware design. Examples include Gajski (1997) and Roth (1998).

Several verification techniques have been developed for synchronous languages. A verification technique using monitors and based on symbolic model checking has been developed for Lustre and described by Halbwachs et al. (1992).

Good surveys of formal verification techniques for hardware and embedded systems have been written by Kern and Greenstreet (1999) and by McFarland (1993). An overview of general formal verification methods is given by Clark and Wing (1996). Clarke et al. (1996) have also written a comprehensible and thorough introduction to model checking.

Monitors, which are sometimes also called *assertions*, are used in many practically oriented validation strategies. Bening and Foster (2001) present an assertion library for the clocked synchronous model and describe a verification methodology covering the usage of assertion with simulation and verification. They distinguish between *combinatorial* and *sequential assertions*, which correspond to stateless and stateful monitors. Sequential assertions can be *time bounded* or *event bounded*. In both cases the activation of the assertion check is triggered by an event. Timed-bounded assertions stop checking after a predefined number of evaluation cycles. Event-bounded assertions stop checking upon the occurrence of a particular event. Similar concepts but from the perspective of testbench design are discussed by Bergeron (2000) under the labels *run time result verification* and *monitoring*.

4.12 Exercises

4.1 The process P in Figure 4-18 is modeled in the clocked synchronous model by means of the processes $P_1 = map2CS(f_1)$ and $P_2 = mapCS(f_2)$. The combinatorial functions f_1 and f_2 are defined as follows:

$$f_1((x,y)) = (x+y, x-y)$$
$$f_2((x,y)) = (x-y, x+y).$$

Calculate the characteristic function f_P of P.

FIGURE 4-18

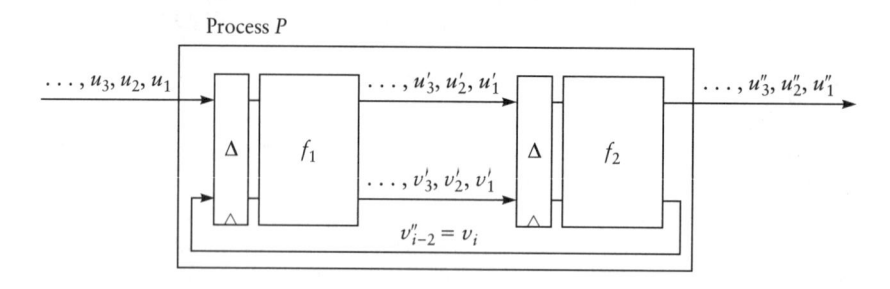

Clocked synchronous process P with an internal feedback loop.

FIGURE 4-19

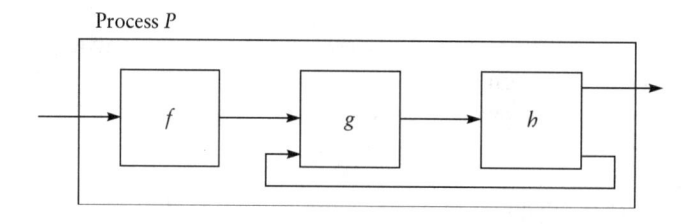

Process with three subprocesses.

4.2 Given is the process P, which consists of three subprocesses characterized by f, g, and h (Figure 4-19):

$$f(x) = 2x$$
$$g(x, y) = 2x + y$$
$$h(x) = (x, -x)$$

Calculate the characteristic function f_P for both the perfectly synchronous and the clocked synchronous model.

4.3 Even in the perfectly synchronous model we need a delay process if we want to deal with processes with internal states. Figure 4-20 shows a symbolic representation and a model of a simple process with an internal state and an initial state value m_0. f is a combinatorial function with two parameters. The characteristic function F_P is defined by the following equations:

$$f_P(x_i) = f(x_i, m_i) \quad \forall i > 0$$
$$m_i = f_\Delta(f(x_i, m_i)) \quad \forall i > 0$$

This process could have been directly instantiated with a process constructor. Which one? Prove it.

FIGURE 4-20

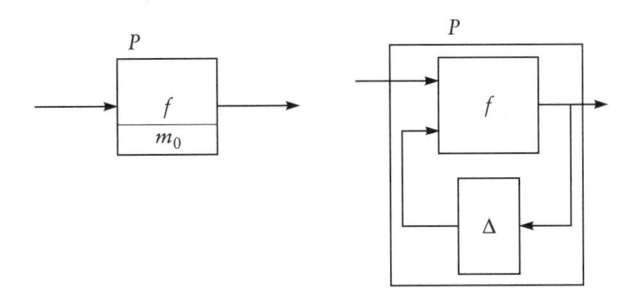

The symbols (left) and the model (right) of a process with state in the perfectly synchronous model.

FIGURE 4-21

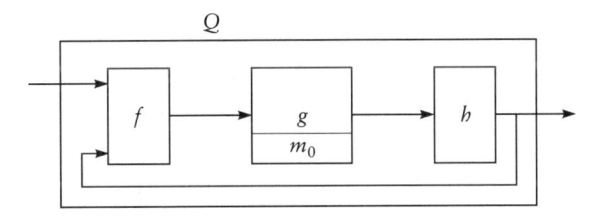

A perfectly synchronous process with feedback loop and internal state.

4.4 Consider process Q in Figure 4-21, which uses the stateful process defined in Exercise 4.3. It consists of three subprocesses defined by the following characteristic functions:

$$f(x, y) = 2x + y$$
$$g(x, m) = x - m$$
$$h(x) = 2x - 1$$

Calculate the characteristic function f_Q.

4.5 Moore and Mealy state machines can be defined in terms of other process constructors.

  a. Instantiate a perfectly synchronous Moore finite state machine p with next-state function g, output encoding function f, and initial state w_0 without using `mooreS`. Instead use only map- and scan-based processes.

  b. Prove that $p = q$ for `mooreS`$(g, f, w_0) = q$.

4.6 To define a Mealy state machine in terms of constructors other than `mealyS` is slightly more involved because the output decoder depends both on the internal state and the input.

   a. Instantiate a perfectly synchronous Mealy finite state machine p with next-state function g, output encoding function f, and initial state w_0 without using `mealyS`. Hint: Use only `zipWithS` and `scanS`.

   b. Prove that $p = q$ for $mealyS(g, f, w_0) = q$.

4.7 Let $[x, y]$ be a list consisting of the two elements x and y. Let P be a process generated by $zipWithSY(g) = P$, with $g(x, y) = [x, y]$.

Model a FIFO process that takes a list of two events as input and emits individual events as output, such that the FIFO input can be connected to the output of P. P merges two signals into one signal s''. s'' consists of events that are lists of two events. These events can be ⊔ events. The FIFO receives lists of events and emits individual events. It does not store ⊔ events. Under the assumption that on average half of the events in s and s' are ⊔ events, the FIFO buffer can be finite. But it has to buffer events if for a short period there are more nonabsent events on its input than it can emit.

Hint: Use the `mooreS` constructor to generate the FIFO process.

FIGURE 4-22

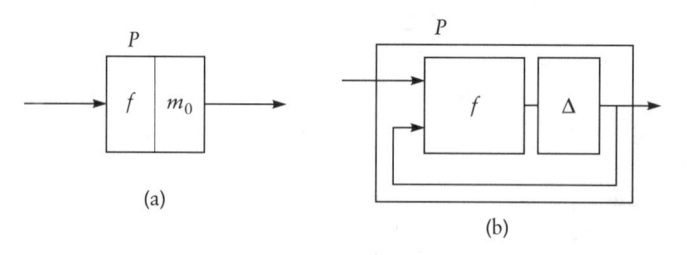

(a) The symbol and (b) the model of another process with state in the perfectly synchronous model. $f_P(x_i) = m_i = f(x_{i-1}, m_{i-1}), \forall i \geq 1$ and $f_P(x_0) = m_0$.

FIGURE 4-23

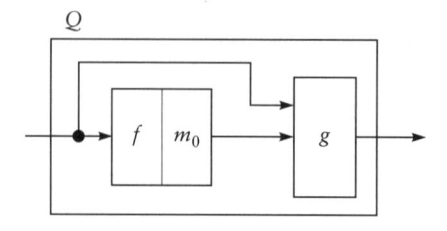

A Mealy state machine.

4.8 The process shown in Figure 4-20 is only one possibility of a simple process with state in the perfectly synchronous model. An alternative is shown in Figure 4-22.

Process Q in Figure 4-23 is using this primitive process. The functions f and g are defined as follows:

$$f(x, m) = x$$
$$m_0 = 0$$
$$g(x, y) = x + y$$

Derive the characteristic function f_P and calculate the output sequence of $P(\langle 1, 2, 3, 4, 5 \rangle)$.

4.9 Model the switch described in Exercise 3.14 in the perfectly synchronous model.

chapter five
The Timed Model of Computation

Time is a first-class property in hardware and embedded software design. On many occasions it must be considered with great accuracy and measured in physical time units. Timing behavior is important in its own right, but it also influences functional behavior.

After discussing the importance of time we introduce a timed MoC. It is based on a generalization of the synchronous MoC from the previous chapter. Timing information is conveyed on the signals by transmitting absent events at regular time intervals. In this way processes always know when a particular event has occurred and when no event has occurred. Their sole sources of information are the input signals; no access to a global state or magic knowledge about a global time is necessary. The differences from the synchronous MoC are the following: (1) The granularity of the timing structure is much finer, and events occur every nanosecond or picosecond rather than every clock cycle. However, the concrete physical time unit is not part of the MoC and is up to interpretaion. (2) Processes can consume and emit any number of events during one evaluation cycle. In this respect the timed MoC is closer to the untimed than to the synchronous MoC. (3) Processes must comply with the causality constraint that output events that are a reaction to input events cannot occur before these input events. When a process consumes a number of input events in a given evaluation cycle, the first output event of this evaluation cycle is emitted not earlier than the latest input event of this cycle. As a consequence, a delay period is associated with each evaluation cycle of a process. (4) The feedback operator is based on the prefix order of signals just like the untimed MoC.

Then we informally discuss variants of the timed MoC. They differ in the degree of separation between the computation of a process and its timing and synchronization behavior. These two issues can be completely separated, resulting in models with nice theoretical properties. The downside is a restricted expressiveness because the functional behavior cannot depend on the timing any more.

We also review three different alternatives for representing a global time and distributing it to all processes. The local timer approach assumes a local timer inside each process that has access to the global time. The time tag approach annotates each event with the time of its occurrence. The absent event approach is the one taken in this chapter. Both the time tag and the absent event approach communicate timing information as part of the signals, while the local timer approach keeps it separate and assumes timing information is communicated by another means to all processes instantaneously.

The δ-delay model is a popular timing model used in languages such as VHDL, Verilog, and SystemC. It has a two-level timing structure. Every two instances of the discrete physical time are separated by potentially infinitely many δ-delays. The δ-delays have no physical correspondence in the real system. Their purpose is to make the simulation behavior deterministic in the presence of simultaneous events. The δ-delay models are foremost simulation devices. However, they are used as input to design and synthesis by interpreting the descriptions according to a clocked synchronous MoC.

5.1 Introduction

The models discussed in the previous chapters do not have a notion of physical time. The untimed MoC is based on data dependences only, just like Petri nets, and neither a transition nor the transport of a token from one place to the next takes a particular amount of time. Synchronous MoCs have a cycled time; that is, the computation of the outputs and the next state is "fast enough," and the communication of events does not take any observable time. This is convenient when we are only concerned with the sequence of states and transitions. But in many cases we want to take into account the precise timing behavior. Consider for instance the problem of determining the highest possible clock frequency of a digital circuit. To do this we need the precise timing behavior of the components in a given target technology, say, a CMOS $0.18\,\mu$ gate library from a particular vendor. From this library we know for each gate the delay from a particular input to a particular output in some physical time unit, say, nanoseconds. Apparently we need to consider the precise timing of the components in the simulation to determine the maximum clock frequency of a given network of gates. But even the functional behavior may depend on the timing behavior.

FIGURE 5-1

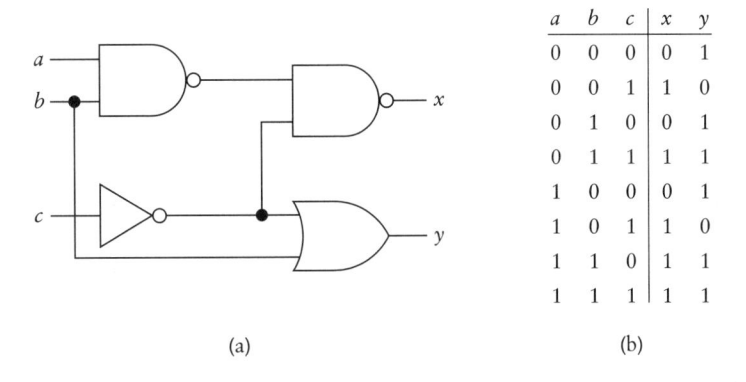

a	b	c	x	y
0	0	0	0	1
0	0	1	1	0
0	1	0	0	1
0	1	1	1	1
1	0	0	0	1
1	0	1	1	0
1	1	0	1	1
1	1	1	1	1

(a) (b)

(a) A simple netlist of gates and (b) its truth table.

TABLE 5-1: *Delays for the gates of Figure 5-1.*

Gate	Delay (ns)
Inverter	1.5
NAND gate	1.8
OR gate	2.1

Consider for instance the netlist in Figure 5-1. Assume it is of utmost importance that, whenever $x = 0$, y holds the value 1. Statically this is guaranteed as the truth table shows. However, as already indicated in the previous chapter, the real answer depends on the exact delays of the gates and the interconnects. Assume we use a gate library with delays as shown in Table 5-1. When an input changes its value, the change will propagate through the network to the outputs and eventually the outputs will settle to new values according to the truth table. However, since the propagation may occur at different speeds on different paths, the outputs may exhibit a transient pattern that is not part of the truth table. Figure 5-2 shows a sequence of values. At time t the network has settled with all output and intermediate values corresponding to the input values. At time t' the value of c changes from 1 to 0. Figure 5-2(b) through (d) shows the propagation of changes through the network until it settles again in Figure 5-2(e) to the new values. We observe that between $t' + 3.3$ ns and $t' + 3.6$ ns the outputs show the pattern $x = 0$, $y = 0$, which is not part of the defining truth table and is undesirable according to our assumption. Essentially, this means that our circuit exhibits a behavior that can only be analyzed if we take the accurate timing of its components into account. None of the previously discussed models can provide this kind of analysis.

The timed model uses a more general concept of time that better reflects physical reality. In particular, it does not make any assumption about the timing of components

FIGURE 5-2

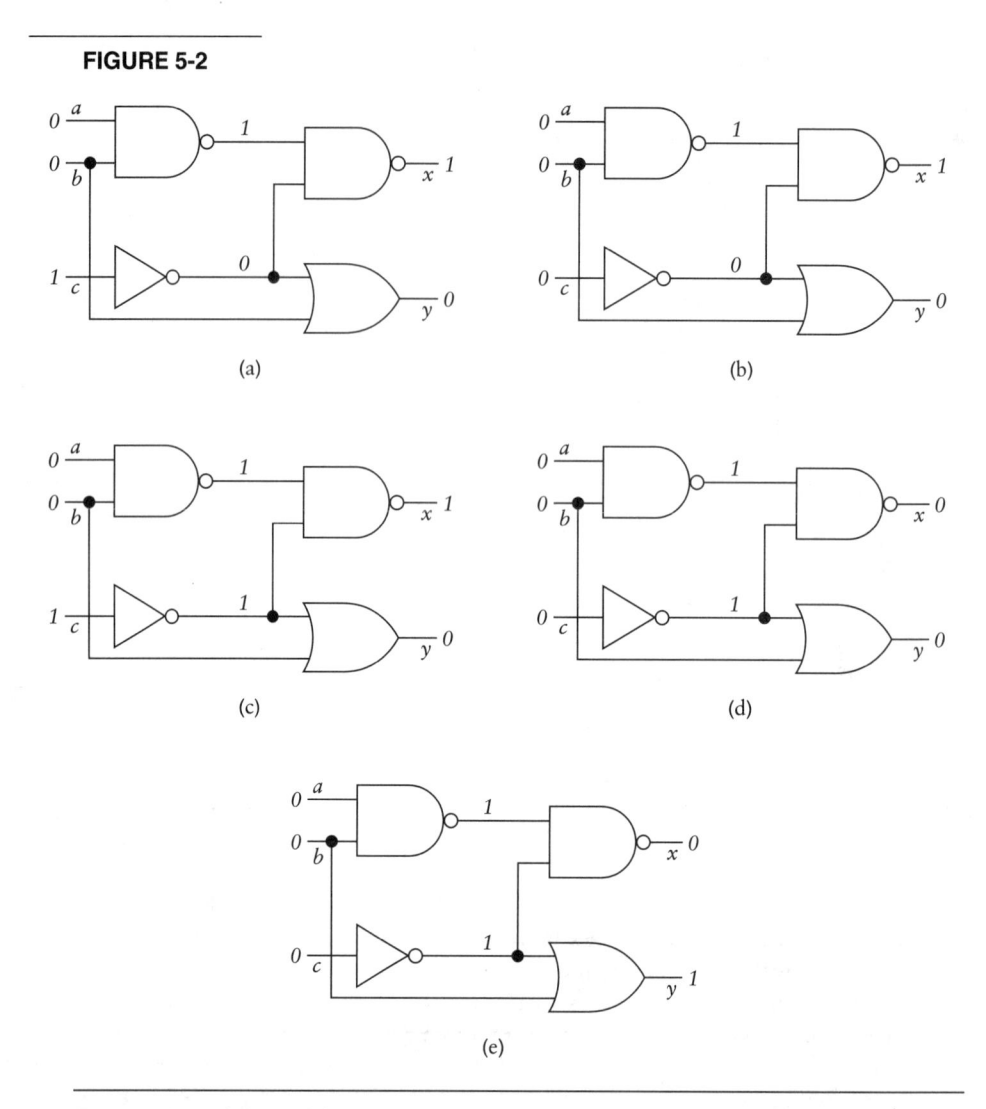

Propagation of changes from the inputs to the outputs: (a) at time t; (b) at time t′; (c) at time t′ + 1.5 ns; (d) at time t′ + 3.3 ns; (e) at time t′ + 3.6 ns.

or computations. On the contrary, it allows for an arbitrary finite assignment of delays to computations.

Using a timed MoC reflects the intention of capturing the timing behavior of physical entities accurately without simplifying assumptions. The only simplification concerns the accuracy of the time representation, in particular when we choose a discrete time representation. However, all physical objects have an identical perception of elapsing time even if they don't interact—at least in Newtonian physics, which is the relevant physical context for the physical objects we have in mind.

5.2 Process Constructors

All timed process constructors and processes operate exclusively on timed signals $\hat{s}$.

Timed processes are a blend of untimed and synchronous processes in that they can consume and produce more than one event per cycle and they also deal with absent events. In addition, they have to comply with the constraint that output events cannot occur before the input events of the same evaluation cycle. This is achieved by enforcing an equal number of input and output events for each evaluation cycle, and by prepending an initial sequence of absent events. Since the signals also represent the progression of time, the prefix of absent events at the outputs corresponds to an initial delay of the process in reacting to the inputs. Moreover, the partitioning of input and output signals corresponds to the duration of each evaluation cycle.

`mealyT` is a process constructor that, given γ, f, g, and w_0 as arguments, instantiates a process $p : \hat{S} \to \hat{S}$ that is defined as follows:

$$\text{mealyT}(\gamma, g, f, w_0) = p \qquad (5.1)$$

where

$$p(\hat{s}) = \hat{s}'$$
$$\pi(v, \hat{s}) = \langle \hat{a}_i \rangle$$
$$\pi(v', \hat{s}') = \langle \hat{b}_i \rangle, \text{ rem}(\pi, v', \hat{s}') = \langle \rangle$$
$$v(i) = \gamma(w_i)$$
$$v'(i) = \gamma(w_i) + K_i$$
$$\hat{b}_i = \langle \sqcup \rangle^{K_i} \oplus \hat{c}_i$$
$$\hat{c}_i = f(w_{i-1}, \hat{a}_{i-1}) \quad \text{for } i > 0$$
$$w_i = g(w_{i-1}, \hat{a}_{i-1}) \quad \text{for } i > 0$$
$$\text{for } i \in \mathbb{N}_0, \hat{s}, \hat{s}', \hat{a}, \hat{b} \in \hat{S}, w_i \in V$$

where $\langle \sqcup \rangle^n$ denotes the sequence of n $\sqcup$ events. The output sequences $\hat{b}_i$ consist of the result of function f and a number of absent events constituting delays whenever necessary. Delays are required to enforce that output events occur not earlier than the input events on which they depend. The number of inserted delay events is determined by the sequence K_i, defined as

$$K_i = \begin{cases} T_i(i) - T_o(i-1) - 1 & \text{if } i \geq 0 \\ T_i(i) - 1 & \text{if } i = 0 \end{cases}$$

$$T_i(i) = \sum_{j=0}^{i} \text{length}(\hat{a}_j)$$

$$T_o(i) = \sum_{j=0}^{i} \text{length}(\hat{b}_j)$$

228 chapter five *The Timed Model of Computation*

TABLE 5-2: *The input and output sequences and inserted delay for the first few evaluation cycles for process* $p = \texttt{mealyT}(\gamma, g, f, 0)$.

i	$T_i(i)$	$T_o(i)$	K_i	$\hat{a}_i$	$\hat{b}_i$	$\hat{c}_i$
0	2	3	1	$\langle \hat{e}_1, \hat{e}_2 \rangle$	$\langle \sqcup, \hat{e}_2, \hat{e}_1 \rangle$	$\langle \hat{e}_2, \hat{e}_1 \rangle$
1	4	5	0	$\langle \hat{e}_3, \hat{e}_4 \rangle$	$\langle \hat{e}_4, \hat{e}_3 \rangle$	$\langle \hat{e}_4, \hat{e}_3 \rangle$
2	6	7	0	$\langle \hat{e}_5, \hat{e}_6 \rangle$	$\langle \hat{e}_6, \hat{e}_5 \rangle$	$\langle \hat{e}_6, \hat{e}_5 \rangle$

Example 5.1 An example may illustrate the concept. Consider a process $p = \texttt{mealyT}(\gamma, g, f, 0)$, $\gamma(x) = 2$, $g(w, x) = 0$, $f(w, \langle x_1, x_2 \rangle) = \langle x_2, x_1 \rangle$, which swaps the order of every two events. In this case an initial delay of one event is required because the second event cannot be emitted before it has been consumed. Given an input signal of only two events, p would produce $p(\langle \hat{e}_1, \hat{e}_2 \rangle) = \langle \sqcup, \hat{e}_2, \hat{e}_1 \rangle$. Input event $\hat{e}_1$ would occur at the same time as the absent event $\sqcup$ at the output, and $\hat{e}_2$ on the input would occur synchronously with $\hat{e}_2$ at the output. The initial delay is the only one introduced by p. Given a longer input sequence, p would react with

$$p(\langle \hat{e}_1, \hat{e}_2, \hat{e}_3, \hat{e}_4, \hat{e}_5, \hat{e}_6 \rangle) = \langle \sqcup, \hat{e}_2, \hat{e}_1, \hat{e}_4, \hat{e}_3, \hat{e}_6, \hat{e}_5 \rangle$$

which is further illustrated in Table 5-2.

In general, $\texttt{mealyT}$-based processes inject additional delays whenever the input sequence in an evaluation cycle becomes longer than it has been previously. Note also that injected delays at the output correspond to buffered events at the input.

Similarly but simpler, the partitioning of all input and output signals for $\texttt{zipT}$-based processes is identical, and there is an initial absent event at the output while consuming the first input events.

Let $\gamma : V \to \mathbb{N}$ be functions. $\texttt{zipT}$ is a process constructor that, given γ as argument, instantiates a process $p : \hat{S} \times \hat{S} \times \hat{S} \to \hat{S}$ that is defined as follows:

$$\texttt{zipT}(\gamma) = p \qquad (5.2)$$

where

$$p(\hat{s}_a, \hat{s}_b, \hat{s}_c) = \hat{s}'$$

$$\dot{e}'_{i+1} = \langle \hat{a}_i, \hat{b}_i \rangle$$

$$\dot{e}'_0 = \sqcup$$

$$\pi(\nu_a, \hat{s}_a) = \langle \hat{a}_i \rangle, \ \nu_a(i) = \gamma(k_i)$$

$$\pi(\nu_b, \hat{s}_b) = \langle \hat{b}_i \rangle, \ \nu_b(i) = \gamma(k_i)$$

$$\pi(\nu_c, \hat{s}_c) = \langle \hat{c}_i \rangle, \ \nu_c(i) = \gamma(k_i)$$

$$\pi(\nu', \hat{s}') = \langle \sqcup \rangle^{\gamma(k_i) - 1} \oplus \langle \dot{e}'_i \rangle, \ \nu'(i) = \gamma(k_i), \ \textsf{rem}(\pi, \nu', \hat{s}') = \langle \rangle$$

$$k_{i+1} = \hat{c}_i[1]$$

5.2 Process Constructors

$$k_0 = 0$$

for $\hat{s}_a, \hat{s}_b, \hat{s}_c, \hat{s}', \hat{a}_i, \hat{b}_i, \dot{c}_i \in \hat{S}, i \in \mathbb{N}_0, k_i \in \hat{E}$

$\mathtt{unzipT}$-based processes perform the reverse operation on a signal. $\mathtt{unzipT}$ is a process constructor that instantiates a process $p : \hat{s} \to \langle \hat{s}, \hat{s} \rangle$ that is defined as follows:

$$\mathtt{unzipT}() = p \qquad (5.3)$$

where

$$p(\hat{s}) = \langle \hat{s}', \hat{s}'' \rangle$$
$$\dot{e}_{i+1} = \langle \hat{a}'_i, \hat{a}''_i \rangle$$
$$\dot{e}_0 = \langle \sqcup, \sqcup \rangle$$
$$\pi(v, \hat{s}) = \langle \dot{e}_i \rangle, \ v(i) = 1$$
$$\pi(v', \hat{s}') = \langle \langle \hat{a}'_i \rangle \rangle, \ v'(i) = 1$$
$$\pi(v'', \hat{s}'') = \langle \langle \hat{a}''_i \rangle \rangle, \ v''(i) = 1$$

for $\hat{s}, \hat{s}', \hat{s}'', \hat{a}'_i, \hat{a}''_i \in \hat{S}, \dot{e}_i \in \hat{E}, i \in \mathbb{N}_0$

Source and sink processes make our MoC complete:

$$\mathtt{sourceT}(g, w_0) = p \qquad (5.4)$$

where

$$p() = \hat{s}'$$
$$w_i = \hat{e}'_i$$
$$g(w_i) = w_{i+1}$$
$$\pi(v', \hat{s}') = \langle \langle \hat{e}'_i \rangle \rangle, v'(i) = 1$$

$$\mathtt{sinkT}(g, w_0) = p \qquad (5.5)$$

where

$$p(\hat{s}) = \langle \, \rangle$$
$$g(w_i) = w_{i+1}$$
$$\pi(v, \hat{s}) = \langle \hat{a}_i \rangle, \ v(i) = 1$$

$$\mathtt{initT}(\hat{r}) = p \qquad (5.6)$$

where

$$p(\hat{s}) = \hat{r} \oplus \hat{s}$$
$$v = v' = 1$$
$$\hat{r}, \hat{s} \in \hat{S}$$

Again, we can now make precise what we mean by the timed model of computation.

Definition 5.1 The *timed model of computation* (timed MoC) is defined as timed MoC $= (C, O)$, where

$$C = \{\mathtt{mealyT}, \mathtt{zipT}, \mathtt{unzipT}, \mathtt{sourceT}, \mathtt{sinkT}, \mathtt{initT}\}$$
$$O = \{\|, \circ, \mathbf{FB_P}\}$$

In other words, a process or a process network belongs to the *timed MoC domain* iff all its processes and process compositions are constructed either by one of the named process constructors or by one of the composition operators. We call such processes *T-MoC processes*.

Process up-rating, down-rating and merging is significantly more involved for timed processes than for the untimed case because additional delays have to be injected during up-rating to preserve a proper causality relation. It is feasible to define up-rating and down-rating operators and merging transformations in a clean and consistent way based on various notions of equality that distinguish between the complete preservation of the behavior and the preservation of only the functionality but not the detailed timing. The practical utility of these techniques, however, depend very much on the modeling style. If functionality and timing aspects are tightly intertwined, the application of up-rating and merging is limited. If functionality and timing aspects are neatly separated in the description of the processes, they allow us to move the functionality from one process to another and merge and split processes while still maintaining the original timing characteristics.

5.2.1 Timed MoC Variants

Without formally introducing them, we briefly discuss three variants of the `mealyT` process constructor that enforce a stricter separation between functionality and timing.

Timer-Based Process Invocation (`mealyPT`)

In `mealyPT`(γ, f, g, w_0)-based processes, the functions f and g never see or return absent events. They are only defined on untimed sequences. The interface of the process strips off all absent events at the input signal, hands over the result to f and g, and inserts absent events at the output as appropriate to provide proper timing for the output signal. The function γ, which may depend on the process state as usual, defines how many events are consumed. Essentially, it represents a timer and determines when the input should be checked the next time.

Event-Count-Based process invocation (`mealyST`)

In `mealyST`(γ, f, g, w_0)-based processes, γ determines the number of nonabsent events that should be handed over to f and g for processing. Again, f and g never see or produce absent events, and the process interface is responsible for providing them with the appropriate input data and for synchronization and timing issues on inputs and outputs. Unlike `mealyPT` processes, functions f and g in `mealyST` processes have no influence on when they are invoked. They only control how many nonabsent events have appeared before their invocation. *f* and *g* in `mealyPT` processes, on the other hand, determine the time instant of their next invocation independent of the number of nonabsent events.

Event Count with Time-out (`mealyTT`)

However, a combination of these two process constructors is `mealyTT`, which allows us to control the number of nonabsent input events and a maximum time period, after which the process is activated in any case independent of the number of nonabsent input events received. This allows us to model processes that wait for input events but can set internal timers to provide time-outs.

These examples illustrate that process constructors and models of computation could be defined that allow us to precisely define to what extent communication issues are separated from the purely functional behavior of the processes. Obviously, a stricter separation greatly facilitates verification and synthesis but may restrict expressiveness.

5.2.2 Representation and Distribution of Global Time

Every timed model has to deal with two conflicting requirements. On the one hand, it is desirable to analyze and understand individual components and processes independently from the rest of the system. This we informally call the *independence property*. It requires that the only information a process receives is through its input ports via input signals. There is no information it receives via some other mechanism such as access to a global state. For instance, if simulation of a process network can be distributed among concurrently active simulators that evaluate their part as long as input is available, parallel simulation would speed up simulation runs very effectively.

On the other hand, a global time has to be maintained centrally, which can only be done based on information about the time status of all processes. Further, the global time has to be communicated to all processes so they know how far they may advance their evaluation.

In general, three different approaches have been taken to represent and distribute a global time among processes in a process network.

Local Timer

The local timer approach places, explicitly or implicitly, a timer that is aware of the global time in every process. The timer can be queried by the process to obtain information about the global time. In a typical usage a time-out period is set by the process and the timer issues an event when the time period has expired.

This approach is typically taken by software specification and programming languages that support modeling of concurrency and real-time features. Prominent examples are SDLC (Ellsberger et al. 1997), Ada (Booch and Bryan 1994), and a number of variants of C. In the latter case the time is often not part of the language but provided by the operating system by means of a system function call. The motivation in these

FIGURE 5-3

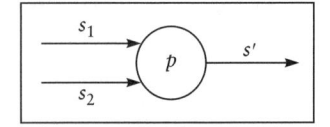

The time-tag approach has difficulties in merging signals in a time-sensitive way.

cases is that the hardware, and based on this the operating system, provides a timer as a primitive that can be set to a specific time-out and that notifies the application software with an interrupt when the time period has expired. This is a natural model when a physical entity (e.g., an oscillating quartz-based clock), provides a reliable source for real-world timing information. Each process can use this resource to obtain information about the current global time. An important condition for this model to work is that the global time perceived by all processes is identical or highly synchronized. This is obviously the case when there is only one timer utilized by all processes such as in a single-processor system. However, in distributed systems different timers used by different processes do not necessarily provide the same global time. Thus, synchronization of the different timers become necessary.

On a conceptual level, this approach means that there is a global state to which all processes have access instantly that can obviously be used for synchronization but also for communication purposes. This violates the independence property and may make it more difficult to analyze the system's behavior or to develop efficient distributed simulations and implementations.

Time Tags

In a second approach, put forward as a theoretical analysis framework (Lee and Sangiovanni-Vincentelli 1998), each event exchanged between two processes is annotated with a time tag that represents the global time of the occurrence of the event. In this approach a process receives information only via explicit events on its input port. There is no explicit or implicit global state shared among all processes. But with each input event received, the process is also informed about the global time. Hence, two processes are synchronized when they receive events with identical time tags.

This approach in some way reconciles with both mentioned requirements: All processes comply with the independence property, and all processes are informed about the global time.

However, there is one difficulty. Consider a process $p = \texttt{zipTs}()$ (Figure 5-3) that consumes time-tagged events on two input ports and merges them into one output signal. For instance, given two signals

$$s_1 = \langle (1, v_1), (2, v_2), (3, v_3) \rangle$$

and

$$s_2 = \langle (2, v_5), (3, v_6), (4, v_7) \rangle$$

FIGURE 5-4

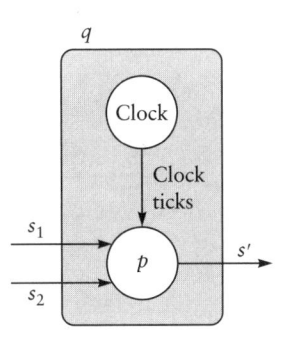

A source process emitting periodic events can emulate a timer allowing q to express p' in (5.7).

the process output becomes

$$p(s_1, s_2) = \langle (2, (v_1, v_5)), (3, (v_2, v_6)), (4, (v_3, v_7)) \rangle$$

But suppose we want a process that does not delay events when there is no matching event on the other input.

$$p'(s_1, s_2) = \langle (1, (v_1, 0)), (2, (v_2, v_5)), (3, (v_3, v_6)), (4, (0, v_6)) \rangle \quad (5.7)$$

Process p' cannot be directly expressed in the time-tagged approach because before the event $(2, v_5)$ arrives, the process does not know if the next event on s_2 will have a tag of 1 or 2. Of course, we can store an event from one input, for example, $(1, v_1)$, until we have received an event from the other input, $(1, v_5)$. Then we can decide that we should emit event $(1, (v_1, 0))$ and store event $(2, v_5)$. However, there are two problems. First, if the next event on the second input has a much higher tag, say, 100,000, we get a huge mismatch between simulated time and real time, making simulation inefficient. Even worse, if the event on the second input never arrives, we cannot output $(1, (v_1, 0))$, even though we should. Imagine there is a feedback loop and the event on the second input depends on the emission of $(1, (v_1, 0))$. The second problem is that we cannot emit an event at time 1 after we have received an event at time 2, which would be necessary for this approach but would violate causality conditions.

On the other hand, process p' can be expressed in the local-timer approach because it does not need to wait for an event on s_2 to know when global time 1 has elapsed. However, a local timer can be emulated by a source process emitting periodic clock events (Figure 5-4). If a process needs to know the global time independent of the arrival of events on its normal input signals such as process p' in equation (5.7), it can be connected to a clock source.

However, both approaches do not mix well with the blocking read semantics that we have adopted in previous chapters. Under this semantics, if process p in Figure 5-4 attempts to read an event from s_2, it is blocked until an event arrives. It cannot, in the meantime, read an event from the clock source and interrupt the reading of signal s_2 if too much time has elapsed.

Absent Events

The approach we have taken in this chapter uses absent events to convey information about the global time. Similarly to the synchronous MoC the time is divided into time slots, and each and every signal in the system carries an event in each and every time slot. If there is no useful event generated by a process for a particular output signal in a time slot, an absent event is emitted instead. Contrary to its name, an absent event is a real event, but its value is distinct from all other possible events in the system. In this way a process always knows the time elapsed in the processes from which it receives events by simply counting the input events. Similarly the process communicates its own time to processes connected to its output signals. Since this happens in a synchronized way for all processes, every process always knows the global time.

The absent-event approach mixes well with the blocking read semantics and can express process p' of equation (5.7) directly because the signals s_1 and s_2 would in this approach be

$$s_1 = \langle v_1, v_2, v_3, \sqcup \rangle$$

and

$$s_2 = \langle \sqcup, v_5, v_6, v_7 \rangle$$

Hence, p_1 would evaluate to

$$p'(s_1, s_2) = \langle (v_1, v_0), (v_2, v_5), (v_3, v_6), (v_0, v_7) \rangle$$

where v_0 is a special value indicating the absence of an input event. In our definition of `zipT` (5.1) and of `zipS` (4.7), we have also used the $\sqcup$ symbol for this purpose.

On the other hand, the absent-event approach does not capture time-tag models where the time tags are not integers but rational numbers or real numbers. Consider two signals (Lee 1998):

$$s_1 = \{(t, v) : t = 0, 1, 2, 3, \dots\}$$
$$s_2 = \left\{(t, v) : t = \frac{1}{2}, \frac{2}{3}, \frac{3}{4}, \dots\right\} \tag{5.8}$$

While both signals individually can be represented in the absent-event approach, we encounter difficulties when modeling both together, for example, when a process takes s_1 and s_2 as inputs $p(s_1, s_2)$. A tuple such as (s_1, s_2) is called a *Zeno tuple* by Lee (1998) because a simulator had to process infinite events from s_2 and could therefore never advance to time instant 1. In the absent-event approach we have to insert infinitely many absent events between events $(0, v_0)$ and $(1, v_1)$ on s_1, which is clearly not possible. Hence, the absent-event approach and the timed MoC that we have presented here require that the tag structure on all signals is statically known and in some way comparable on all signals. If it were not known but dynamically dependent on input values, some global agent had to communicate the information about the tag structure dynamically during evaluation to all processes. In other words, as elaborated by Lee (1998), the absent-event approach can only represent a time structure where the tags on all signals are *order-isomorphic* to a subset of the integers. Order isomorphism requires that the set of tags can be mapped on the integers (and vice versa) while preserving the order. Since the preservation of order is important, the rational numbers are not order isomorphic with the integers. Although both sequences 1/2, 2/3, 3/4,... and 0, 1, 2,... are order isomorphic with the integers, their union

$$\left\{\frac{1}{2}, \frac{2}{3}, \frac{3}{4}, \ldots\right\} \cup \{0, 1, 2, \ldots\}$$

is not.

Discussion

Thus, comparing the three approaches, we observe that the processes in the time tag approach have no complete knowledge of the global time as long as they are not connected to a clock source process, and situations where this is required cannot be modeled as a single process as equation (5.7) has illustrated. In the presence of a clock-tick-providing process, this approach becomes equivalent to the local timer approach, where every process essentially receives the ticks of a global clock source. However, both approaches lose significant expressiveness when a blocking read semantics is used because processes are hindered from using the clock information when occupied with reading an event. The absent-event approach, however, requires a fixed and inflexible time structure that also restricts expressiveness (5.8). Finally, in the local-timer approach the time structure is very flexible and processes are always ideally synchronized, but the independence property is sacrificed.

Another interesting point to notice is that both the local-timer and time tag approaches, due to their flexible time structure, can be naturally extended to continuous time models; this would be very unnatural or even impossible for the absent-event approach.

As we discussed at the end of the previous section when introducing constructor alternatives (`mealyT`, `mealyPT`, `mealyST`, `mealyTT`), there are different degrees of entanglement between these synchronization and timing issues and the behavior proper of a process in this approach. Also, the concepts of modeling these issues should be kept separate from the way they are implemented. Absent events should be considered foremost as a modeling concept. The implementation can be based on a shared-timer resource, as is common in single-processor multitasking systems, or on local timers and a synchronization scheme that regularly synchronizes the local timers to provide a consistent view of the global time. The latter approach is often taken in distributed systems, sometimes with simple synchronization schemes and other times with more sophisticated approaches.

Although it is illuminative to keep modeling concepts and implementation techniques separate, it should be noted that some modeling concepts can be more effectively implemented by some techniques than by others. For instance, `mealyPT`-based processes can be naturally and efficiently implemented by means of a shared-timer resource. `mealyST`-based processes can be implemented in a distributed system with minimal synchronization overhead because the behavior proper of the processes (functions *f* and *g*) does not depend on the timing information on the input signals and does not need to know the global time. Thus, the selection of the right MoC with the right set of process constructors can be very important for cost-effective and high-performance implementations. In general we can observe that the more we separate synchronization and timing issues from the behavior proper of processes, the more efficient implementations we will be able to develop. However, higher degrees of separation, as exemplified by `mealyST` processes, restrict expressiveness because the processes have no access to timing information.

5.3 Discrete Event Models Based on δ-Delay

Many important languages and so-called discrete time modeling systems are based on a mixture of the local-timer and the time tag approaches rather than the absent-event approach that we have followed so far in this chapter. To elaborate this approach further, we present a particular variant of discrete event systems that has become popular for hardware modeling. It maintains a global time centrally and makes it instantly available to all processes, as in the local-timer approach, and thus sacrifices the independence property. Moreover, it annotates each event with a time tag.

Discrete event models are usually, but not always, based on a timed model of computation. They are used in a variety of applications, from the simulation of transistor networks to stochastic performance analysis of communication systems. A comprehensive account of discrete event modeling techniques and applications is given by Cassandras (1993).

We will discuss discrete event simulation as one prominent application of a timed MoC. In particular, we will focus on the simulation of digital circuits through hardware description languages like VHDL, Verilog, or SystemC.

5.3.1 The Two-Level Time Structure

The central concept of the discrete event simulation is the *event* occuring at a particular time instance. The occurrence of an event triggers computations, such as the evaluation of a primitive component like a gate, which in turn may generate new events. For instance, the value change on input c from 1 to 0 in Figure 5-2(b) is an event that occurs at time t'. This event causes the evaluation of the inverter, which after its assigned delay of 1.5 ns generates a new event. This prompts the evaluation of the NAND and OR gates, which after appropriate delays generate new events that change the outputs x and y.

A straightforward formulation of a discrete event simulation is nondeterministic due to the simultaneous occurrence of events. Consider the netlist in Figure 5-5. Component A generates two events $\hat{e}$ and $\hat{e}'$ at the same time t. Component B reacts to the input event with an output event $\hat{e}''$ without delay. Consequently, both events, $\hat{e}$ and $\hat{e}''$, occur concurrently at time t. The network is nondeterministic because there is no rule that decides in which order component C processes its input events. Should it process event $\hat{e}$ first and then $\hat{e}''$, or $\hat{e}''$ first and then $\hat{e}$, or both events simultaneously? Suppose the rule demanded both events to be processed simultaneously. That would be very difficult to implement because a simulator does not know after the evaluation

FIGURE 5-5

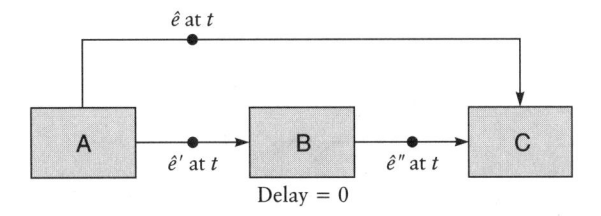

A nondeterministic network.

FIGURE 5-6

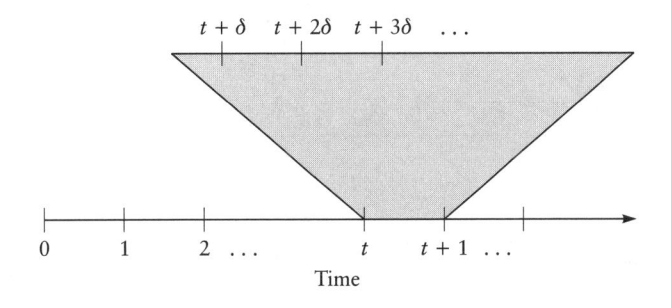

The δ-delay-based time structure.

of A if any of the ready components will generate a new event at time t. It has to evaluate all ready components in a trial step. If it finds a component, such as B in our example, it has to check again if other components, sensitised by B's output, would produce an event at time t.

Thus, this would be a very complicated simulation algorithm, and it would not even terminate in all cases. One can devise models with feedback loops for which such an algorithm would not find a stable solution but oscillate between two or several solutions. In fact, there exists no simple, intuitively appealing simulation scheme that would be deterministic.

This nondeterminism in naive discrete event simulation schemes has led to the invention of the δ-delay-based discrete event simulation. The δ-delay mechanism essentially introduces a two-level hierarchical time structure, as illustrated in Figure 5-6. Normal time, as it is used to describe delays of all components and computations, is denoted as $0, 1, 2, \ldots t, t+1, \ldots$, that is, the positive integers, with 0 denoting the time instance when the simulation starts. Between every two time instances, t and $t+1$, there is a potentially infinite number of δ time instances, ordered as $t+\delta, t+2\delta, t+3\delta, \ldots$. To avoid zero-delay computations, every computation exhibits at least a δ-delay. In this way every zero-delay computation has a finite positive delay that is shorter than the

FIGURE 5-7

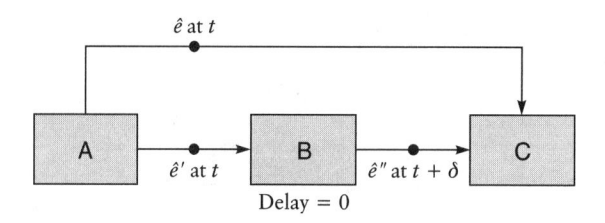

A netlist simulated with δ-delay is always deterministic.

FIGURE 5-8

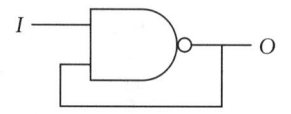

A netlist with feedback loop.

shortest possible real delay. Also, an event $\hat{e}$ that triggers the generation of another event $\hat{e}'$ can never have the same time tag as the resulting event $\hat{e}'$; cause and effect are at least 1δ apart.

Figure 5-7 shows the same example as Figure 5-5; the only difference is that component B generates its output with a delay of at least 1δ, even though it is modeled as having a delay of zero. Consequently, there is a well-defined order in which component C will process its inputs: first $\hat{e}$, then $\hat{e}''$.

The concept of δ-delay also alleviates the problem of zero-delay feedback loops. Consider the example in Figure 5-8 and assume that the NAND gate has zero delay. If the input I is 0, the output is well defined and 1. But when the input switches from 0 to 1, the output is contradictory, whatever its value is. However, with a δ-delay of the NAND gate, this contradiction is resolved, and the output would oscillate between 0 and 1 with a period of 2δ. This is a perfectly consistent situation even though it might not be the desired behavior. In fact, it points to another problem that δ-delay-based discrete event simulation models may have. Feedback loops can lead to an infinite sequence of δ cycles without ever increasing the real time. Thus, the simulation might get stuck at a time t, advancing forever in the δ time axis without ever reaching time $t + 1$. This is also known as *Zeno*[1] *behavior*.

In summary, we can conclude that the introduction of the δ-delay concept solves two problems: It makes the discrete event simulation deterministic and it gives arbitrary

---

[1] Zeno of Elea (495–430 BC) was a Greek philosopher and mathematician who is known particularly for his paradoxes (e.g., Achilles and the tortoise) that address the effect of infinitely many ever shorter time periods.

models including feedback loops a consistent behavior. But at the same time it leaves us with the possibility of infinite loops in the δ time axis.

It can be claimed that the δ-delay discrete event model is at odds with the timed MoC that we introduced earlier in this chapter. It is more closely related to the local-timer or the time tag approach, as discussed in Section 5.2.3. Thus, it does not use absent events for communicating timing information, it can hardly be used with a blocking read semantics, and it needs a central agent for maintaining and distributing global time. For these reasons we do not attempt to formally express it in our MoC framework. In fact, developing a formal semantics for δ-delay-based languages such as VHDL is no simple task, as demonstrated by Kloos and Breuer (1995), who collected several alternative approaches to the problem.

5.3.2 The Event-Driven Simulation Cycle

We use VHDL as an example of an event-driven simulation (Lippsett et al. 1993). The simulator maintains a central event list, which contains all the *pending events*. A pending event is an event that has been generated but not yet applied to the model to update signal and port values. In general an event is a change of a value of a signal. Hence, if a computation such as a signal assignment changes the value of the signal, an event is generated. An event is associated with a time stamp to mark its occurrence. The event list is ordered according to the time stamps of the events. Thus, the first element in the list represents the next event to occur. A time stamp consists of two components: the "real" time t plus the number of δ's; the event list is ordered according to both components.

Figure 5-9 shows the flow of the event-driven simulation cycle. First, the current time t_c is set to the time of the first element in the event list. Then all events with the same time stamp are identified. They should come right after each other in the event list. The events are "applied" to the model, that is, the values of signals are updated. Due to the advance of time or to new values on signals and ports, elements in the model are activated, for example, a process that is sensitive to a particular signal or that waits for a particular amount of time. All of these activated elements are evaluated and generate new pending events with a time stamp at least 1δ greater than t_c. The new events are inserted into the event list appropriately sorted.

The simulation ends when the event list becomes empty or when t_c exceeds a predefined maximum simulation time.

5.4 Rugby Coordinates

In the terminology of the Rugby metamodel, the timed MoC is denoted as ⟨[Alg-LB], [IPC-Top], [Sym-LV], PhysicalTime⟩. Thus, the timing abstraction is its characteristic feature. In fact there are many timed models, with computational abstractions from the Algorithm level to the LogicBlock level. VHDL is a prominent example that covers both these levels. At the Transistor level, differential equations are the natural modeling mechanism, requiring both continuous time and continuous state variables, which is

FIGURE 5-9

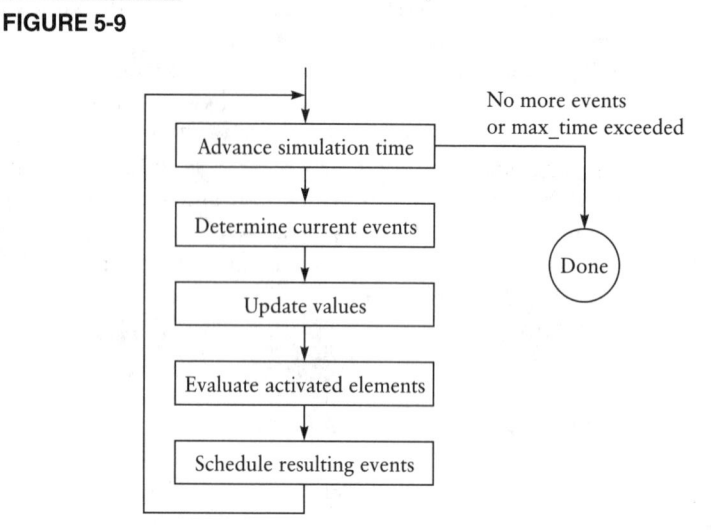

```
         ┌──────────────────────────┐
         │  Advance simulation time │────── No more events
         └──────────────────────────┘        or max_time exceeded
                      ↓                              │
         ┌──────────────────────────┐                ↓
         │ Determine current events │             ( Done )
         └──────────────────────────┘
                      ↓
         ┌──────────────────────────┐
         │      Update values       │
         └──────────────────────────┘
                      ↓
         ┌──────────────────────────┐
         │ Evaluate activated elements │
         └──────────────────────────┘
                      ↓
         ┌──────────────────────────┐
         │ Schedule resulting events │
         └──────────────────────────┘
```

Event-driven simulation cycle.

beyond the scope of timed models as discussed in this chapter. At the SystemFunctions and the RelationConstraints level, events are only partially ordered with respect to their causal dependences. This is in contrast to the total order of events in the timed models.

In the other two domains, the timed MoC is also positioned in the middle. Neither the lower levels (Layout, ContValue) nor the higher levels (InterfaceConstraints, DataTypeConstraints) are covered.

5.5 Applications

The timed model is foremost a simulation tool. The low timing abstraction defies any attempts at synthesis. Synthesis requires some inaccuracy of the input description in order to make design decisions and find a good solution. But if delays of individual components and computations are defined as, say, 2 ns, not 1.9 ns and not 2.1 ns, then it is hopeless for the synthesis tool to match this specification exactly. This is true for the general timed model, but it is even more so for the δ-delay-based variant. The concept of δ-delay is a means to sequentialize the simulation and to make it unambiguous in the presence of simultaneous events. To achieve exactly the same behavior in an implementation, the synthesis tool essentially has to implement the simulation cycle as well, which of course is prohibitively expensive. For the very same reasons formal verification cannot be applied to discrete event models. Since no implementation will ever match exactly the timing behavior as specified in a discrete event model, a formal proof of equivalence is never possible.

However, due to the popularity of timed models for system as well as for implementation modeling, researchers have been hard pressed to find a way to apply synthesis and formal verification to them as well. As a result it has been proposed to *interpret* VHDL and Verilog descriptions at a different timing abstraction, that is, at the ClockedTime level, for the purpose of synthesis and verification. Thus, by following certain modeling conventions (e.g., to identify the clock signal), it is possible to consistently interpret VHDL and Verilog descriptions according to the ClockedTime semantics, even though the same description is simulated according to the δ-delay timed semantics based on discrete physical time. This approach is not without its problems, but in general it has been very successful and is employed today as a standard technique.

We will come back to these issues in Chapter 9, where we discuss the requirements and needs of particular applications.

5.6 Further Reading

A framework for representing untimed and timed MoCs based on a time tag approach has been introduced by Lee and Sangiovanni-Vincentelli (1998). There the properties of the time tags determine the MoC. A partially ordered set of time tags can represent an untimed MoC, a totally ordered discrete set can represent a discrete timed model, and a continuous set can represent a continuous timed model. Lee (1998) also discusses the concept of order-isomorphic sets for time tags and Zeno tuples.

δ-delay-based discrete event systems and the corresponding simulation mechanism have been described many times. For example, Lippsett et al. (1993) and Zwoliński (2000) discuss it in the context of VHDL, and Grötker et al. (2002) for SystemC.

Also, the clocked synchronous interpretation of VHDL and Verilog models has been thoroughly covered, for instance, by Gajski et al. (1993) and by Bhatnagar (2002) for synthesis, and by Kloos and Breuer (1995) and by Rashinkar et al. (2001) for formal verification.

5.7 Exercises

5.1 Model an inverter, a NAND, and an OR gate by means of timed MoC processes with the delays given in Table 5-1. Then model the netlist of Figure 5-1 and show by simulation that the output pattern can be $x = 0, y = 0$.

5.2 Model the switch described in Exercise 3.14 in the timed Moc.

5.3 Model the traffic light controller of Section 4.8 in the timed MoC.

5.4 Give precise definitions of the following process constructors:

   a. *mealyPT, zipdPT, unzipPT, sourcePT, sinkPT, initPT*

   b. *mealyST, zipdST, unzipST, sourceST, sinkST, initST*

   c. *mealyTT, zipdTT, unzipTT, sourceTT, sinkTT, initTT*

5.5 Model round-robin scheduling by means of `mealyPT` process constructors:

   a. Model three processes with arbitrary functionality. Only one of them is active at any particular time and they perform a round-robin scheduling. After 10,000 event ticks the active process hands over the control to the next process.

   b. Use a scheduler process for control arbitration. The active process hands over control to the scheduler after 10,000 event ticks, which in turn hands over control to the next process.

   c. Modify the three processes such that they run forever but allow them to be interrupted by a separate interrupt input signal. Model a timer process that interrupts the active process after 10,000 event ticks and hands over control to the scheduler.

chapter six
MoC Interfaces

Embedded systems are usually heterogeneous in that they combine different types of applications. Signal-processing subsystems often have to coexist with control-dominated parts and reactive real-time parts. This means for our discussion that different MoC domains have to be integrated in the system model.

We start this chapter (Section 6.1) with the observation that even two domains, that is, separate process networks, of the same MoC type may have a different time structure. For example, two S-MoC domains that are not connected can have different durations of their evaluation cycles. When connecting them, the relation between their respective time structures has to be defined. This relation can be constant and simple (e.g., 1:1 or 1:3), or it can vary dynamically. The interface processes between the MoC domains define this relation.

Next we define interface processes between different MoC types (Section 6.2). We distinguish between interfaces that add timing information and those that remove timing information. In general, both can be done in many different ways, but we describe only one possible solution.

Based on these interface processes we introduce an integrated MoC *(Section 6.3), consisting of different MoC domains and interface processes. Several profound issues emerge when connecting MoC domains, which go beyond the technical insertion and suppression of time information, and we discuss some of them. We propose distinguishing between (1) the time structures of different MoC domains and their relations to each other, (2) the protocol used to model communication between different domains, and (3) the modeling of delays on the channels between different domains (Section 6.4).*

After the illumination of some important aspects of interfacing MoC domains, we consider moving functionality between domains (Section 6.5). For the untimed, the synchronous, and the timed MoCs we discuss all possible cases of migrating a process from one domain to another, although some cases are treated only superficially. There are many possible ways to formulate these migrations, and sensible and efficient solutions depend on the application context, which we do not elaborate on further.

Finally (Section 6.6), we take up some possible applications of an integrated MoC. One example is the integration of different front-end languages, such as SDL and Matlab, in a common modeling and design environment. Each front-end language is mapped on a particular MoC, and the integrated MoC is used to define the relation between the different MoCs and thus between the front-end languages.

6.1 Interfaces between Domains of the Same MoC

We have defined three major models of computation and discussed some of their fundamental properties and applications. Since the design of an electronic system usually requires the deployment of different computational models, integration is desirable. The main difficulty obviously is the representation and conversion of timing information. This is an issue even for disconnected process networks of the same MoC. Consider the two process networks in two MoC domains in Figure 6-1. MoC A and MoC B may be the same or different. Only the case where both are the untimed MoC (i.e., MoC A = MoC B = untimed MoC) is unambiguous and unproblematic. Since in both domains there is no timing information present, the two process networks can be directly connected with no need for a special interface. However, if both networks form a perfect synchronous domain (i.e., MoC A = MoC B = synchronous MoC) the elementary timing unit may be different. Since there is no reference to an absolute time in the synchronous MoC, the duration of the basic evaluation cycle may be quite different in MoC A and

FIGURE 6-1

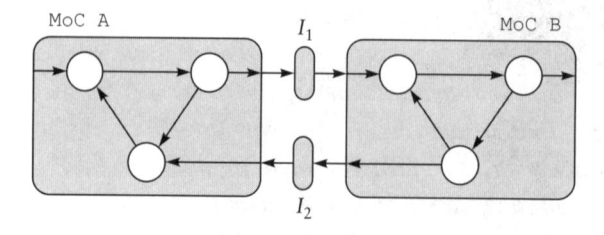

The interfaces I_1 and I_2 define the relation between different MoC domains. All interfaces between two domains must be consistent with each other.

MoC B. MoC A's evaluation cycle could have a duration r times that of MoC B's cycle. r could be a positive integer, but it could also be a rational or real number. In fact, r does not even have to be a constant since the evaluation cycles in a synchronous MoC can vary arbitrarily. This does not become apparent as long as we deal with an isolated domain because there the duration of each evaluation cycle is the same for all processes and it cannot be observed from within the domain whether consecutive cycles vary in length. It can only be observed when compared to another, disconnected domain or some "absolute time reference." Thus, in the general case, interfaces I_1 and I_2 in Figure 6-1 could be completely asynchronous even if both domains are perfectly synchronous MoC domains.

A practically useful case, however, is when r is a constant positive integer, $r \in \mathbb{N}$. The following interface constructors can be used to interface synchronous domains in this case:

$$intSup(r,f) = mapU(1,f) \tag{6.1}$$

with

$$\mathsf{length}(f(\bar{e})) = r$$
$$\bar{e} \in \bar{E}, r \in \mathbb{N}$$

$$intSdown(r,f) = mapU(r,f) \tag{6.2}$$

with

$$\mathsf{length}(f(\bar{a})) = 1$$
$$\bar{a} \in \bar{S}, r \in \mathbb{N}$$

$intSup$ emits r events for each input event, and function f defines the values of the emitted events. $intSdown$ emits one event for r input events. Hence, assuming MoC A operates at a rate two times the rate of the MoC B domain in Figure 6-1, we could define the interface processes as follows:

$$I_1 = intSdown(2,f_1) \quad \text{with } f_1(\langle \bar{e}_1, \bar{e}_2 \rangle) = \bar{e}_1$$
$$I_2 = intSup(2,f_2) \quad \text{with } f_2(\bar{e}) = \langle \bar{e}, \sqcup \rangle$$

It is obviously crucial that all interfaces between two domains are consistent, which for the special case of constant integer factors r is straightforward to check. Figure 6-2 shows an example of three MoC domains. Their relative cycle period ratio, r_1, r_2, r_3, annotates the connecting arcs. Obviously different arcs connecting two domains must express the same ratio. Hence, the two arcs connecting MoC A and MoC C have the same ratio r_3, with one being based on $intSup$ and the other on $intSdown$. It is interesting to note that the MoC domain graph can be viewed as a synchronous dataflow graph (Section 3.15), and SDF techniques can be used to check consistency in the graph and to determine buffer requirements on domain boundaries.

The situation for connecting two separate clocked synchronous MoC domains or two timed MoC domains is exactly the same as for the perfectly synchronous case. Even

FIGURE 6-2

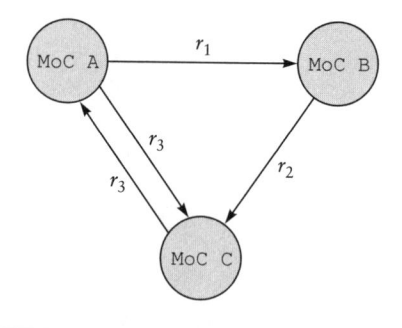

An MoC domain graph.

though the time granularity or assumptions about computation delays are different, the main issue again is to relate the relative time base in different domains to each other. Thus we can define

$$intTup = intSup \qquad (6.3)$$

$$intTdown = intSdown \qquad (6.4)$$

Without going into more details, it is obvious that this is closely related to the problem of designing a hardware circuit with multiple clock domains.

6.2 Interfaces between Different Computational Models

If MoC A and MoC B in Figure 6-1 are different MoCs, the interfaces have to bridge domains with different information content concerning time. Consequently interfaces have to filter out or insert timing information. This can be done in many different ways depending on the modeling objectives. Unfortunately we cannot give a systematic presentation of the choices, but we present only selected, simple solutions.

Interface processes that connect processes with different timing regimes either remove or add timing information. Interface constructors that add timing information are labeled with the prefix `insert`; those that remove timing information have the prefix `strip`. Table 6-1 lists the names of process constructors that instantiate interface processes. We define them in the following sections.

6.2.1 Strip-Based Interface Constructors

The strip-based processes remove the timing information that they receive on their input signals. This is straightforward when the output signal is an untimed signal because then

6.2 Interfaces between Different Computational Models

TABLE 6-1: *Interface processes between timing regimes.*

From/to	Timed	Synchronous	Untimed
Timed	—	`stripT2S`	`stripT2U`
Synchronous	`insertS2T`	—	`stripS2U`
Untimed	`insertU2T`	`insertU2S`	—

only the ⊔ symbols have to be removed and other events are passed to the output in the same order they appear at the input.

Definition 6.1   `stripT2U` is a process constructor that instantiates a process $p: \hat{S} \to \dot{S}$, which is defined as follows:

$$stripT2U() = p$$

where

$$p(\hat{s}) = \dot{s}$$
$$\pi(v, \hat{s}) = \langle \hat{e}_i \rangle, \ v(i) = 1$$
$$\pi(v', \dot{s}) = \langle \dot{a}_i \rangle$$

$$\dot{a}_i = \begin{cases} \langle \rangle & \text{if } \dot{e}_i = \sqcup \\ \langle \dot{e}_i \rangle & \text{otherwise} \end{cases}$$

$$v'(i) = \begin{cases} 0 & \text{if } \dot{e}_i = \sqcup \\ 1 & \text{otherwise} \end{cases}$$

Definition 6.2   `stripS2U` is a process constructor that instantiates a process $p: \hat{S} \to \dot{S}$, which is defined as

$$stripS2U = stripT2U$$

However, `stripT2S`-based processes need additional information to determine the time relation between timed and synchronous events. We use a parameter $\lambda \in \mathbb{N}$ that defines the number of events in the timed input signal for each synchronous output event.

Definition 6.3   `stripT2S` is a process constructor that instantiates a process $p: \hat{S} \to \bar{S}$, which is defined as follows:

$$stripS2U(\lambda) = p$$

where

$$p(\hat{s}) = \bar{s}$$
$$\pi(v, \hat{s}) = \langle \hat{a}_i \rangle, \ v(i) = \lambda$$
$$\pi(v', \bar{s}) = \langle \bar{e}_i \rangle, \ v'(i) = 1$$

$$\bar{e}_i = \begin{cases} \sqcup & \text{if strip}(\hat{a}_i) = \langle\rangle \\ \text{lastt}(\hat{a}_i) & \text{otherwise} \end{cases}$$

$$\text{for } \lambda \in \mathbb{N}, \hat{s}, \hat{a} \in \hat{S}, \bar{s} \in \bar{S}, \bar{e}_i \in \bar{E}, i \in \mathbb{N}_0$$

where $\text{lastt}(\hat{s})$ denotes the last nonabsent event in signal $\hat{s}$.

This is not the only way to define interface processes between the timed and the synchronous domains. It is easy to imagine the usefulness of processes with a control input that determines the time relation between input and output signals.

6.2.2 Insert-Based Interface Constructors

For insert-based processes we have to decide on the ratio of input events to output events. This ratio may vary and may be context dependent. However, here we adopt the simple solution with a constant ratio λ.

Definition 6.4 $\mathit{insertU2S}$ is a process constructor that, given a natural number λ, instantiates a process $p : \dot{S} \to \bar{S}$, defined as follows:

$$\mathit{insertU2S}(\lambda) = p$$

where

$$p(\dot{s}) = \bar{s}$$
$$\pi(\nu, \dot{s}) = \langle \dot{e}_i \rangle, \nu(i) = 1$$
$$\pi(\nu', \bar{s}) = \langle \bar{a}_i \rangle, \nu'(i) = \lambda$$
$$\bar{a}_i = \langle \dot{e}_i \rangle \oplus \langle \sqcup \rangle^{\lambda-1}$$
$$\text{for } \lambda \in \mathbb{N}, \dot{s} \in \dot{S}, \bar{s}, \bar{a}_i \in \bar{S}, \dot{e}_i \in \dot{E}, i \in \mathbb{N}_0$$

Definition 6.5 $\mathit{insertU2T}$ is a process constructor that, given a natural number λ, instantiates a process $p : \dot{S} \to \hat{S}$, defined as follows:

$$\mathit{insertU2S}(\lambda) = \mathit{insertU2S}(\lambda)$$

Definition 6.6 $\mathit{insertS2T}$ is a process constructor that, given a natural number λ, instantiates a process $p : \bar{S} \to \hat{S}$, defined as follows:

$$\mathit{insertS2T}(\lambda) = p$$

where

$$p(\bar{s}) = \hat{s}$$
$$\pi(\nu, \bar{s}) = \langle \bar{e}_i \rangle, \nu(i) = 1$$
$$\pi(\nu', \hat{s}) = \langle \hat{a}_i \rangle, \nu'(i) = \lambda$$
$$\hat{a}_i = \langle \bar{e}_i \rangle \oplus \langle \sqcup \rangle^{\lambda-1}$$
$$\text{for } \lambda \in \mathbb{N}, \bar{s} \in \bar{S}, \hat{s}, \hat{a} \in \hat{S}, \bar{e}_i \in \bar{E}, i \in \mathbb{N}_0$$

6.3 Integrated Model of Computation

We now have the means to reason about all three computational models in a unifying framework. Before we formulate an MoC that contains several MoC domains, we generalize the MoC concept.

Definition 6.7 A *hierarchical model of computation* (HMoC) is a 3-tuple HMoC = (M, C, O), where

M is a set of HMoCs, each capable of instantiating processes;

C is a set of process constructors, each of which, when given constructor-specific parameters, instantiates a process;

O is a set of process composition operators, each of which, when given processes as arguments, instantiates a new process.

By "process" we mean either an elementary process or a process network.

This allows us to formulate hierarchical MoCs with specialized MoC domains in a structured way.

Definition 6.8 The *integrated model of computation (integrated MoC)* is defined as integrated HMoC $= (M, C, O)$, where

$$M = \{\text{U-MoC, S-MoC, T-MoC}\}$$
$$C = \{\mathtt{intSup, intSdown, intTup, intTdown,}$$
$$\mathtt{stripT2S, stripT2U, stripS2U,}$$
$$\mathtt{insertS2T, insertU2T, insertU2S}\}$$
$$O = \{\|, \circ, \mathbf{FB_P}\}$$

In other words, a process or a process network belongs to the *integrated MoC domain* iff all its processes and process compositions are constructed either by one of the named MoCs, one of the process constructors, or by one of the composition operators. We call such processes *I-MoC processes*.

We use the feedback operator $\mathbf{FB_P}$, which is based on the prefix order of signals, to define the semantics of feedback loops. In the synchronous MoC we have used the $\mathbf{FB_S}$ operator based on the Scott order of values. As a result we have different semantics in the different subdomains. This is not a problem, however, because the process network in a subdomain is viewed from the outside as just another process. The interior of this process is of no concern to the outside as long as it complies at the interfaces with the expectations of the surrounding processes. Although it is not formally required in Definition 6.8, we have to constrain models such that proper interfaces are always used for connecting subdomains, as is the case in Figure 6-3. Thus, if a S-MoC domain is surrounded by proper interface processes, its internal semantics does not cause inconsistencies with other subdomains.

FIGURE 6-3

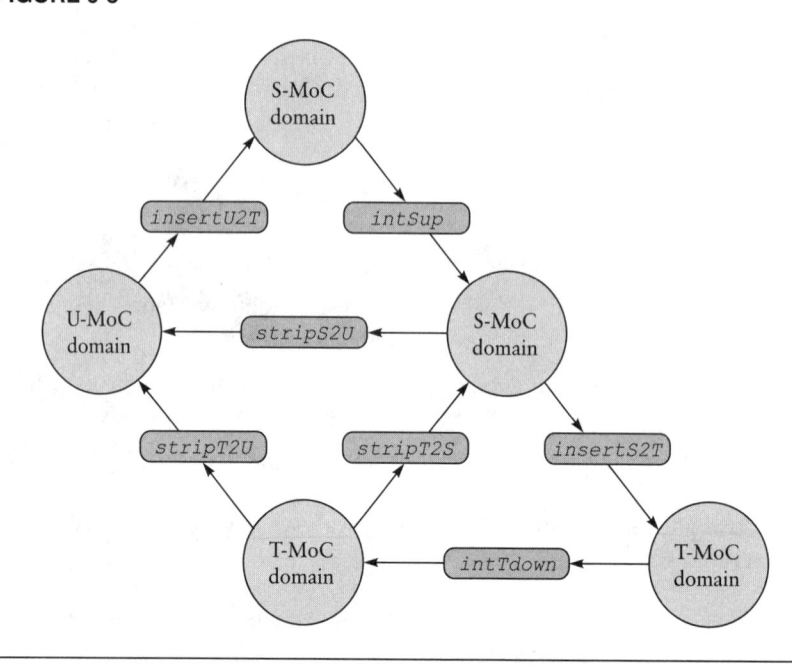

A system model with multiple subdomains and proper interfaces.

Example 6.1 As an example with two MoC domains, we consider a digital equalizer system (Bjuréus and Jantsch 2000), illustrated in Figure 6-4. We only discuss the highest hierarchical level, consisting of four processes. The Filter process reads the primary input, an audio signal, in chunks of 4096 data points. The Filter contains high-pass, band-pass, and low-pass filters and amplifiers. The amplifiers are controlled by the Button control process. Button control receives control inputs from a user interface, which sets the amplification for volume, bass, and treble of the output audio signal. The output of Filter is analyzed by the Analyzer process with the goal of detecting harmful output signals to avoid damaging the loud speakers. The Distortion control process uses the result of the Analyzer and provides input to the Button control, which in turn sets the parameters for the Filter.

Figure 6-4 shows the process graph. The inputs and outputs of each process are annotated by the number of events consumed and emitted in each evaluation cycle. The interfaces between the two domains are simple. They establish a one-to-one correspondence between the untimed and the synchronous events on their inputs and outputs. The interface from S-MoC to the U-MoC domain uses a process I in addition to $stripS2U$, defined as follows:

$$I = mapS(f)$$

6.3 Integrated Model of Computation 251

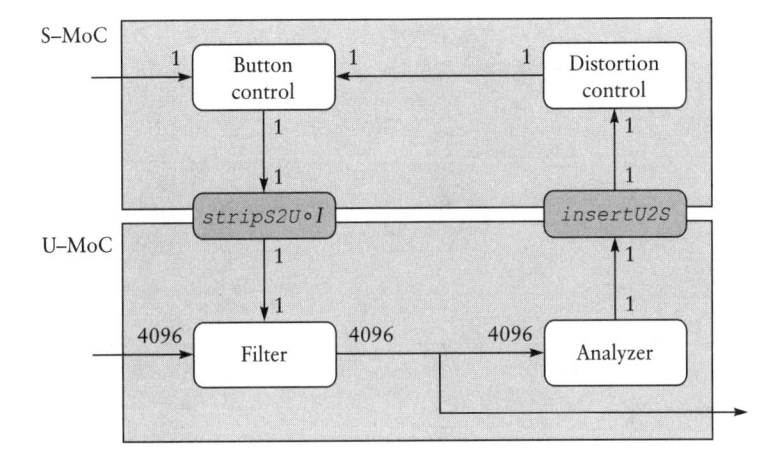

FIGURE 6-4

The digital equalizer system consists of a control part, modeled in S-MoC, and a dataflow part, modeled in U-MoC.

with

$$f(\bar{e}) = \begin{cases} 0 & \text{if } \bar{e} = \sqcup \\ \bar{e} & \text{otherwise} \end{cases}$$

I replaces absent events with 0, meaning that the control values in the Filter should not change.

This digital equalizer is quite typical: it has a datapath with a regular dataflow and a control part with relatively few control events occurring at irregular points in time. Input to the Distortion control arrives periodically triggered by the activity in the datapath. But its output and the inputs and the output of the Button control process will occur very rarely and irregularly depending on user inputs and on the need to change Filter parameters.

As this is a very typical situation, it is worth while to carefully examine the choices we have to link the two domains. First, we observe that, by connecting the two domains, the untimed processes take on a notion of time. Suddenly, the activities of the untimed processes can be related to the time of the synchronous domain. But could the interfaces, which are very simple in our example, be designed differently to avoid this coupling? The answer is no for the modeling devices we have introduced so far. Since the overall equalizer system is deterministic, we have to decide and describe precisely how the output of the Analyzer is merged into the time fabric of the synchronous domain. Similarly, the activity of the Filter process is pressed into the time fabric originating from the Button control process. This effect propagates to all untimed processes, even those not directly connected to the S-MoC domain.

FIGURE 6-5

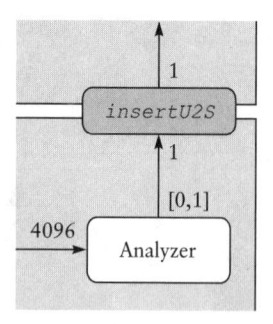

The Analyzer *emits data only when there is a change in the analysis result.*

Is this coupling effect the result of all the untimed processes imposing constant partitionings on their input and output signals? Consider the case where the Analyzer emits data only when the current analysis result is different from the previous one (Figure 6-5). The consequence would be that the entire time structure of the S-MoC domain would be aligned with the output of the Analyzer. This would be an unpleasant situation because it would make the operation of the Button control and of the Filter processes dependent on Analyzer output. Recall that the timing structure in the S-MoC is determined by the primary inputs. Consequently, all the inputs to a synchronous domain must have a consistent timing structure, which is not the case if the timing structure on one input is completely dependent on the datapath activity, and the other input, coming from a user interface, is rooted in the time structure of the real world.

Hence, when untimed processes have a variable and dynamic partitioning of their signals, the timing structure can radically change, but the coupling effect to the synchronous domain is not decreased. The conclusion is that we must interpret the activities of the untimed processes with respect to a time structure if we want to connect them to a synchronous or timed MoC domain.

This consequence may contradict the ambition of keeping considerations of time out of the untimed domain and restricting time effects to the synchronous or timed domain. Perhaps at this point you are thinking that an asynchronous interface, which reads data from the untimed domain whenever it arrives, is called for. However, an asynchronous interface can only connect two domains that both have a time structure, albeit two different and uncorrelated time structures. If one domain has no time structure whatsoever, an asynchronous interface does not help. Its introduction would only link the two domains and establish correlated, but not necessarily synchronized, time structures in both domains.

We can make the coupling more flexible by adding nondeterminism into the interface. Figure 6-6 indicates this possibility.[1] Process *R* inserts a nondeterministic number of absent events between every two events from the Analyzer. Similarly, we can

---

[1] Our modeling concepts do not allow us to express nondeterminism. However, we will introduce stochastic processes in Chapter 8, which can be used to the same effect.

FIGURE 6-6

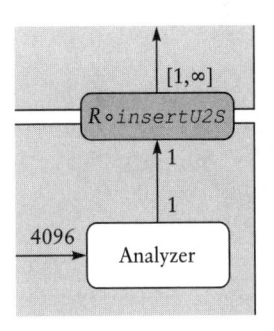

The interface process R inserts a random number of absent events between two events from the Analyzer.

add a nondeterministic element to I, the other domain interface process in Figure 6-4. I could emit a nondeterministic number of zeroes for each consumed absent event. Doing so would nondeterministically vary the delay between the control part and the datapath. It would not avoid, however, imposing a timing structure on the untimed domain.

Assume R adds 100 absent events between two particular data events, e_1 and e_2, from the Analyzer. The effect of e_2 would appear at the input of the Filter 100 S-MoC evaluation cycles after the effect of e_1. However, in the meantime Filter would have received 100 zero-events, one for each absent event, from the controller. It would have continued its operation, that is, evaluating 100 chunks of the input signal. Thus, the obvious interpretation is that there is a clear and unambiguous time relation between the untimed and the synchronous domains. By adding delays in the interfaces nondeterministically, we can model nondeterministic delay, but we would still import a timing structure from the synchronous into the untimed domain.

In summary, we can conclude that by connecting two MoC domains, we relate their time structure to each other. Untimed MoC domains import a time structure from synchronous or timed domains. It is crucial that the correlation between the involved time structures, which is determined by the interfaces, is in fact the one desired by the designer. Moreover, it is imperative that all the interfaces are consistent with each other. We do not yet have a systematic method to support the designer in achieving all this. But techniques and tools exist that address specific instances of this general problem. For example, Jantsch and Bjuréus (2000) have developed a modeling environment and a methodology to integrate the timed MoC defined by SDL (Olsen et al. 1995) with the untimed model of the Matlab (Hanselman and Littkefield, 1998) language.

6.4 Asynchronous Interfaces

As elaborated above, when two MoC domains are connected, their time structures are set in a particular relationship to each other. How shall we model the situation where we do not make assumptions about this relationship, but we still want to design a reliable communication? We propose the following three-step procedure:

1. *Add a time interface:* To connect two MoC domains, whether of the same or of different types, we have to use an interface that defines the relationship of the time structure. Examples are the interface constructors `intSup`, `stripT2U`, `insertU2S`, and so on, introduced earlier. They usually establish a constant and fixed relationship. This is not a necessity, however. Consider the following interface constructor:

$$intSups(r,s,f',f'') = mealyU(1,f,g,0)$$

with

$$f(\text{state}, \bar{e}) = \begin{cases} f'(\bar{e}) & \text{if state} < s \\ f''(\bar{e}) & \text{if state} = s \end{cases}$$

$$\text{length}(f'(\bar{e})) = r$$

$$\text{length}(f''(\bar{e})) = r+1$$

$$g(\text{state}, \bar{e}) = (\text{state} \mod s) + 1$$

$$\bar{e} \in \bar{E}, r, s \in \mathbb{N}, \text{state} \in \mathbb{N}_0$$

`intSups` is just like `intSup` except that every s cycles it emits $r+1$ events rather than r. For these two different situations, the two functions f' and f'' are used. This can capture the case where the frequencies in the two MoC domains are not a simple multiple of each other. s could even be modeled as a stochastic variable to account for fluctuations in the relationship of the two time structures. Hence, there is considerable flexibility in defining the time relationship between the two domains, but nevertheless it should be defined precisely.

2. *Refine the protocol:* Model the asynchronous protocol that will provide the reliable communication between the two MoC domains.

3. *Model the channel delay:* If the channel is subject to delay variations, use a stochastic process to model the channel delay. Stochastic processes will be introduced in Chapter 8.

Example 6.2 To illustrate this procedure, consider the communication between processes P and Q in two different timed MoC domains, MoC A and MoC B, respectively (Figure 6-7). Both P and Q are simple identity processes, passing their inputs unchanged to their outputs. The time structure relationship, as defined by the interface I_1, is such that two evaluation cycles in MoC A correspond to three cycles in MoC B. f_1 and f_2 are defined as

$$f_1(\hat{e}) = \langle \hat{e}, \sqcup, \sqcup \rangle$$
$$f_2(\langle \sqcup, \sqcup \rangle) = \sqcup$$
$$f_2(\langle \sqcup, \hat{e} \rangle) = \hat{e}$$

6.4 Asynchronous Interfaces 255

FIGURE 6-7

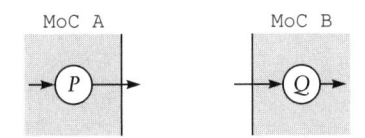

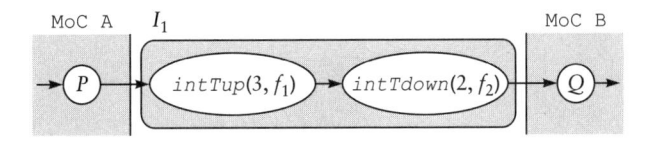

Processes P and Q in two different timed domains are connected via the interface $I_1 = intTdown(2, f_2) \circ intTup(3, f_1)$.

FIGURE 6-8

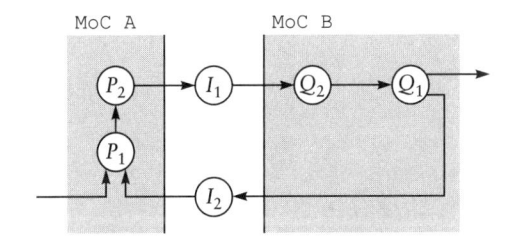

Processes P and Q are refined to model a handshake behavior.

$$f_2(\langle \hat{e}, \sqcup \rangle) = \hat{e}$$

$$f_2(\langle \hat{e}_1, \hat{e}_2 \rangle) = \hat{e}_1$$

In the second step of the procedure we model a protocol that is robust to changes in the time structure relationship. Figure 6-8 shows how P is refined into two processes, P_1 and P_2, and Q is refined into Q_1 and Q_2. We can reuse the interface I_1 for the communication from MoC A to MoC B. For the opposite direction we have to design a corresponding interface process I_2:

$$I_2 = intTdown(3, f_4) \circ intTup(2, f_3)$$

with

$$f_3(\hat{e}) = \langle \hat{e}, \sqcup \rangle$$

$$f_4(\langle \sqcup, \sqcup, \sqcup \rangle) = \sqcup$$

FIGURE 6-9

$P_1 = \texttt{zipUs}(1,1)$

$P_2 = \texttt{mealyT}(1, f, g, \texttt{idle})$

with

$$
\begin{aligned}
f(\texttt{idle}, (\hat{e}, _)) &= (\hat{e}, \texttt{valid}) \\
f(\texttt{idle}, (\sqcup, _)) &= \sqcup \\
f(\texttt{waiting}, (\hat{e}, \sqcup)) &= \sqcup \\
f(\texttt{waiting}, (\sqcup, \texttt{ack})) &= \sqcup \\
f(\texttt{waiting}, (\hat{e}, \texttt{ack})) &= (\hat{e}, \texttt{valid}) \\
g(\texttt{idle}, (\hat{e}, _)) &= \texttt{waiting} \\
g(\texttt{idle}, (\sqcup, _)) &= \texttt{idle} \\
g(\texttt{waiting}, (\hat{e}, \sqcup)) &= \texttt{waiting} \\
g(\texttt{waiting}, (\sqcup, \texttt{ack})) &= \texttt{idle} \\
g(\texttt{waiting}, (\hat{e}, \texttt{ack})) &= \texttt{waiting}
\end{aligned}
$$

$Q_1 = \texttt{unzipT}()$

$Q_2 = \texttt{mapT}(1, f)$

with

$$
\begin{aligned}
f(\sqcup) &= \sqcup \\
f((_, \sqcup)) &= \sqcup \\
f((\hat{e}, \texttt{valid})) &= (\hat{e}, \texttt{ack})
\end{aligned}
$$

P_1 simply zips the data input and the control signal from Q_1 together. P_2 waits for an acknowledgment before sending new data to Q_2.

$$
\begin{aligned}
f_4(\langle \hat{e}, _, _ \rangle) &= \hat{e} \\
f_4(\langle \sqcup, \hat{e}, _ \rangle) &= \hat{e} \\
f_4(\langle \sqcup, \sqcup, \hat{e} \rangle) &= \hat{e}
\end{aligned}
$$

The "_" symbol indicates a wild card matching any value or event including the absent event.

I_2 implements a time structure relationship consistent with I_1, but care has to be taken because events may get lost in I_2, while events can never be lost in I_1.

Process P and Q are refined into processes P_1, P_2, Q_1, and Q_2 as defined in Figures 6-8 and 6-9. The main idea is as follows: P_1 packs the primary input data and a control signal from Q_1 into an event that is then processed by P_2. The only possible value of the control event from Q_1 is an acknowledgment `ack`. After P_2 has sent a valid data to Q_2, it awaits an `ack` event before sending new data. P_2 does not buffer any data. Thus, if new data arrives while P_2 is waiting for an acknowledgment, it is silently dropped. A buffer of any size could, of course, be added to avoid loss of data, but we have suppressed it to keep the model simple. When P_2 sends valid data, it also sends a control event, `valid`.

Q_1 is a simple unzip process separating the primary data output signal from the acknowledgment signal for the handshake procedure. Q_2 is a stateless process that emits an acknowledgment event for every valid data event it receives.

FIGURE 6-10

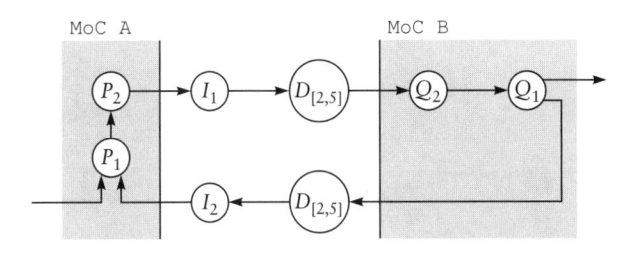

Stochastic processes can be used to model channel delay.

The handshake protocol, as simple as it may be, illustrates step 2, the refinement of an abstract channel into a concrete protocol implementation.

Step 3 introduces a delay model of the channel between the two domains. In Figure 6-10 two processes, labeled $D_{[2,5]}$, are inserted into the interface between the two MoC domains. They are stochastic processes and delay events by two to five cycles. Since we will introduce stochastic processes only in Chapter 8, we will not provide further details here. A deterministic delay could be easily modeled with `initT`-based processes.

If you have been observant, you will have certainly noted that the handshake protocol does not work correctly because the interface I_2 drops some events since it receives more events than it emits. See Exercise 6.1 to rectify this error.

In summary, we advocate separating channel delay modeling, or delay modeling in general, from the relationship of time structures. Whenever separate MoC domains are connected, their time structure is related. It is worth while defining this relationship explicitly and carefully. Doing this, however, should not be confused with defining and modeling the delays between the MoC domains.

6.5 Process Migration

One of our objectives is to capture the different computational models in a uniform way, such that the characteristics in each domain are preserved but both optimizations and verification can be done across domain boundaries. A necessary requirement for this, in addition to well-defined interface processes, is that processes can be moved from one domain into another.

Figure 6-11 shows one example of a process moving from the untimed domain into the synchronous domain. Apparently, the process must change in order to preserve the overall system behavior.

Based on the three computational models and the interface processes we have defined, we briefly discuss all the possible migrations for processes with one input and one output signal. All these cases are listed in Table 6-2.

FIGURE 6-11

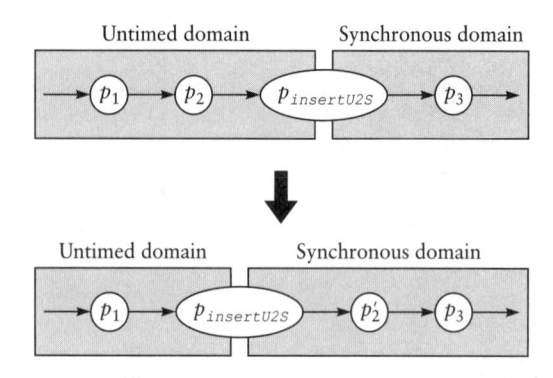

Migration of a process from the untimed into the synchronous domain.

TABLE 6-2: *Migration of processes between timing regimes.*

1a.	$P_{insertU2S} \circ P_U$	$\Rightarrow$	$P_S \circ P_{insertU2S}$
b.	$P_S \circ P_{insertU2S}$	$\Rightarrow$	$P_{insertU2S} \circ P_U$
2a.	$P_{insertU2T} \circ P_U$	$\Rightarrow$	$P_T \circ P_{insertU2T}$
b.	$P_T \circ P_{insertU2T}$	$\Rightarrow$	$P_{insertU2T} \circ P_U$
3a.	$P_{insertS2T} \circ P_S$	$\Rightarrow$	$P_T \circ P_{insertS2T}$
b.	$P_T \circ P_{insertS2T}$	$\Rightarrow$	$P_{insertS2T} \circ P_S$
4a.	$P_{stripT2S} \circ P_T$	$\Rightarrow$	$P_S \circ P_{stripT2S}$
b.	$P_T \circ P_{stripT2S}$	$\Rightarrow$	$P_{stripT2S} \circ P_T$
5a.	$P_{stripT2U} \circ P_T$	$\Rightarrow$	$P_U \circ P_{stripT2U}$
b.	$P_U \circ P_{stripT2U}$	$\Rightarrow$	$P_{stripT2U} \circ P_T$
6a.	$P_{stripS2U} \circ P_S$	$\Rightarrow$	$P_U \circ P_{stripS2U}$
b.	$P_U \circ P_{stripS2U}$	$\Rightarrow$	$P_{stripS2U} \circ P_S$

Before we discuss the individual cases, however, we need some supporting processes to describe the migrations concisely.

The process par_c is a synchronous process that takes c consecutive events and packs them into one event as a record or sequence:

$$par\_c = mealyS(f, g, \langle\rangle)$$

where

$$f(e,w) = \begin{cases} w \oplus e & \text{if length}(w) < c \\ \langle\rangle & \text{otherwise} \end{cases}$$

$$g(e,w) = \begin{cases} w \oplus e & \text{if length}(w) = c \\ \sqcup & \text{otherwise} \end{cases}$$

For par_c the number of events to be packed into a record is constant. But for par it is variable and given by a second input. Whenever the first event of a new record arrives at the first input signal, it is expected that the other input signal provides the number that defines the size of this record. par consists of two processes: the first zips the two inputs together, and the second performs the packing.

$$par = p_1 \circ p_2$$
$$p_1 = \text{zipS}()$$
$$p_1 = \text{mealyS}(f, g, (\langle\rangle, 0))$$

where

$$f((e,c),w,d)) = \begin{cases} (w \oplus e, d) & \text{if length}(w) < d \\ (\langle\rangle, c) & \text{otherwise} \end{cases}$$

$$g((e,c),(w,d)) = \begin{cases} w \oplus e & \text{if length}(w) = d \\ \sqcup & \text{otherwise} \end{cases}$$

The ser process is the inverse operation. It serializes sequences that come in a single event. Recall that tail returns all elements of a sequence except the first one and head returns only the first element of a sequence.

$$ser = \text{mooreS}(f, g, \langle\rangle)$$

where

$$f(e,w) = \begin{cases} \text{tail}(w) & \text{if } e = \sqcup \\ \text{tail}(w) \oplus e & \text{otherwise} \end{cases}$$

$$g(w) = \begin{cases} \text{head}(w) & \text{if } w \neq \langle\rangle \\ \sqcup & \text{otherwise} \end{cases}$$

In the following, we discuss 12 cases that should be considered as examples rather than a systematic methodology. As will be seen, the best solution on how to transform the migrated processes and the interfaces depends on the details of the processes and the concrete context. Therefore, a designer would like to have a rather large library of migration patterns available from which specific solutions can be selected and parameterized.

FIGURE 6-12

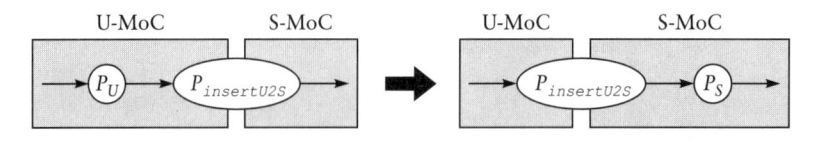

Downstream migration of a process from the untimed to the synchronous domain.

Case 1a: $(P_{insertU2S} \circ P_U \Rightarrow P_S \circ P_{insertU2S})$

Consider two processes connected together: $p_I \circ p$ (Figure 6-12). First, we discuss the migration of a stateless process. Let p be a `mapU`-based untimed process and p_I be an interface process from the untimed to the synchronous domain.

$$p = mapU(c, f_1)$$
$$p_I = insertU2S(1)$$

Then the two processes can be transformed into $p_I \circ q$, with q being defined as follows:

$$q = q_3 \circ q_2 \circ q_1$$
$$q_1 = par\_c$$
$$q_3 = ser$$
$$q_2 = mapS(f_2)$$

$$f_2(e) = \begin{cases} \sqcup & \text{if } e = \sqcup \\ \sqcup & \text{if } f_1(e) = \langle \rangle \\ f_1(e) & \text{otherwise} \end{cases}$$

Because the original process p takes c events during each execution cycle, but a synchronous process can only take one, we have to pack c event values into one record event (by q_1) before we can apply the original function f_1 on these records (done by q_2 and f_2). Similarly, because the untimed process p can generate any number of output events, we have to serialize the resulting value records into individual synchronous events, which is done by q_3.

If p is a process with state, the migration is more involved. Let $p = mealyU(\gamma, f, g, w_0)$. Process p and p_I are, as above, connected together, $p_I \circ p$.

$$p = mealyU(\gamma, f, g, w_0)$$
$$p_I = insertU2S(1)$$

Then, they can be transformed into processes $q \circ p_I$, where q is a process consisting of five subprocesses as depicted in Figure 6-13.

Process q_1 parallelizes the incoming events. q_2 represents the scan part of the process p; thus it contains the internal state and updates it. q_3 is the output encoder.

6.5 Process Migration

FIGURE 6-13

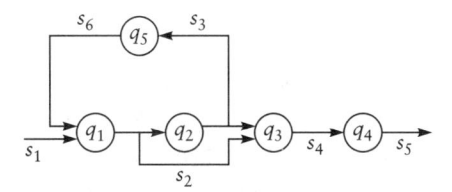

The process resulting from the migration of a Mealy-based process from the untimed into the synchronous domain.

Next-state processing and output encoding had to be separated because we need the information concerning how many events should be assembled into the next record by q_1. To infer this we need access to the internal state, which is now visible at signal s_3. Process q_5 infers this information and provides q_1 with the record sizes. Finally, q_4 serializes the packed events.

More formally, the process q is defined as follows:

$$q(s_1) = s_5$$

where

$$s_5 = q_4(s_4)$$
$$s_4 = q_3(s_3, s_2)$$
$$s_3 = q_2(s_2)$$
$$s_2 = q_1(s_1, s_6)$$
$$s_6 = q_5(s_3)$$
$$q_1 = \text{par}$$
$$q_2 = \text{scanS}(f_2, w_0)$$

where

$$f_2(e, w) = \begin{cases} \sqcup & \text{if } e = \sqcup \\ f(e, w) & \text{otherwise} \end{cases}$$

$$q_3 = \text{mapS}(g_3)$$

where

$$g_3(e, w) = \begin{cases} \sqcup & \text{if } e = \sqcup \\ g(e, w) & \text{otherwise} \end{cases}$$

$$q_4 = \text{ser}$$
$$q_5 = \text{mapS}(\gamma_5)$$

where

$$\gamma_S(e) = \begin{cases} \sqcup & \text{if } e = \sqcup \\ \gamma(e) & \text{otherwise} \end{cases}$$

Case 1b: $(P_S \circ P_{insertU2S} \Rightarrow P_{insertU2S} \circ P_U)$

In general it is difficult to fully emulate a synchronous process in the untimed domain because its behavior may depend on timing information that is not available for untimed processes. It is much simpler to move a process the opposite way from the synchronous into the untimed domain. As elaborated below for case 2b, we have two options to deal with this problem. However, for the special case $\lambda = 1$, the following transformation fully preserves the behavior.

Consider two processes connected together:

$$p \circ p_I$$

with

$$p_I = insertU2S(1)$$
$$p = mealyS(f, g, w_0)$$

We can easily transform them into

$$p_I \circ q$$

with

$$q = mealyU(1, f, g, w_0)$$

Case 2a: $(P_{insertU2T} \circ P_U \Rightarrow P_T \circ P_{insertU2T})$

Migrating a process from the untimed into the timed domain is easier than moving it into the synchronous domain because both the timed and untimed processes can consume an arbitrary number of events in each firing cycle. Again we only consider the special case where the interface process provides a time ratio of $\lambda = 1$.

Two processes, $p = mealyU(\gamma, g, f, w_0)$ and $p_I = insertU2T(1)$, which are connected together ($p_I \circ p$), can be transformed into two other processes $q \circ p_I$ if q is defined as $q = mealyT(\gamma, g, f, w_0)$. If λ is not 1, we could still have the new interface process with $\lambda = 1$ and process q generate absent events in the same way as the interface process does. This exactly preserves the behavior and the timing. In many practical cases the preservation of the timing may not be required, allowing transformations that result in simpler processes.

Case 2b: $(P_T \circ P_{insertU2T} \Rightarrow P_{insertU2T} \circ P_U)$

In contrast to case 2a this transformation is a difficult issue. The reason is that we simply cannot emulate the full behavior of a timed process by an untimed process. The timed process's behavior may depend on the timing of the input events, and the timing of the output events may be an essential part of the behavior. There are two approaches to address this situation. First, we could encode and preserve the timing information in the untimed domain, for example, by representing absent events with a special data value that is not used otherwise or by annotating each event with a time tag. This would allow us to fully emulate the timed process's behavior in the untimed domain at the expense of increased complexity of the untimed process. Second, we could ignore the timing information and only emulate the timed process's behavior as far as it depends on the values and sequence of events but not on their timing. This will only be possible if these two aspects are already separated to some degree in the timed domain.

However, we will abstain from developing these ideas and leave as future work the development of a set of practically useful transformations that allow designers to choose the most appropriate approach for a given problem.

Case 3a: $(P_{insertS2T} \circ P_S \Rightarrow P_T \circ P_{insertS2T})$

Two processes, $p = mealyS(g,f,w_0)$ and $p_I = insertS2T(\lambda)$, which are connected together ($p_I \circ p$), can be transformed into two other processes $q \circ p'_I$ if q and p'_I are defined as follows:

$$p'_I = insertS2T(1)$$
$$q = mealyT(1,g,f',w_0)$$

where

$$f'(w,\hat{e}) = \langle f(w,\hat{e}) \rangle \oplus \langle \sqcup \rangle^{\lambda-1}$$

We have simply integrated both p and the p_I into the new process q while simplifying the new interface process.

Case 3b: $(P_T \circ P_{insertS2T} \Rightarrow P_{insertS2T} \circ P_S)$

Similarly to case 1a, we have to deal with the situation where the timed process can consume and produce any number of events, while the constructed synchronous process can consume and produce only one event at a time. Hence, we must parallelize events before the synchronous process and serialize them after the synchronous process. Again, we consider only the special case with a time ratio $\lambda = 1$.

Let $p = mealyT(\gamma,f,g,w_0)$ and $p_I = insertS2T(1)$ be connected together, $p \circ p_I$. They can be transformed into processes $p_I \circ q$, where q is a process consisting of five subprocesses as depicted in Figure 6-13 and derived in exactly the same way as in case 1a.

Case 4a: $(P_{stripT2S} \circ P_T \Rightarrow P_S \circ P_{stripT2S})$

We again have a similar situation as in cases 1a and 3b where we had to parallelize the input in the synchronous domain to emulate the behavior of the timed and untimed processes. The same scheme as we employed in the earlier cases can be used here for the situation of a time ratio of $\lambda = 1$.

Case 4b: $(P_S \circ P_{stripT2S} \Rightarrow P_{stripT2S} \circ P_T)$

The $stripT2S$ process filters out all events except the last in a clock cycle as determined by λ. When we move a synchronous process upwards against the data stream across a $stripT2S$ process and transform it into a timed process, the new process will potentially see many more events. However, since the interface process filters events based only on their timing, we can make the data processing also in the timed domain provided that the timed process does not rely on detailed timing information. Hence, this transformation is fairly simple but may lead to the unnecessary processing of many events in the timed domain.

Given are two processes, $p = \mathtt{mealyS}(g, f, w_0)$ and $p_I = \mathtt{stripT2S}(\lambda)$, which are connected together, $p \circ p_I$. These two processes can be transformed into two other processes $p'_I \circ q$, defined as follows:

$$q = \mathtt{mealyT}(\lambda, g', f', w_0)$$

where

$$g'(w, \hat{a}) = g(w, \mathsf{lastt}(\hat{a}))$$
$$f'(w, \hat{a}) = f(w, \mathsf{lastt}(\hat{a}))$$
$$p'_I = \mathtt{stripT2S}(1)$$

Case 5a: $(P_{stripT2U} \circ P_T \Rightarrow P_U \circ P_{stripT2U})$

Again, as in case 1b and 2b, this transformation is in general not possible because the necessary information (i.e., the time information) is not available to the untimed process.

Case 5b: $(P_U \circ P_{stripT2U} \Rightarrow P_{stripT2U} \circ P_T)$

Since the untimed process operates only on the nonabsent events, we can emulate its behavior in the timed domain by ignoring all absent events. Conveniently, we filter absent events in a separate process that resembles the interface process.

Let $p = \mathtt{mealyU}(\gamma, g, f, w_0)$ and $p_I = \mathtt{stripT2U}()$ be connected together, $p \circ p_I$. They can be transformed into processes $p_I \circ q$ with

$$q = q_1 \circ q_2$$
$$q_2 = \mathtt{mapT}(1, f)$$

where

$$f(\hat{e}) = \begin{cases} \langle\rangle & \text{if } \hat{e} = \sqcup \\ \langle \hat{e} \rangle & \text{otherwise} \end{cases}$$

$$q_1 = \mathtt{mealyT}(\gamma, g, f, w_0)$$

Case 6a: $(P_{stripS2U} \circ P_S \Rightarrow P_U \circ P_{stripS2U})$

As in cases 1b, 2b, and 5b, the full behavior of a synchronous process is difficult to emulate in the untimed domain due to a lack of timing information.

Case 6b: $(P_U \circ P_{stripS2U} \Rightarrow P_{stripS2U} \circ P_S)$

We can completely emulate the behavior of the untimed process in the synchronous domain, but as in case 1a we have to properly parallelize the incoming events first and serialize them after processing. We can use the same scheme as in case 1a, but in addition we have to filter out all absent events before the parallelization.

The process migration techniques described here are merely examples and are not developed in a systematic way for two reasons. First, the details of such a transformation often depend on the application and on the context. To make a process migration useful and feasible, context- and application-specific information has to be used, which may not always be explicit in the model at all. Second, we just do not have the understanding and practical experience with inter-domain process migration. The systematic development of a set of techniques and tools to support process migration and cross-domain formal verification of functionality and performance figures is still an area open for research.

6.6 Applications

Traditionally, the term "model of computation" has been used to intuitively describe different concepts of communication and synchronization between concurrent processes. Occasionally, interfaces between different models of computation have been defined to allow the joint simulation of processes in different MoC domains. Ptolemy is the most prominent example of a framework that integrates a number of different models of computation for the purpose of simulation and modeling.

Our formal framework attempts to study the integration of different MoCs on a more fundamental level—to go beyond simulation and eventually address integrated synthesis, optimization, and verification across various MoCs. Since this is a long-term research program, we only want to sketch some possible concrete applications.

First, the framework is a means to study the relationships between different MoCs. Technically we have introduced the framework as a dataflow process network without timing information. Other MoCs have been embedded in this framework by introducing a particular data type, the absent event, which is interpreted by processes as timing information. Hence, all MoCs are formulated in the same semantic framework and can therefore be easily studied together. For instance, we have formulated the dataflow process network with firing rules (Lee 1997) in our framework. It can then be shown that the U-MoC contains a strictly larger set of processes because the partitioning function for the input signal need not be bounded in the number of events consumed in an evaluation cycle while the firing rules require bounded input partitionings.

The composite signal flow model (Jantsch and Bjuréus 2000) is another example of the integration of MoCs. It allows us to model both untimed and timed processes and their interaction with the same formalism. It has been successfully applied to integrating the design languages Matlab and SDL into a common modeling environment (Bjuréus and Jantsch 2001). The same can be achieved in our framework since both formalisms are very similar. Again, the composite signal flow can be formulated in our framework as a particular MoC, and its expressive power can be compared with the timed MoC. The composite signal flow is also a good example of how a formal foundation for a computational model can be used to facilitate the practical task of integration of different design tasks and languages that have no formal relationship to each other per se. As illustrated in Figure 6-14, each of the design languages has to be mapped onto a particular MoC. Then interface processes between the resulting MoCs have to be designed, and the communication between the design languages has to be implemented in compliance to the semantics of the interface processes. The implementation of the communication is typically glue code using the foreign language interfaces of the involved simulators. In the SDL-Matlab case, both the Matlab execution

FIGURE 6-14

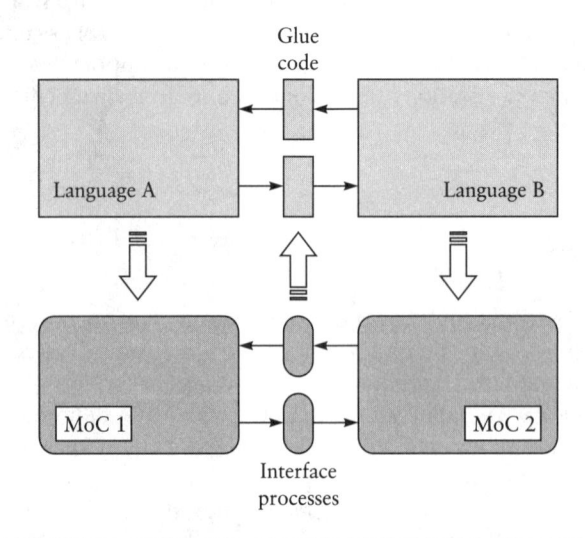

Integration of two design languages by means of an MoC framework.

environment and the SDL simulator allow the inclusion of arbitrary C code, which is used to establish a connection between the two environments and implement the communication.

The mapping of a design language onto an MoC may not be complete either because some language features are not relevant for the task at hand or because the language contains features that cannot easily be expressed in an MoC framework. This may lead to a modeling technique restricting the use of the design languages. For example, Matlab allows functions to communicate arbitrarily via shared variables. The Mascot modeling technique (Bjuréus and Jantsch 2001) prohibits this for Matlab functions residing in different processes because the interprocess communication is governed by the composite signal flow. Similarly, the interface processes connecting the two different MoCs may not allow the expression of all communication mechanisms available in the design languages. Again, this leads to a restriction of expressiveness that can be justified by a simpler implementation or by stronger formal properties.

Beyond connecting different languages and MoCs on the modeling and simulation plane, the proposed framework has the potential to perform synthesis and formal verification across MoC domain boundaries. By embedding different MoCs in the same formal framework, well-defined interfaces can be designed that are the basis for moving functionality from one domain to another. In this way the system can be optimized globally across all present MoC domains. Moreover, formal reasoning and verification techniques can be used to prove or disprove the equivalence of process networks in different MoC domains. Not much research has been done on these issues, but some first steps have been taken toward developing design transformations.

The *ForSyDe* methodology (Sander and Jantsch 1999, 2002) uses the S-MoC for specifying the system functionality. Design transformations gradually add more details

FIGURE 6-15

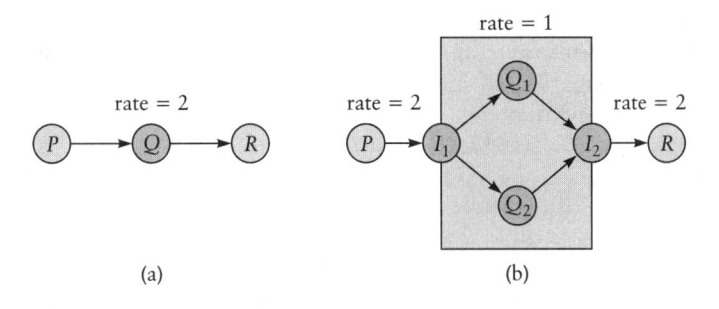

Transformation of (a) process Q into (b) two processes Q_1 and Q_2 with half the execution speed each.

and design decisions until the resulting model can be directly mapped to an efficient implementation expressed in VHDL or C. The transformations, which include process merging and splitting, pipelining, and resource sharing, influence the cost and performance of the implementation but affect the functionality either not at all or in only a well-defined way based on different definitions of equivalence. For instance, pipelining incurs an initial delay and a latency but does not change the functionality otherwise. A particularly interesting transformation is the introduction of a different clock domain. Figure 6-15(a) shows three processes operating all with the same relative rate; that is, all processes consume and produce the same number of events in every evaluation cycle. In Figure 6-15(b) process Q has been transformed into two parallel processes Q_1 and Q_2, which operate at half the rate; that is, they need to execute half as fast as process Q. The interface processes I_1 and I_2 adapt the different rates and they split and merge the data stream. In this way one fast and expensive (or infeasible) processing unit could be replaced with two slower and cheaper units.

The MoC used in *ForSyDe* with different clock domain rates is a generalization of the S-MoC, but it is more restricted than the T-MoC in that a process cannot consume and emit a varying number of events in different evaluation cycles.

The MoC framework can prove useful even if the proposed formalism and notation is not directly visible to the designer. It can be used to define appropriate MoCs that are than realized and embedded in a design language such as VHDL or SystemC. As observed earlier, this would lead to a restriction of expressiveness, with the benefit that analysis, synthesis, and verification tools could exploit the stronger properties to perform their objectives more efficiently with better results. By choosing the MoCs properly the restrictions can be negligible or acceptable while the gains can be extraordinary.

6.7 Further Reading

The Ptolemy project (Davis et al. 2001) studies systematically different MoCs and their interaction. A modeling and simulation framework has been developed in which a number of important MoCs have been implemented and integrated. The Ptolemy project

Web page (*ptolemy.eecs.berkeley.edu*) is a rich source of documentation, reports, and publications about the project and many facets of the research on MoCs and their integration.

The composite signal flow (Jantsch and Bjuréus 2000; Bjuréus and Jantsch 2001) maps the languages Matlab and SDL on a separate MoC to integrate them in a simulation and modeling environment.

Other recent work (Sander and Jantsch 2002; Sander et al. 2003) develops transformations of a synchronous MoC domain into subdomains with a different time structure.

A very similar intention is pursued by Burch et al. (2001a; 2001b). Their approach is based on *trace algebras*, where traces correspond to our input and output signals of processes. Processes and their behavior are solely defined by traces, which is in contrast to our framework that defines processes by describing how processes consume, process, and emit events. However, they also define three different abstraction levels that are also distinguished by their representation of time. Their *metric time* corresponds loosely to our timed MoC, but also includes continuous time models. *Non-metric time* imposes only a partial order on events and thus corresponds to our untimed MoC. *Pre-post time* has no direct correspondence to one of our discussed MoCs and represents only initial and final states of behaviors but abstracts away intermediate states. It is intended to model noninteractive, terminating systems.

6.8 Exercises

6.1 The handshake protocol as defined in Figures 6-8 and 6-9 does not work properly because the interface I_2 drops one out of three events. Thus an `ack` event could also be dropped and P_2 would never be notified that the data has been successfully received by Q. Then P would never send any more data, resulting in a deadlock. Design an improved protocol based on the observation that out of three consecutive events process I_2 drops exactly one.

6.2 Improve the asynchronous protocol of Exercise 6.1 further by implementing a buffer in process P_2 such that a maximum of five input events can be buffered while waiting for an acknowledgment from Q.

chapter seven
Tightly Coupled Process Networks

Models of computation are not restricted to those we have discussed so far. As a matter of fact, practically important MoCs fall outside the frame we have developed. Our framework enforces three properties in all its concrete MoCs: blocking read, decoupled sender/receiver, and determinism. In this chapter we discuss more informally MoCs that violate these properties.

In particular we investigate the effects of a nonblocking read semantics for a message receiver (Section 7.1). Together with nondeterministic communication delays it leads to nondeterministic behavior.

A blocking read/blocking write semantics of the communication (Section 7.2) leads to a very tight coupling between the processes in a process network because the delay or deadlock of one process can lead to delay and deadlock of processes that do not need data from the blocked process. Thus, this particular communication semantics incurs control dependences in addition to the data dependences among processes.

This phenomenon is generalized and elaborated on under the term oversynchronization (Section 7.3) to understand its cause and effects. It turns out that it can also be viewed as a resource-sharing problem. Oversynchronization occurs when processes have to wait for the availability of communication buffers that are the shared resources.

In Chapters 3 through 5 we have dealt with three different computational models under a uniform framework. The three models are distinguished by the way they represent time. The unifying framework enforces three important properties:

- *Blocking read:* The reader of a channel cannot probe the channel for the presence of data, but it always waits until an event has arrived. In the untimed MoC an arbitrary, possibly infinite, amount of time may elapse until data arrives and the reading process resumes its activity. In the synchronous and the timed MoCs the special events ⊔ make sure that events are regularly received. Thus, the blocking read notwithstanding, processes will never block arbitrarily long.

- *Decoupled sender/receiver:* The consumer and producer processes of a channel are decoupled from each other as far as their causal relationship allows. There is no unnecessary inter-dependence, for instance, due to resource constraints. The communication channels are in principle unbounded FIFO buffers. The writing of a data item to a channel is entirely decoupled from the reading of that data from the channel. In the untimed and the timed models, this may have the consequence that channel buffers may grow beyond any limit, and the channel reader and writer are only synchronized when the buffer becomes empty and the reader expects more data. In the synchronous MoC every process emits and consumes exactly one event in each and every evaluation cycle. Consequently, all channel buffers can be limited to size 1. Sender and receiver are still decoupled in the sense that the sender always can write to the output channel and the receiver can always read from its input channel and no blocking can ever occur.

- *Determinism:* The process network is determinate if all processes are determinate; that is, the connecting network does not introduce nondeterminism.

These three properties are not independent. A blocking read is an essential requirement for a determinate process network, but nondeterminism can also be introduced by other means. Bounded channel capacity can, depending on the channel policy, either introduce nondeterminism or have unexpected coupling effects between seemingly unrelated parts of the process network. If the channel drops data items or the writer can sense the fullness of the channel, the process network is nondeterminate. If the writer blocks upon fullness of the channel, fewer output data may be computed as compared to the unbounded case.

In this chapter we briefly review computational models that violate some of these properties. Violation of the blocking-read property leads in Section 7.1 also to nondeterminate behavior in the presence of nondeterministic timing. The violation of the decoupled-send-receiver property in Section 7.2 leads to a much tighter coupling between the processes. The effects of this can be far-reaching, as further elaborated on in Section 7.3. In the next chapter (Chapter 8) we investigate thoroughly the concept of nondeterminism.

The blocking or nonblocking of read and write operations is the feature we use in this chapter to distinguish between computational models. A blocking read and nonblocking write leads to the models described in previous chapters (Chapters 3 through 5). The two cases of nonblocking read/nonblocking write and blocking read/blocking write are discussed in the following three sections. The case of nonblocking read/blocking write is a mixture of the other models and is not described separately.

7.1 Nonblocking Read

There are popular languages that use a model similar to our untimed model of computation but with the difference that a process can check for the presence of data at an

FIGURE 7-1

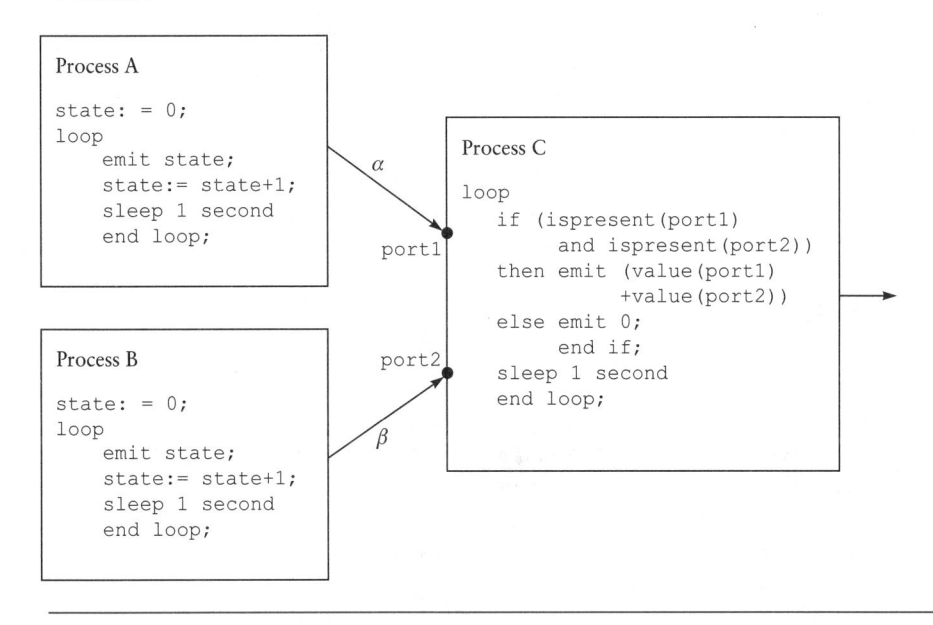

```
Process A

state: = 0;
loop
    emit state;
    state:= state+1;
    sleep 1 second
end loop;
```

```
Process C

loop
    if (ispresent(port1)
        and ispresent(port2))
    then emit (value(port1)
                +value(port2))
    else emit 0;
        end if;
    sleep 1 second
end loop;
```

```
Process B

state: = 0;
loop
    emit state;
    state:= state+1;
    sleep 1 second
end loop;
```

Channels: α from Process A to port1, β from Process B to port2.

Communicating processes.

input channel. For instance in SDL (Olsen et al. 1995), processes are modeled as finite state machines and they communicate via unbounded FIFO channels. When a process wants to read an input from a channel, it first tests the channel for the presence of data. If a data item is available, it is consumed and processed; if no data is present, the process may decide to take some other action. The reason why this is considered to be a useful language feature is the lack of knowledge about the channel. In particular we may not know how long it takes for a data item to be transported from a sending process over a channel to the receiving process. Moreover, the channel may not be reliable and it may lose data. With the possibility of checking for the presence of data at a particular time we can model a process that behaves robustly and reliably even if the channel imposes arbitrary delays or is not reliable. The disadvantage is that we introduce the possibility of nondeterministic behavior in our process network.

Consider Figure 7-1, where one process C continuously receives messages from two other processes A and B and emits messages. If messages are available on both input ports, it emits the sum of the input values on its output port; otherwise it emits 0. Assume that both channels α and β impose zero delay on messages. Process C would then emit the sequence $0, 2, 4, 6, \ldots$. However, for instance, if channel β imposes a delay of 1 second on its messages, process C would emit the sequence $0, 1, 3, 5, \ldots$, as illustrated in Table 7-1. In general an infinite number of different output sequences are possible if the channel delay varies nondeterministically during the computation.

Hence, not only the timing of the output events of process C is changed but also their values. In this example the nondeterministic channel delay was the cause of nondeterminism in the behavior of process C. In general, nondeterministic timing

TABLE 7-1: *Process behavior depends on channel delay.*

	Process ports	Output sequence				
delay(α) = 0	A output	0	1	2	3	4
delay(β) = 0	B output	0	1	2	3	4
	C input 1	0	1	2	3	4
	C input 2	0	1	2	3	4
	C output	0	2	4	6	8
delay(α) = 0	A output	0	1	2	3	4
delay(β) = 1	B output	0	1	2	3	4
	C input 1	0	1	2	3	4
	C input 2	—	0	1	2	3
	C output	0	1	3	5	7

of either channels or process computations may result in nondeterministic behavior because the time when an event *e* appears on an input port depends on the entire process network that contributes to the generation of that event. If any of the processes or channels in this network exhibits nondeterministic timing, the time when *e* appears may be nondeterministic.

To understand this phenomenon better, we first observe that in the synchronous and the timed computational models we can test for the absence of an event at a particular time without introducing nondeterminism. Both the synchronous and the timed models of computation are as deterministic as the untimed model. The reason is that in these cases we model time explicitly and make it in some way an integral part of the functional behavior. In both the synchronous and the timed model we divide the time into slots, and when we process all events in a particular time slot, we know precisely if there is an event in this slot on an input channel or not. The ⊔ symbol explicitly models absent events.

Thus, the problem of nondeterminism arises only if we do not model time explicitly, like in the untimed model and SDL, but we still would like to test the presence of data at a particular time. This desire is questionable, but there are important languages like SDL (Olsen et al. 1995) and Ada (Booch and Bryan 1994) that fall into this category. The resulting nondeterminism has been accepted deliberately with the argument that nondeterminism is an inherent feature of real systems and should therefore also be an integral part of system models. We argue that it is preferable to use a better suitable computational model with the right timing abstraction if issues of time are to be taken into account.

In practice we often deal with computational models and languages that are not ideal from a theoretical perspective. We have to acknowledge that a common practical task is to devise a globally determinate behavior in the presence of nondeterministic features of the language, the underlying computational model, or the physical environment of the implemented system. A straightforward technique to achieve fully determinate behavior is an obvious consequence of our discussion so far. We always implement a blocking read. Whenever we attempt to read data from an input channel, we wait until it becomes available without doing anything else in between. This may or may not be efficient depending on the concrete situation.

FIGURE 7-2

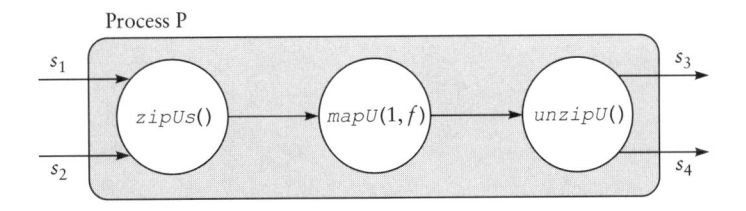

Process P couples two streams $s_1 - s_3$ and $s_2 - s_4$, although they are not functionally dependent on one another, if f is defined as $f((x,y)) = (f_1(x), f_2(y))$.

In addition to this simple and effective technique, which can always be applied, there is a special and important case that can be handled in a more sophisticated way. Consider a process that deals with two or more different inputs in functionally different ways. Since there is only one process, the operations on the different inputs have to be interleaved even though they are not data dependent on each other. A common example is a server that serves requests arriving on different input signals. The server is a shared resource. The service time may depend on the number of competing requests, but the functionality of the processing of each request should not be mutually dependent.

Consider as a simple example process P in Figure 7-2. P receives data on two input signals, zips them together into pairs, processes them with the function f, and unzips the result into separate output signals. If function f uses both inputs to compute both outputs, each output signal functionally depends on both input signals. However, if f is defined as $f((x,y)) = (f_1(x), f_2(y))$, output signal s_3 only depends on s_1, and s_4 only depends on s_2.

Process P suffers from the problem that a lack of input on signal s_1 will also block output on s_4 and vice versa. As a remedy, P can be defined in the untimed domain as P′ in Figure 7-3. The output s_3 is computed independently from s_2, and s_4 is computed independently from s_1.

P′ can be refined in the synchronous computational model to reflect the intention to implement only one process (i.e., the shared server), but still avoid blocking one stream by another. A proper refinement is P″ = $unzipS() \circ mapS(f) \circ zipS()$ if f is defined as follows:

$$f(\sqcup) = \sqcup$$
$$f(\sqcup, y) = (\sqcup, f_2(y))$$
$$f(x, \sqcup) = (f_1(x), \sqcup)$$
$$f(x, y) = (f_1(x), f_2(y))$$

The same intention can be captured in a variant of the untimed computational model, where the presence of data on an input channel can be probed with a primitive function present, as illustrated in Figure 7-4.

The general rule can be formulated as follows. If a set of output signals R_o does not depend on a set R_i of input signals, then the consumption of, processing of, and

FIGURE 7-3

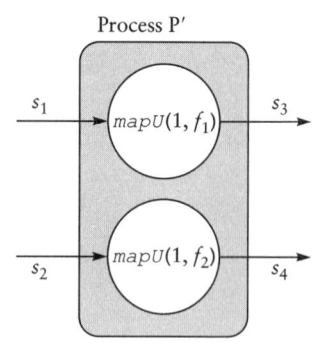

Process P′ treats the two streams $s_1 - s_3$ and $s_2 - s_4$ separately.

FIGURE 7-4

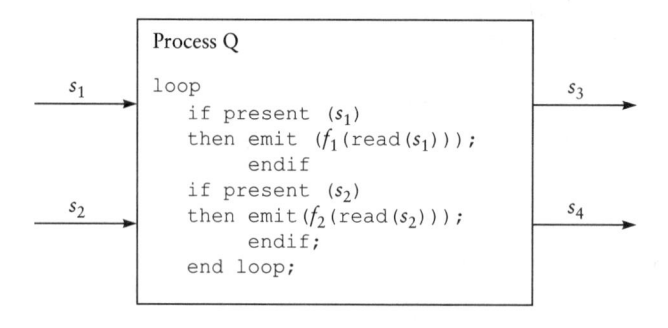

```
Process Q
loop
    if present (s₁)
    then emit (f₁(read(s₁)));
        endif
    if present (s₂)
    then emit (f₂(read(s₂)));
        endif;
end loop;
```

Process Q uses the atomic function present, *which probes an input signal for the availability of data.*

waiting for data from input signals $s \in R_i$ must not change or block the generation of data for the output signals in R_o. If this rule is observed, the blocking read rule can be relaxed while determinism is maintained.

7.2 Blocking Read and Blocking Write

If communication channels are bounded and data are discarded upon overflow of the buffer, the resulting process network becomes nondeterminate, since the relative processing speed of sender and consumer determines if and when data are discarded. The only way to avoid this is to block the sending process on an attempt to write on a full channel.

FIGURE 7-5

```
┌─────────────────────────┐                    ┌─────────────────────────┐
│ Process A               │                    │ Process B               │
│                         │       port1        │                         │
│ loop                    │◄───────────●       │ loop                    │
│     read data;          │                    │     do something;       │
│     do something;       │                    │     write port1;        │
│ end loop;               │                    │ end loop;               │
└─────────────────────────┘                    └─────────────────────────┘
```

(a)

```
┌─────────────────────┐           ┌─────────────────────┐           ┌─────────────────────┐
│ Process A           │           │ Process B           │           │ Process C           │
│                     │  port1    │                     │  port2    │                     │
│ loop                │◄────●     │ loop                │◄────●     │ loop                │
│     read data;      │           │     do something;   │           │     faulty();       │
│     do something;   │           │     write port1;    │           │     read data;      │
│ end loop;           │           │     do something;   │           │ end loop;           │
│                     │           │     write port2;    │           │                     │
│                     │           │ end loop;           │           │                     │
└─────────────────────┘           └─────────────────────┘           └─────────────────────┘
```

(b)

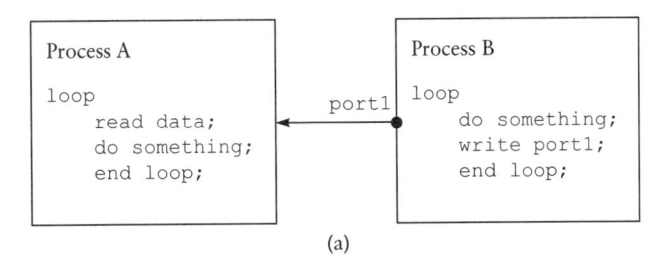

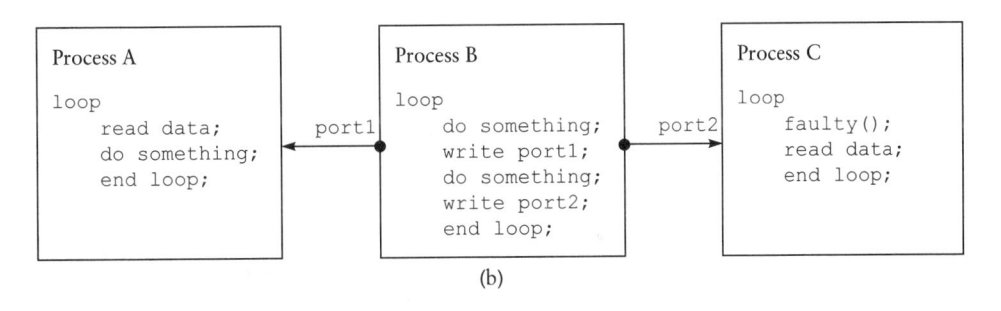

Blocking synchronization between A and C due to rendezvous-based communication.

As a special case of a bounded channel with blocking write, we discuss the case of a channel with zero buffer capacity. Since both reading from and writing to such a channel is blocking, each communication becomes also a synchronization between the involved processes. This is sometimes called a *rendezvous*. It leads to a much stronger coupling of processes, which makes it easier to analyze the behavior of an entire system, but makes it more difficult to design individual processes independently from other processes because processes proceed in a synchronized, lock-step manner.

Consider the example of two processes in Figure 7-5(a), where one process generates data and the other consumes it. If communication is rendezvous based, adding another process C, which consumes data from B, might have an impact for process A. Unfortunately, subroutine faulty () never returns and thus process C is not reading any data. Process B could send only one data item to A and would then remain in a blocking write state trying to write to port2. Consequently, process A could receive only one data item and would also be blocked when trying to receive the second item. This kind of synchronization between processes A and C is called *blocking synchronization* because it occurs due to blocking communication mechanisms. It is in general undesirable to intertwine communication and synchronization in this way since it may lead to unintentional and unnecessary synchronization.

Consider the same scenario with nonblocking write/blocking read and unbounded FIFOs on the channels separating the processes. Process B would continuously send data on both ports, and process A would not be affected by C. If process C is working incorrectly, the effects would be confined locally.

However, since the rendezvous channel does not introduce nondeterminism, its effect as compared to unbounded channels with nonblocking write is at most the production of less data; the values of the data produced cannot differ. This is an interesting and important result for the refinement of unbounded, ideal channels to finite channels with a fixed buffer capacity.

In practice the rendezvous communication paradigm has not been popular, and very few languages implement it in a strict way. Ada's (Booch and Bryan 1994) communication mechanism is called "rendezvous," but it has features for probing and accepting messages selectively, which makes it essentially equivalent to a nonblocking read model.

7.3 Oversynchronization

Blocking synchronization due to rendezvous is in fact a special case of a more general phenomenon called *oversynchronization*. Rendezvous can be considered a special case of FIFO-based communication, where the FIFO is finite, the read blocks upon an empty FIFO, and the write blocks upon a full FIFO. Let's call this form of communication channel a *blocking FIFO*. If the FIFO has length 0, we obtain a rendezvous mechanism. Further, we can observe the blocking synchronization with any size of the FIFO. If we had FIFOs of size N in Figure 7-5(b) and process C would not consume any data, process B and, consequently, process A would block after N loop iterations.

As in the case of the rendezvous, a blocking FIFO does not introduce nondeterminism but may result in the production of fewer data tokens as compared to a model with infinite FIFOs. In fact, all the signals in a blocking FIFO model are prefixes of the corresponding signals in the otherwise equivalent infinite FIFO model, and, by increasing the FIFO sizes, the behavior of the infinite FIFO model is gradually approximated.

Hence, in summary we can say that blocking FIFOs may lead to oversynchronization between otherwise independent processes, which in turn may lead to system behavior where the signals are merely strict prefixes of the corresponding signals in an infinite-FIFO-based model. However, we can also consider the finite FIFO as a resource that is shared among the data tokens communicating from one process to another. Shared resources require, in general, arbitration mechanisms that may cause processes to wait and stop until the resource is available. In this interpretation it is the arbitration mechanism, controlling the access to the shared resource, that causes the oversynchronization. As a consequence, every shared resource may potentially cause oversynchronization. It is indeed easy to see that shared servers, shared buses, shared memories, and so on, regularly give rise to this phenomenon, which all too often provokes unpleasant surprises when a well-behaved model is implemented on an architecture where resources are shared and arbitrated.

Conversely, it is justified to view a blocking, finite FIFO as a less abstract, refined model of an ideal, infinite FIFO. Replacing an infinite FIFO with a finite one is of course a necessary step toward an implementation, and thus it is in interesting question how we can guarantee that the refined model behaves identically to the ideal model. In general

FIGURE 7-6

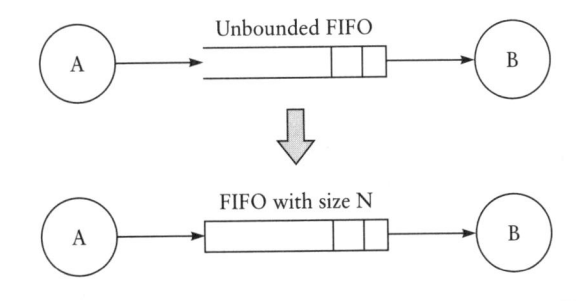

Refinement of an infinite to a finite FIFO communication channel.

this will not be the case for arbitrary models and for arbitrary inputs, but we can give conditions that guarantee identical behavior. We derive such a condition for the refinement of a FIFO channel, but similar conditions can be formulated for the refinement toward any kind of shared resource.

Consider the two U-MoC processes A and B in Figure 7-6. In each evaluation cycle of A, it emits tokens into the FIFO, and in each evaluation cycle of B, it consumes tokens from the FIFO. Our intention is to formulate a condition under which process B never tries to write into a full FIFO of size N. However, this depends how often A evaluates in relation to B. Moreover, since A may be implemented on a different processor with very different evaluation times than B, we have to establish some absolute global time. Many different factors of the actual implementation, such as processor performance and scheduling policy, will influence the precise timing of data emission and consumption, some of which we have already discussed from a different perspective in Section 3.15. Since we want to formulate a constraint on any kind of implementation, we denote the time instant of data emission and consumption in terms of an absolute reference time that can be considered the real time. But we make the simplifying assumption that an evaluation cycle for a process takes zero time, and all data consumed or emitted in the same evaluation cycle is consumed or emitted at the time instant of the evaluation cycle.

Let ev-time(p) be the sequence of time instants when process p is evaluated. Thus, if ev-time(A) = $\langle t_0, t_1, t_2, \ldots \rangle$, then t_0 would be the time of the first evaluation of process A, t_1 would be the time of the second evaluation, and so on. Further, let emitted(p, i) and consumed(p, i) be the number of tokens emitted and consumed, respectively, by process p in evaluation cycle i. And finally, let $E_{p,t}$ and $C_{p,t}$ be the sum of all data tokens emitted and consumed, respectively, by process p before time t.

$$E_{p,t} = \sum_{j=0}^{J} \text{emitted}(p,j)$$

$$C_{p,t} = \sum_{j=0}^{J} \text{consumed}(p,j)$$

FIGURE 7-7

$A = \mathtt{zipUS}(1, 1)$

$B = \mathtt{mealyU}(1, f, g, 0)$

where $g(0, (x_1, x_2)) = 1$

$g(1, (x_1, x_2)) = 2$

$g(n, (x_1, x_2)) = 2$ for $n > 2$

$f(0, (x_1, x_2)) = \langle x_1, x_1, x_2 \rangle$

$f(1, (x_1, x_2)) = \langle x_1, x_2 \rangle$

$f(2, (x_1, x_2)) = \langle x_1 \rangle$

$C = \mathtt{initU}(\langle x_0, x_2 \rangle)$

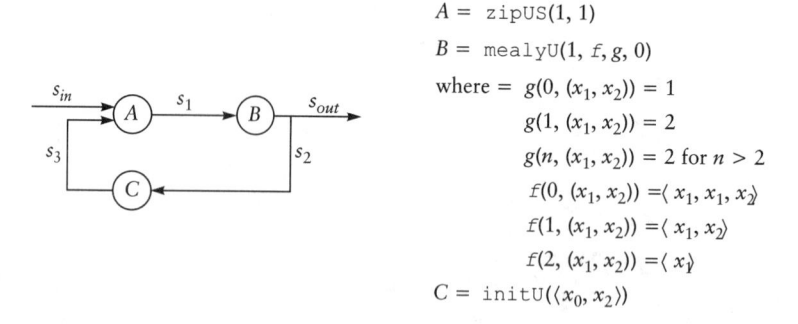

All processes consume and emit one token in every cycle except B, which emits three tokens in the first cycle, two tokens in the second, and one thereafter.

where

$$J = \max(k : t_k \leq t)$$
$$\text{ev-time}(p) = \langle t_0, t_1, t_2, \ldots \rangle$$

Then, process A will never be blocked due to a full FIFO of size N if

$$E_{A,t} \geq C_{B,t} \geq E_{A,t} - N \quad \text{for all } t \in \text{ev-time}(A) \tag{7.1}$$

For a rendezvous (i.e., $N = 0$), the number of emitted and consumed tokens are always identical. An implementation that always meets condition (7.1) will exhibit the same behavior as the corresponding infinite FIFO model. A similar analysis can be done for other types of resource sharing (see, for example, Exercise 7.2).

Violating condition (7.1) is not necessarily harmful. If there is no feedback loop in our system, the implementation model will behave identically to the abstract model. There may be a performance penalty such that tokens are produced later, but eventually all signals in the implementation will be identical to the corresponding signals of the abstract model.

However, consider the process network in Figure 7-7. A is a simple zip process, and C initializes signal s_2 with the sequence $\langle x_0 \rangle$. Process B emits a sequence of three events in the first evaluation round, two events in the second, and one event thereafter. Table 7-2 shows the tokens on the signal that have been emitted but not yet consumed for consecutive evaluation cycles for the input $s_{in} = \langle x_1, x_2, x_3 \rangle$. Note that signal s_2 contains more than one token, emitted by B but not yet consumed by C, several times after evaluation step 3. Assuming unbounded FIFOs for all signals as Table 7-2 does, the final value of the output signal is $\langle x_0, x_1, x_0, x_2, x_1 \rangle$.

However, if we assume blocking FIFOs of length 1, we observe an evaluation sequence as depicted in Table 7-3. B cannot emit all three tokens of its first evaluation cycle at once due to the limited output buffer. Hence, it emits it in three parts in steps 3, 5, and 8. However, after step 8, the feedback loop is blocked altogether. All three

TABLE 7-2: *Evaluation sequence and signal states for the model in Figure 7-7 with infinite buffers.*

Evaluation step	Evaluated process	s_{in}	s_1	s_2	s_3	s_{out}
—	—	$\langle x_1, x_2, x_3 \rangle$	$\langle\rangle$	$\langle\rangle$	$\langle\rangle$	$\langle\rangle$
1	C	$\langle x_1, x_2, x_3 \rangle$	$\langle\rangle$	$\langle\rangle$	$\langle x_0 \rangle$	$\langle\rangle$
2	A	$\langle x_2, x_3 \rangle$	$\langle (x_0, x_1) \rangle$	$\langle\rangle$	$\langle\rangle$	$\langle\rangle$
3	B	$\langle x_2, x_3 \rangle$	$\langle\rangle$	$\langle x_0, x_0, x_1 \rangle$	$\langle\rangle$	$\langle x_0, x_0, x_1 \rangle$
4	C	$\langle x_2, x_3 \rangle$	$\langle\rangle$	$\langle x_0, x_1 \rangle$	$\langle x_0 \rangle$	$\langle x_0, x_0, x_1 \rangle$
5	A	$\langle x_3 \rangle$	$\langle (x_0, x_2) \rangle$	$\langle x_0, x_1 \rangle$	$\langle\rangle$	$\langle x_0, x_0, x_1 \rangle$
6	B	$\langle x_3 \rangle$	$\langle\rangle$	$\langle x_0, x_1, x_0, x_2 \rangle$	$\langle\rangle$	$\langle x_0, x_0, x_1, x_0, x_2 \rangle$
7	C	$\langle x_3 \rangle$	$\langle\rangle$	$\langle x_1, x_0, x_2 \rangle$	$\langle x_0 \rangle$	$\langle x_0, x_0, x_1, x_0, x_2 \rangle$
8	A	$\langle\rangle$	$\langle (x_0, x_3) \rangle$	$\langle x_1, x_0, x_2 \rangle$	$\langle\rangle$	$\langle x_0, x_0, x_1, x_0, x_2 \rangle$
9	B	$\langle\rangle$	$\langle\rangle$	$\langle x_1, x_0, x_2, x_0 \rangle$	$\langle\rangle$	$\langle x_0, x_0, x_1, x_0, x_2, x_0 \rangle$
10	C	$\langle\rangle$	$\langle\rangle$	$\langle x_0, x_2, x_0 \rangle$	$\langle x_1 \rangle$	$\langle x_0, x_0, x_1, x_0, x_2, x_0 \rangle$

TABLE 7-3: *Evaluation sequence and signal states for the model in Figure 7-7 with unit sized FIFOs.*

Evaluation step	Evaluated process	s_{in}	s_1	s_2	s_3	s_{out}
—	—	$\langle x_1, x_2, x_3 \rangle$	$\langle\rangle$	$\langle\rangle$	$\langle\rangle$	$\langle\rangle$
1	C	$\langle x_1, x_2, x_3 \rangle$	$\langle\rangle$	$\langle\rangle$	$\langle x_0 \rangle$	$\langle\rangle$
2	A	$\langle x_2, x_3 \rangle$	$\langle (x_0, x_1) \rangle$	$\langle\rangle$	$\langle\rangle$	$\langle\rangle$
3	B blocked	$\langle x_2, x_3 \rangle$	$\langle\rangle$	$\langle x_0 \rangle$	$\langle\rangle$	$\langle x_0 \rangle$
4	C	$\langle x_2, x_3 \rangle$	$\langle\rangle$	$\langle\rangle$	$\langle x_0 \rangle$	$\langle x_0 \rangle$
5	B resumed	$\langle x_2, x_3 \rangle$	$\langle\rangle$	$\langle x_0 \rangle$	$\langle x_0 \rangle$	$\langle x_0, x_0 \rangle$
6	A	$\langle x_3 \rangle$	$\langle (x_0, x_2) \rangle$	$\langle x_0 \rangle$	$\langle\rangle$	$\langle x_0, x_0 \rangle$
7	C	$\langle x_3 \rangle$	$\langle (x_0, x_2) \rangle$	$\langle\rangle$	$\langle x_0 \rangle$	$\langle x_0, x_0 \rangle$
8	B resumed	$\langle x_3 \rangle$	$\langle (x_0, x_2) \rangle$	$\langle x_1 \rangle$	$\langle x_0 \rangle$	$\langle x_0, x_0, x_1 \rangle$

processes are blocked because their output FIFOs are full. As a result, the process network produces as output the sequence $\langle x_0, x_0, x_1 \rangle$, which is clearly different than the result in Table 7-2. If we assume the processes have internal memory, such that they can consume tokens from their inputs even though their output FIFOs are full, our example would not block. However, unless the processor internal memories are infinite, the example can be easily modified such that a mutual blocking still occurs.

Note that this is a deadlock situation as we have discussed it in Section 2.3.4 and Figure 2.34. Thus, oversynchronization can lead to deadlocks if two or more processes compete for two or more resources. In our case we have three processes, A, B, and C, that compete for the FIFO memories s_1, s_2, and s_3.

In summary we can conclude that the behavior of a process network can be different due to oversynchronization in the presence of a feedback loop. Note, however, that feedback loops can affect remote and seemingly unrelated parts of the network. Consider the two branches of a process network in Figure 7-8. The feedback loop in

FIGURE 7-8

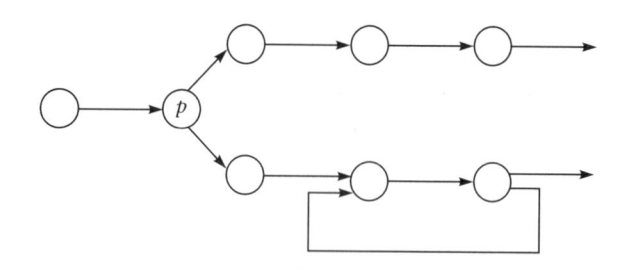

A feedback loop can affect remote parts of a process network in the presence of oversynchronization.

the lower branch can change the behavior in the upper branch if they are coupled via an oversynchronization mechanism in process p.

7.4 Rugby Coordinates

The MoC variants discussed in this chapter can be denoted as ⟨Alg, IPC, [Caus-PhyT], [Sym-LV]⟩. They are defined as communicating processes that determine the levels in the computation and communication domain. The role of the time domain is less obvious. Often processes are formulated at the **Causality** level based only on data dependences and a partial order of events. However, some of them are additionally equipped with a concept of totally ordered physical time like the languages Ada and SDL. In practical models we observe therefore a mix of both levels. Although not common in practice, these process models can be combined with the concept of **ClockedTime**. Consequently the Rugby coordinate covers the range [Caus-PhyT].

7.5 Further Reading

There are several important languages that implement variants and blends of the concepts we have discussed. SDL (Specification and Design Language) is a language standardized by the ITU (International Telecommunication Union). Introductory books for this language are Ellsberger et al. (1997) and Olsen et al. (1995). A shorter summary is given by Saraco and Tilanus (1987).

Ada has been developed under the guidance of the U.S. Department of Defense (DoD). An introduction to the language together with a systematic method on how to use it is provided by Booch and Bryan (1994).

Erlang (Armstrong et al. 1993), a functional language used for programming telecom systems, has a nonblocking read communication mechanism.

POLIS is a recently developed research framework based on Codesign Finite State Machines (CFSMs), which have features similar to those discussed in this chapter.

A summary of its concepts and the many techniques available in this framework is provided by Balarin et al. (1997).

The problem of finding minimal buffer sizes for process networks has been addressed in the context of the scheduling and synthesis of synchronous dataflow graphs. A complete account of the underlying theory and many practical algorithms is given by Bhattacharyya et al. (1996).

7.6 Exercises

7.1 Reformulate process C in Figure 7-1 such that the process network becomes deterministic independent of the delays on the channels α and β. You may change C's behavior such that it does not output anything if it does not receive inputs but make sure that C only consumes an input with `value(port)` if an event has in fact arrived.

7.2 Consider the refinement of process A into B in Figure 7-9 where f is defined as follows:

$$f(x,y) = (f_1(x), f_2(y))$$

FIGURE 7-9

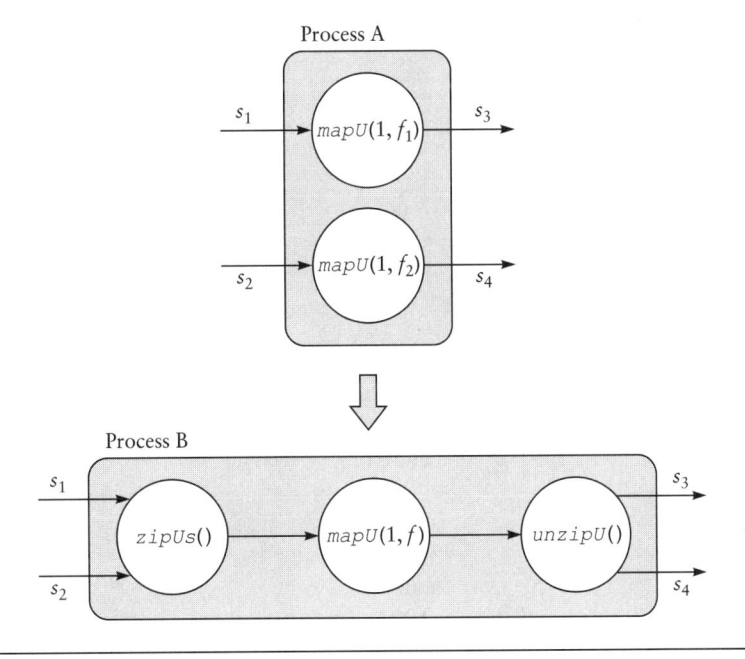

Transformation of two processes into one, reflecting a shared resource in the implementation.

In analogy to the FIFO channel refinement in Section 7.3, derive the condition under which no oversynchronization occurs in process B and both processes behave identically.

7.3 Refine the unbounded FIFO model of Figure 7-7 into a bounded FIFO model such that no deadlock can occur while minimizing the FIFO sizes for each signal.

chapter eight
Nondeterminism and Probability

Nondeterminism, its expressiveness, and its difficulties have played a prominent role in the history of computer science and in the development of computational models. It is symptomatic that we have touched upon it in almost every one of the preceding chapters.

Nondeterminism as a modeling concept is used with two distinct objectives: In its descriptive role (Section 8.1) nondeterminism is used to capture uncertainties in the behavior and timing of a system. In particular it has been used to model the unknown delays of interprocess communication. We review briefly several deterministic (e.g., Kahn process networks and synchronous languages) and nondeterministic (e.g., history relations and process algebras) attempts to deal with these uncertainties.

The second main purpose of nondeterminism is to constrain product implementations (Section 8.2). The intention of having several nondeterministic alternatives in a model is to allow implementations to select one of them. Thus, the design process is given freedom to select the most efficient solution. It is obvious that these two purposes of nondeterminism have to be handled very differently. In its descriptive role the interpretation of nondeterminism is that the modeled entity can potentially behave according to any of the nondeterministic choices. It can even change its behavior dynamically at will. In the constraining role of nondeterminism the model entity will behave according to only one of the choices, and this selection if fixed.

Due to the inherent difficulties with nondeterminism, we propose using stochastic processes to achieve the same objectives (Section 8.3). Sigma processes generate output signals with a given probability distribution. These stochastic processes can then be used to formulate stochastic variants of all previously introduced processes such as map-based and

Mealy-based processes. We distinguish between sigma bar, $\bar{\sigma}$, and sigma tilde, $\tilde{\sigma}$, processes to accommodate the two distinct objectives of nondeterminism. The sigma bar process type aims at the constraining purpose and allows the implementation to select exactly one of the possible behaviors within the stochastic range. The sigma tilde process type aims at the descriptive purpose and requires the implementation to be faithful to the stochastic properties.

Nondeterminism as a modeling concept has been used with two different objectives. One objective is to capture an aspect of the world that is not completely known and that behaves in an unpredictable manner. We call this usage the *descriptive purpose*. The second objective is to designate different possibilities for implementation, which we call the *constraining purpose*.

We use the terms "deterministic" and "determinate" in the following sense. An entity is *deterministic* if its entire internal mechanism is fully functional, that is, in each part and in each step the same output is produced for the same input. An entity is *determinate* if its externally visible behavior is functional, that is, if the entity always produces the same output for the same input. Thus, a nondeterministic system may be determinate, but a nondeterminate system cannot be deterministic.

8.1 The Descriptive Purpose

We review the history briefly because it is instructive to see that the question of determinism has been a dividing line between major tracks of research.

8.1.1 Determinate Models

In theoretical computer science nondeterminism has received continuous attention over decades because of the difficulties in dealing with it in a satisfactory manner. Nondeterminism has been considered mandatory as a modeling concept when writing distributed programs. When these programs are compiled and executed on a particular machine, the delays of computation and communication depend on the details of the target machine. If the different delays may potentially lead to different behavior, the abstract program is nondeterministic. Hence, nondeterminism is used to capture the timing behavior of the target machine.

However, the inclusion of nondeterminism severely complicates attempts to define a precise semantics for a computer program. One track of research has therefore excluded nondeterminism by defining the semantics of a language in such a way that its behavior is independent of the execution delays of the target machine. In Kahn's language for parallel programs (Kahn 1974) both the individual processes as well as an arbitrary composition of processes are determinate functions. The functions are defined over the entire history of inputs and outputs. In this way functions can have an internal state because, given an initial value of the state and a history of inputs, the function will always generate the same sequence of outputs. This is in fact the approach that we followed in Chapter 3. Our untimed MoC is essentially a Kahn process network model. Kahn's semantic is very elegant and useful and had long-lasting influence on various research directions and application fields. But the restriction that he imposed for the sake of deterministic behavior sometimes impedes the programmer from formulating more efficient solutions to a problem. For instance one restriction in Kahn's language is that processes cannot test for the emptiness of an input channel, a feature known as a "blocking read." Often it is obvious to a programmer that resources are better utilized if a process may check if input data is available and do something else if it is not.

Many variants of Kahn's process network model have been proposed: synchronous dataflow (Lee and Messerschmitt 1987b), Boolean Dataflow (Buck 1993), cyclo-static dataflow (Bilsen et al. 1995), and composite signal flow (Jantsch and Bjuréus 2000), among others. In addition to being determinate, all of them impose further restrictions in exchange for some formal property that facilitates efficient analysis and synthesis methods. Synchronous dataflow is a case in point, as we saw in Section 3.15. They all belong to the family of untimed models.

While Kahn process networks and their descendants took the approach of defining a behavior that is independent of timing properties, the synchronous MoC of Chapter 4 and the perfectly synchronous languages (Benveniste and Berry 1991b) such as Esterel (Berry et al. 1988) and Lustre (Halbwachs et al. 1991) impose on any implementation the constraint that it has to be "fast enough." For programs and their implementations that fulfill this assumption, the behavior is determinate, again by separating timing properties from the behavior.

To the same end, clocked synchronous models have been used in hardware design. A circuit behavior can be described determinately independent of the detailed timing of gates, by separating combinatorial blocks from each other with clocked registers. An implementation will have the same behavior as the abstract circuit description under the assumption that all combinatorial blocks are "fast enough." This assumption has been successfully used for the design, synthesis, and formal verification of circuits.

Fully timed models, as we discussed them in Chapter 5, are determinate by accepting delays as an integral and equally important part of the behavior. They are not vague about the effect of timing on the overall behavior.

In summary, determinate models achieve determinate behavior by either separating timing properties from behavioral properties (untimed and synchronous models) or giving them equal prominence (timed models). Variants of Kahn process networks define the semantics such that any behaviorally correct implementation is acceptable independent of its timing. Perfectly synchronous and clocked synchronous models, however, divide possible implementations into two classes: those that are "fast enough" are acceptable, and those that are "too slow" are not acceptable.

8.1.2 Nondeterminate Models

Nondeterminism has been studied in dataflow networks with asynchronous, infinitely buffered communication and in process algebras with synchronous, unbuffered communication.

FIGURE 8-1

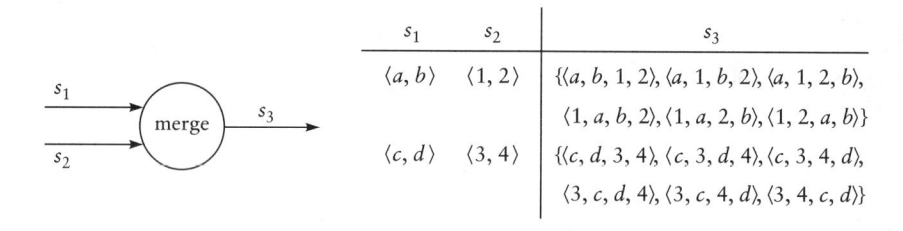

	s_1	s_2	s_3
	$\langle a, b \rangle$	$\langle 1, 2 \rangle$	$\{\langle a, b, 1, 2 \rangle, \langle a, 1, b, 2 \rangle, \langle a, 1, 2, b \rangle,$ $\langle 1, a, b, 2 \rangle, \langle 1, a, 2, b \rangle, \langle 1, 2, a, b \rangle\}$
	$\langle c, d \rangle$	$\langle 3, 4 \rangle$	$\{\langle c, d, 3, 4 \rangle, \langle c, 3, d, 4 \rangle, \langle c, 3, 4, d \rangle,$ $\langle 3, c, d, 4 \rangle, \langle 3, c, 4, d \rangle, \langle 3, 4, c, d \rangle\}$

$s_1 \rightarrow$ merge $\rightarrow s_3$
$s_2 \rightarrow$

The merge process relates two input sequences to a set of possible output sequences.

Dataflow Networks

One approach to generalize Kahn's theory to nondeterminate process networks is to use history relations rather than history functions. A history relation maps an input stream onto a set of possible output streams instead of a single determinately defined output stream. For instance, a merge process maps two input streams to a set of possible output streams:

$$\text{merge}(s_1, \langle\rangle) = s_3$$
$$\text{merge}(\langle\rangle, s_2) = s_3$$
$$\text{merge}(e_1 \oplus s_1, e_2 \oplus s_2) = \{e_1 \oplus s_3 : s_3 \in \text{merge}(s_1, e_2 \oplus s_2)\}$$
$$\cup \{e_2 \oplus s_3 : s_3 \in \text{merge}(e_1 \oplus s_2, s_2)\}$$

Figure 8-1 shows the set of possible output signals for the inputs $\langle a, b \rangle, \langle 1, 2 \rangle$.

However, history relations are not sufficient to model a nondeterministic process network in a satisfactory way. Keller (1977) and Brock and Ackerman (1981) showed that two components with identical history relations, if placed in the same context of a bigger system, may cause the system to behave in different ways; that is, the system has different history relations. This means that history relations are not sufficient to capture all relevant information about a component. In particular causality information between events must be included.

To see why, we follow the example given by Brock and Ackerman (1981). Two process networks, C_1 and C_2 (Figure 8-2), are defined as

$$C_1(s_1, s_2) = B_1(\text{merge}(A(s_1), A(s_2)))$$
$$C_2(s_1, s_2) = B_2(\text{merge}(A(s_1), A(s_2)))$$

and consist of the following processes:

$A(\langle\rangle) = \langle\rangle$
$A(e \oplus s) = \langle e, e \rangle$

$B_1(\langle\rangle) = \langle\rangle$
$B_1(\langle e \rangle \oplus s) = \langle e \rangle \oplus f_1(s)$
$f_1(\langle\rangle) = \langle\rangle$
$f_1(\langle e' \rangle \oplus s) = \langle e' \rangle$

$B_2(\langle\rangle) = \langle\rangle$
$B_2(\langle e \rangle \oplus s) = f_2(e, s)$
$f_2(e, \langle\rangle) = \langle\rangle$
$f_2(e, \langle e' \rangle \oplus s) = \langle e, e' \rangle$

8.1 *The Descriptive Purpose*   287

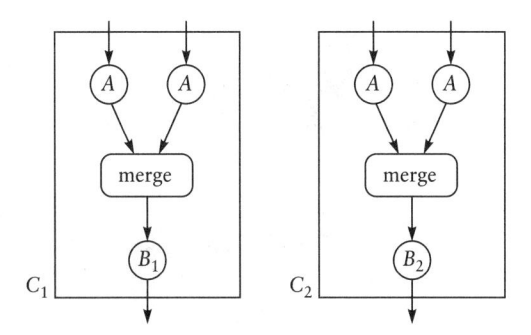

FIGURE 8-2

The two processes C_1 and C_2 are defined differently but have identical history relations.

Processes B_1 and B_2, and hence also C_1 and C_2, emit at most two output events. The difference between B_1 and B_2 is that the former already produces an output after the first input, while the latter needs two input events to emit two output events. However, because process A always emits two events for each input event, the B_i processes are always guaranteed two input events, thus masking the difference in the B_i processes. Consequently, both processes C_1 and C_2 behave identically and can be described by the same history relations:

$$C_i(\langle\rangle, \langle\rangle) = \langle\rangle$$
$$C_i(e \oplus s, \langle\rangle) = \langle e, e \rangle$$
$$C_i(\langle\rangle, e' \oplus s', \langle\rangle) = \langle e', e' \rangle$$
$$C_i(e \oplus s, e' \oplus s') = \{\langle e, e\rangle, \langle e, e'\rangle, \langle e', e\rangle, \langle e', e'\rangle\}$$

If the inherent information in history relations would be sufficient to describe nondeterministic process networks, we could put two processes with identical history relations (such as C_1 and C_2) into identical environments and expect that the resulting systems are indistinguishable. This is not the case, however. Consider the process networks D_1 and D_2 in Figure 8-3.[1] Although they provide the same environment for C_1 and C_2, they have different history relations. Suppose D_1 receives a single event with value 5. It passes through the left A and the merge processes and becomes the first output event. It is also the first input to the plus 1 process, which outputs a 6. The event 6 would be doubled in the right A process, and the result would be merged with the second 5 from the left A. Consequently, the history relation of D_1 is

$$D_1(\langle 5 \rangle) = \{\langle 5, 5\rangle, \langle 5, 6\rangle\}$$

---

[1] We use here the feedback operator **FBp** as in the untimed MoC.

FIGURE 8-3

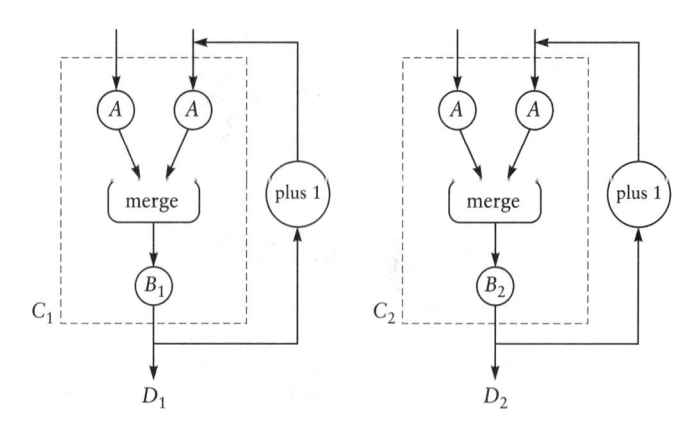

Processes D_1 and D_2 provide an identical context to C_1 and C_2 but still have different history relations.

In contrast, process B_2 in the network D_2 would not emit anything until it has consumed both 5s, and the history relation of D_2 would thus contain only one possible output sequence:

$$D_2(\langle 5 \rangle) = \{\langle 5, 5 \rangle\}$$

The difference is due to the fact that B_1 emits one event after its first input and one event after its second input while B_2 emits no event after its first input and two events after its second input. This difference is not captured by history relations but may still have effects in feedback loops. Note that in the terminology of Chapter 3 B_2 is an up-rated version of B_1, and we have noted that, in the untimed MoC, up-rated processes in feedback loops may lead to the computation of fewer outputs but never to contradictory outputs.

Based on the insight of the insufficiency of history relations, (Brock and Ackerman 1981; Brock 1983) gave a formal semantics based on history relations augmented with causality information in terms of so-called scenarios. The processes C_1 and C_2 above are then distinguished by different scenarios that reflect the internal mechanics of how the processes react to incoming events.

A different approach has been taken by Kosinski (1978), who described a semantic for nondeterminate dataflow programs based on the idea of annotating each event with the sequence of nondeterminate choices that leads to that event. In this way the causality information is also preserved, but the result is a very complex formalism that cannot be applied except to the smallest systems.

Park's formal semantics of dataflow (Park 1983) models nondeterminism with oracles rather than with history relations (Figure 8-4). All processes are determinate with the exception of oracles, which are pure data sources. They output a nondeterministic sequence of numbers. Each merge operator, which is itself determinate, is provided with an extra control argument that is connected to an oracle. It controls from which

FIGURE 8-4

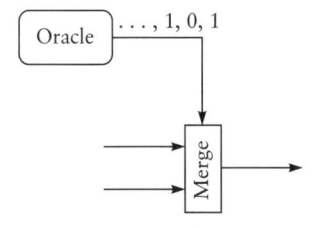

The merge selects the input based on the oracle values.

FIGURE 8-5

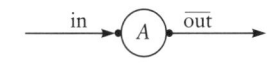

Process A with one input and one output port.

input stream the next token for the output stream is selected. Hence, together with the oracle the determinate merge becomes a nondeterminate merge operator. Below we will take up the idea of oracles but with a stochastic rather than a nondeterminate behavior.

Process Algebras

Hoare's CSP (Communicating Sequential Processes) (Hoare 1978) and Milner's CCS (Calculus of Communicating Systems) (Milner 1980) have been developed in response to two difficulties with dataflow models. First, it appeared difficult to find elegant solutions to define the formal semantics of nondeterminate dataflow languages. Second, dataflow models require unbounded buffers for communication, which lead to difficulties in implementation.

In CCS a process is determined by its acts of communication with other processes. "Communication" means either receiving a token via an input port or sending a token to an output port. For instance, a process A can be defined as follows:

$$A \stackrel{\text{def}}{=} \text{in}.\overline{\text{out}}.A$$

A receives a token on its input port in and then it emits a token on its output port $\overline{\text{out}}$ (Figure 8-5). After having performed these two acts of communication, it behaves like process A. Thus, this is a recursive definition of a process with exactly one state, in which it can receive an input and emit an output. CCS defines a number of operators for describing processes and for combining them into process networks. For instance,

FIGURE 8-6

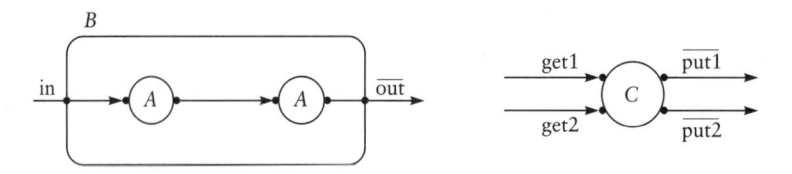

B *is a sequential composition of two* A *processes, and* C *is the union of two alternative behaviors.*

FIGURE 8-7

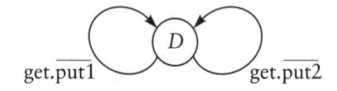

The transition diagram for process D.

the $+$ operator defines alternative behaviors, and $|$ connects two processes.

$$B \stackrel{\text{def}}{=} A|A$$

$$C \stackrel{\text{def}}{=} \overline{\text{get1}}.\overline{\text{put1}}.C + \overline{\text{get2}}.\overline{\text{put2}}.C$$

Process B is a simple sequential composition of two A processes (Figure 8-6). When process C receives an input on port $\overline{\text{get1}}$, it responds with an output on port $\overline{\text{put1}}$ and it responds with an output on port $\overline{\text{put2}}$ to an input on get2. C is determinate, but we can easily define a nondeterminate process:

$$D \stackrel{\text{def}}{=} \overline{\text{get1}}.\overline{\text{put1}}.D + \overline{\text{get1}}.\overline{\text{put2}}.D$$

Upon reception of an input on port get1, process D can either respond with an output on port $\overline{\text{put1}}$ or $\overline{\text{put2}}$.

A transition diagram shows the possible sequences of actions (Figure 8-7). Process D has only one state, denoted as D, and two possible transitions.

Milner defines the concept of *weak determinacy* (Milner 1989), which is based on observational equivalence. A system in a given state with given inputs can enter a set of different successor states nondeterministically. If the system in all successor states behaves identically, as far as it can be observed from the outside, the system is weakly determinate. Process D above is not weakly determinate because, depending on the choice of transition, it will emit different output tokens. But process D′, defined as follows,

$$D' \stackrel{\text{def}}{=} \text{get1}.D_1 + \text{get1}.D_2$$

FIGURE 8-8

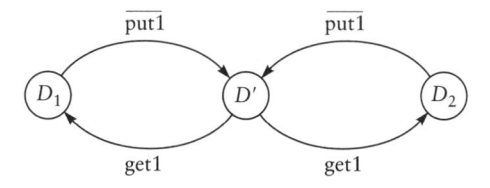

The transition diagram for process D′.

FIGURE 8-9

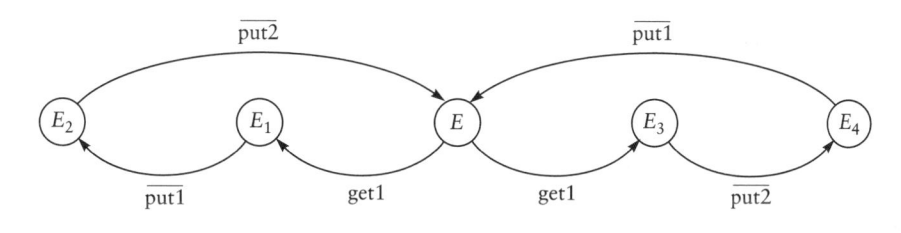

The transition diagram for process E.

$$D_1 \stackrel{\text{def}}{=} \overline{\text{put1}}. D'$$
$$D_2 \stackrel{\text{def}}{=} \overline{\text{put1}}. D'$$

is weakly determinate because it will always exhibit the same behavior as it can be observed from the outside. Figure 8-8 shows its transition diagram.

A somewhat broader concept discussed by Milner is *weak confluence*. A system is confluent if for every two possible actions, the occurrence of one can never preclude the other. Thus, even though one of the two is selected nondeterministically, the other will eventually also occur.

$$E \stackrel{\text{def}}{=} \text{get1}. E_1 + \text{get1}. E_3 \qquad \begin{array}{ll} E_1 \stackrel{\text{def}}{=} \overline{\text{put1}}. E_2 & E_3 \stackrel{\text{def}}{=} \overline{\text{put2}}. E_4 \\ E_2 \stackrel{\text{def}}{=} \overline{\text{put2}}. E & E_4 \stackrel{\text{def}}{=} \overline{\text{put1}}. E \end{array}$$

Process E emits the two outputs $\overline{\text{put1}}$ and $\overline{\text{put2}}$. However, the order of emission depends on the nondeterministic choice of the transition taken after receiving get1. Figure 8-9 depicts the corresponding decision diagram.

A realistic example for the usefulness of this concept is the service of interrupts by operating systems. If two interrupts are pending, the operating system may select one interrupt first. However, no matter which interrupt is served first, the other must also be processed eventually. Hence, we do not require a determinate behavior by the operating system, but we do require weak confluence.

Milner then gives a set of construction rules that preserve confluence (Milner 1989, Chapter 11). For example, a restricted form of process composition, where the types of input events consumed by the individual processes are disjoined, preserves confluence. The main idea is that determinate or confluent behavior is desirable when designing systems. In the presence of nondeterminism, the designer needs (1) techniques to show that a given model is determinate or confluent and (2) rules that allow the construction of determinate or confluent models from simpler determinate or confluent components.

This can be compared to the idea of fully determinate models: restrict the models to be deterministic and define conditions under which the implementations, the presence of nondeterministic mechanisms notwithstanding, still behave determinate and equivalent to the abstract models.

Both approaches fight nondeterminism but in different ways. From the designer's perspective, nondeterministic MoCs such as CCS are richer in terms of expressiveness, but the designer has to be more careful in how to model and how to construct the system. Deterministic MoCs are more restrictive but make the task of modeling and designing easier.

8.2 The Constraining Purpose

For the purpose of describing requirements on a system, two techniques related to nondeterminism have been used. Relations divide the possible responses of a system to a given input into two categories: acceptable and not acceptable. `execution time (Program)` $< 5\,\text{ms}$ and `size(Chip)` $< 1\,\text{cm}^2$ are two relations constraining the nonfunctional properties of a system. The relation defining a sorted integer array has been used numerous times as an example of a functional requirement. Dennis and Gao (1995) describe the example of a transaction server that accepts requests at several inputs and processes them (Figure 8-10). The merge process, which decides the order in which requests are served, is subject to several functional and nonfunctional constraints. Apparently each request should eventually be served. Perhaps we require that the average response time is similar for requests on all input lines. And most likely we would like to have a high performance of the merge operator itself while minimizing its implementation cost. Thus, there is a significant freedom for the implementation to determine the way the requests are merged into a sequential stream, which should be used to meet the nonfunctional requirements as well as possible. Clearly, the functional specification should not define a determinate merge mechanism, but it should rather

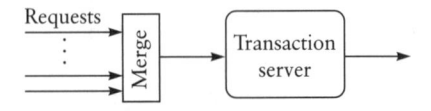

FIGURE 8-10

A transaction server with request merger.

8.3 The σ Process

On the one hand, as already indicated, nondeterminism brings many difficulties. Formal analysis and verification are significantly harder because they must take into account all the possible behaviors rather than only one definite behavior. Simulation cannot really achieve nondeterministic behavior but typically resorts to stochastic models. On the other hand, there are more options to explore in synthesis and design when a high-level specification model employs nondeterminism because nondeterminism is a way to avoid overspecification.

To reconcile these different views we develop in the following sections stochastic processes that are used to approximate nondeterminate behavior. First we introduce the basic processes, which are called σ processes. They provide a desired stochastic distribution. Then we introduce the process constructors that instantiate stochastic processes in line with the notation we use in the rest of the book.

The σ process is not much more than a pseudorandom generator with a defined stochastic distribution. We use it in two ways.

First, we use it to constrain the implementation of the system with respect to behavior. Depending on the precise kind of σ process, the implementation may or may not be required to respect the statistical properties of the specifying process.

Second, environment elements can be modeled with σ processes when we cannot or do not want to represent their exact behavior or timing. We think a statistical distribution is more appropriate than plain nondeterminism. Consider an ATM switch that receives ATM cells from the environment. The type of ATM cells, user cells, alarm cells, maintenance cells, erroneous cells, and so on, follow a statistical distribution. To generate ATM cells according to given probabilities is both more accurate and more useful for the design and validation of the ATM switch.

Different kinds of σ processes generate integers with different random distributions. Figure 8-11 shows two of the possibilities. $\sigma^u_{<s,r>}$ generates a uniform distribution of integer values in the range r, with the initial seed s for the random number generator. $\sigma^n_{<s,r,m,d>}$ generates a normal distribution of integer values in the range r with the mean value m and the standard deviation d. In the following we restrict the discussion to uniform distributions. The `sigmauC` constructor generates a σ^u process. Let $p_v(e)$ be

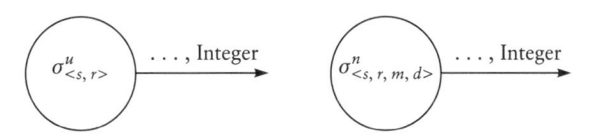

FIGURE 8-11

The σ processes for uniform and normal distributions.

the probability that the value of event e is v.

$$\texttt{sigmauC}(min, max, w_0) \rightarrow \sigma^u_{<w_0, [min, max]>}() = s'$$

with

$$min \leq \text{val}(e') \leq max \quad \text{for all } e' \in s'$$

and

$$p_v(e') = \frac{1}{max - min + 1} \quad \text{for all } v \in [min, max]$$

The σ process is based on a pseudorandom generator that is initialized with a specific seed value. Different seed values define different σ processes. But two σ processes with the same instantiation parameters will generate two identical sequences of output values.

We assign different interpretations to the σ processes, depending on the activity under consideration. The description that we have given above is identical to the interpretation for simulation.

8.4 Synthesis and Formal Verification

The interpretation of the sigma processes is the same for synthesis and formal verification.

First, we have to keep in mind that these stochastic processes are often used to model environment components, which will not be implemented. But σ processes can be part of the system and then they must be implemented. We distinguish two variants.

First, synthesized *sigma bar process* $\bar{\sigma}$ can generate any of the possible outputs of a σ process without restriction. The output of the synthesized process may or may not have the statistical properties of the specification process. For instance a synthesized $\bar{\sigma}^u$ could be implemented in such a way that it generates always a 0 for each request with a range (0,10).

Second, *sigma tilde processes* $\tilde{\sigma}$ have to be implemented such that the statistical properties are preserved. This distinction is similar in intention to, but formally quite different from, Park's "tight nondeterminism" and "loose nondeterminism" (Park 1979).

As a consequence the outputs generated by the implementation will be different from the outputs generated during a simulation run for both σ process variants.

The merge of the transaction server in Figure 8-10 definitely requires a $\tilde{\sigma}$ process. However, consider the bar merge operation illustrated in Figure 8-12. In each processing step the bar merge receives one token from each of its two inputs, I1 and I2, and emits the two tokens in arbitrary order. Since we do not want to specify the order of the two output tokens deterministically, we use a $\bar{\sigma}$ process to drive the third input of the bar merge. If that token is a 0, we emit first the x token and then the y; if it is a 1, we emit the tokens in reverse order. For the implementation we are free to select any order; we may select a hardwired solution to always emit x before y.

Although the sequence of output values of the σ processes is generated deterministically, formal verification and analysis tools are not allowed to use the concrete values

FIGURE 8-12

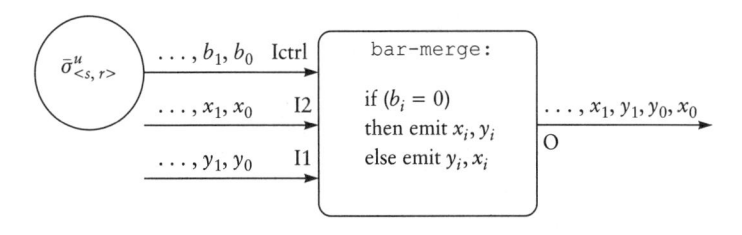

A merge that uses a $\bar{\sigma}$ process.

of these sequences for reasoning. They are only allowed to use the statistical properties that are guaranteed by any implementation. Hence, the interpretation by formal verification follows the synthesis interpretation by distinguishing bar and the tilde variants of σ processes.

8.5 Process Constructors

Based on σ processes we can define variants of the familiar process constructors that instantiate stochastic processes. In the following constructors for only the synchronous model are introduced, but they can be generalized to the untimed and timed models.

We discuss two different approaches. First, we use a select operator that applies one out of two functions on an input value depending on the result of a σ process. Second, we internalize the choice into the combinatorial functions.

8.5.1 select-Based Constructors

The function select takes three parameters and returns a function. The first one, δ, can be 0 or 1, and it decides which of the functions f or g is returned by select.

$$\text{select}(\delta, f, g) = \begin{cases} f & \text{if } \delta = 0 \\ g & \text{if } \delta = 1 \end{cases}$$

Select-based processes resemble the oracle mechanism described earlier.

selMapS (Figure 8.13) is similar to mapS, but it has two functions that are selectively applied depending on the output of an internal $\sigma^u_{w_0,[0,1]}$ process. The third argument of selMapS is the seed value for the random number generator.

$$\text{selMapS}(f_1, f_2, r_0) = p \tag{8.1}$$

with

$$p(\bar{s}) = p_1(p_2(\sigma^u_{<r_0,[0,1]>}, \bar{s}))$$

$$p_1 = \text{mapS}(f)$$

FIGURE 8-13

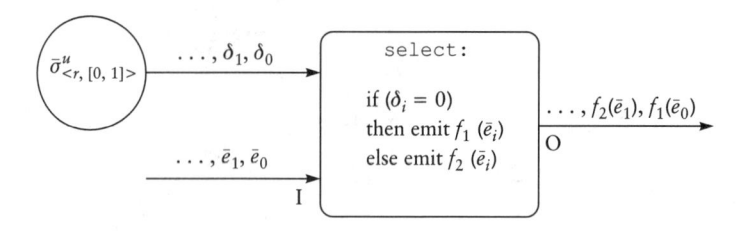

A process instantiated by `selMapS`.

$$p_2 = \text{zipS}()$$
$$f((\delta, \bar{e})) = \text{select}(\delta, f_1, f_2)(\bar{e})$$

Similarly, we define `selMooreS`, which contains two internal σ processes and has two different next-state functions and two output encoding functions to choose from in each computational step.

$$\text{selMooreS}(g_1, g_2, f_1, f_2, r_0, r'_0, w_0) = p \qquad (8.2)$$

with

$$p(\bar{s}) = p_1(p_2(s_f, p_3(p_4(s_g, \bar{s}))))$$
$$s_f = \sigma^u_{<r'_0, [0,1]>}$$
$$s_g = \sigma^u_{<r_0, [0,1]>}$$
$$p_1 = \text{mapS}(f)$$
$$p_2 = \text{zipS}()$$
$$p_3 = \text{scanS}(g, w_0)$$
$$p_4 = \text{zipS}()$$
$$f((\delta, \bar{e})) = \text{select}(\delta, f_1, f_2)(\bar{e})$$
$$g(w, (\delta, \bar{e})) = \text{select}(\delta, g_1, g_2)(w, \bar{e})$$

Other select-based process constructors can be defined accordingly for the synchronous as well as for the timed and untimed computational models.

8.5.2 Consolidation-Based Constructors

An alternative to the usage of the select operator is to supply the combinatorial functions directly with the output of the σ process. In this case these functions have to understand the additional input as a stochastic process. Because the basic idea is simple, we only describe the `conMooreS` constructor, which results in the same internal process netlist as `selMooreS` processes (Figure 8-14). It again contains two internal σ^u processes with

FIGURE 8-14

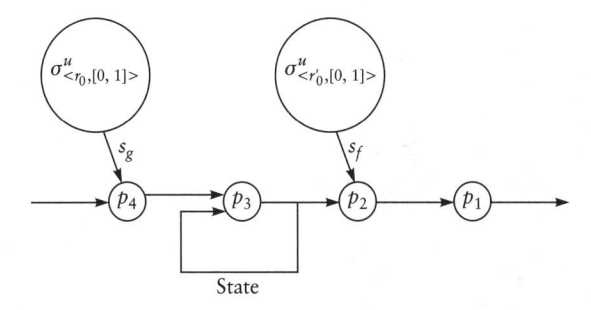

Both `selMooreS`*- and* `conMooreS`*-based processes have the same internal process structure with two stochastic processes. In* `selMooreS` *processes* p_1 *and* p_3 *apply one out of two possible functions depending on the values from* s_f *and* s_g*, respectively. In* `conMooreS` p_1 *and* p_3 *each have only one function, which directly interpret the inputs from* s_f *and* s_g*, respectively.*

two different seed values r_0 and r'_0.

$$conMooreS(g, f, r_0, r'_0, w_0) = p \qquad (8.3)$$

with

$$p(\bar{s}) = p_1(p_2(s_f, p_3(p_4(s_g, \bar{s}))))$$
$$s_f = \sigma^u_{<r'_0,[0,1]>}$$
$$s_g = \sigma^u_{<r_0,[0,1]>}$$
$$p_1 = mapS(f)$$
$$p_2 = zipS()$$
$$p_3 = scanS(g, w_0)$$
$$p_4 = zipS()$$

This approach is preferable if there are not two distinct functions from which we stochastically choose. The functions f and g have to understand the meaning of the input coming from the stochastic processes. In principle the two types of stochastic skeletons are equivalent, and it is mainly a matter of convenience which one is chosen for a particular modeling task.

8.6 Usage of Stochastic Skeletons

In previous sections we have indicated that we want to use the stochastic skeletons for similar purposes as nondeterminism has been used for. We have also discussed briefly how the different activities (simulation, synthesis, and verification) should interpret these skeletons. Now we summarize our objectives and delineate them from issues that should not be addressed in this way.

8.6.1 The Descriptive Purpose

One important application of stochastic skeletons is to model environment components. Very often we cannot know the exact behavior of the environment, or we do not care to model the environment in all its details. The uncertainty about the environment concerns both the functional as well as the timing behavior. When we use traditional nondeterminism for this, we enumerate the possible environment behaviors, and we give no further information about how likely the different cases are. If we add a notion of fairness, we exclude a few specifically undesirable possibilities. However, very often we know a bit more, and we would like to assign probabilities to the different possible behaviors. The simulations that we can perform based on probabilistic environment behavior will be more realistic in many cases. If we do not know anything about the probabilities, we can use a uniform distribution to give each case equal probability. Even if we do not know how realistic this is, it has the advantage that we will trigger each of the cases in the simulation, provided we can run it long enough.

We have to acknowledge, however, that the notion of nondeterminism is broader than any specific probability distribution can capture because it encompasses all possible probabilistic distributions. A nondeterministic process may generate sequences of numbers that exhibit any possible probability distribution, while a stochastic process can only generate sequences with a given probability distribution. However, for two reasons we doubt that this difference is of practical importance. First, a stochastic process can generate any sequence that a nondeterministic process can, although it may be very unlikely. Second, all implementations of simulators that allow the simulation of nondeterministic behavior use in fact some stochastic process for this. Our computer technology does not allow us to simulate nondeterministic behavior that does not follow a specific probability distribution.

While this argument allows us to substitute a stochastic process for a nondeterministic process even on the ideal level, we can even go further by postulating that it is in fact an advantage to do so. If a user includes a nondeterministic process in her model, she has no control over what the implementation of the simulator is in fact doing. Since a nondeterministic process may generate any sequence of numbers, the implementor of the simulator typically selects one of the possibilities which is most convenient for her. But this decision is usually unknown to the end user who uses the simulator. On the other hand, if we specify a stochastic process with a specific probability distribution, the simulator must respect the statistical properties. Thus, the end user who writes the model has better control over the behavior of the model and knows exactly what she is simulating. From this practical argument we conclude that it is in fact an advantage to use stochastic processes rather than nondeterministic processes for modeling environment behavior.

While we advocate the usage of stochastic processes for the description of the environment, we do not propose to use them for describing uncertainties of the system under design. The notion that we do not know the exact behavior with respect to timing or function is not satisfactory. We design, implement, and manufacture the system.

So in principle we have full control and knowledge about it if we decide to dedicate the necessary effort. Sometimes we may not want to spend the effort because we do not care as long as the behavior falls into a given class or range of acceptable behaviors. Consequently, we prefer the notion that we *constrain the system*.

We should understand, however, that this point of view may not be appropriate for general software systems. We mainly discuss the specification and modeling of embedded systems, hardware, and embedded software, where we always are concerned with timing and performance issues and where we very often face hard timing constraints. Therefore we find the concept of constraining the timing behavior of the implementation more appealing than the assumption that the timing behavior could be arbitrary.

8.6.2 The Constraining Purpose

The functional and timing behavior of a system implementation can be constrained in a variety of ways, and stochastic processes should only be used in some specific cases. A perfectly synchronous timing model is very well suited to constraining the timing behavior. Both in hardware and in embedded software design, a synchronous design style has been used with great success. It effectively separates functionality from timing issues. Static timing analysis can be done independently from functional validation and verification. A rich set of pipelining and retiming techniques has been developed that allow us to tune the timing behavior while keeping the functionality unmodified.

The general method to express constraints on the functionality is by means of relations. As we discussed in Section 1.5, in the early phase of system development, the requirements analysis phase, general requirements and constraints are formulated in terms of relations. However, because relations allow for a huge design space, efficient synthesis techniques for both hardware and software are still out of reach. A system specification model, which captures most of the high-level design decisions, is therefore a necessity. This model should be determinate because nondeterminism greatly complicates synthesis and validation.

However, as we tried to illustrate in several examples, there are occasions when we would prefer to leave several options open in order to give the later design phases more opportunities to find optimal implementations. The merging of data streams is a good example (Figure 8-10) of a situation where a specification model should avoid a decision. The stochastic processes, as introduced in the previous sections, are a good way to address this issue. For simulation they allow us to exercise all possibilities that might occur in a concrete implementation. Synthesis can exploit the possibilities that a stochastic process exposes. Validation can use their statistical properties to verify system properties and to establish that a specific implementation complies with the specification model.

8.7 Further Reading

Several approaches to equip nondeterminate dataflow process networks with a formal semantics have been reported (Staples and Nguyen 1985; Park 1983; Kosinski 1978; Brock 1983; Brock and Ackerman 1981).

Process algebras as developed by Hoare (1978) and Milner (1980, 1989), and their variants, are inherently nondeterministic, and much of the theory on nondeterminism has been developed in these frameworks. An early formulation of nondeterminism, which is still worth reading, was written by Dijkstra (1968).

The issue of fairness is a central aspect of nondeterminism. It has been discussed in depth several times (Parrow 1985; Park 1983, 1979; Panangaden and Shanbhogue 1992), and a book is even devoted to the subject (Francez 1986).

The theory and application of stochastic processes to system analysis and performance modeling is highly developed. Good books on these subjects are by Cassandras (1993) and by Severance (2001). Stochastic processes have not often been used for specifying and constraining a system as we discussed in this chapter. One approach similar to what we developed here is described in Jantsch et al. (2001).

8.8 Exercises

8.1 A telecommunication network has many traffic sources and sinks (Figure 8-15). Let's assume we have to deal with a packet-switched network and the average traffic source emits three different kinds of packets. *Operation administration and maintenance* (OAM) packets are used for operating and controlling the network. They are, for instance, used to set up and close connections, measure performance, and identify and locate errors. *Real-time* (RT) packets have real-time performance requirements such as a maximum delay. They are used to implement telephone connections and audio and video streams. *Best-effort* (BE) packets have no hard

FIGURE 8-15

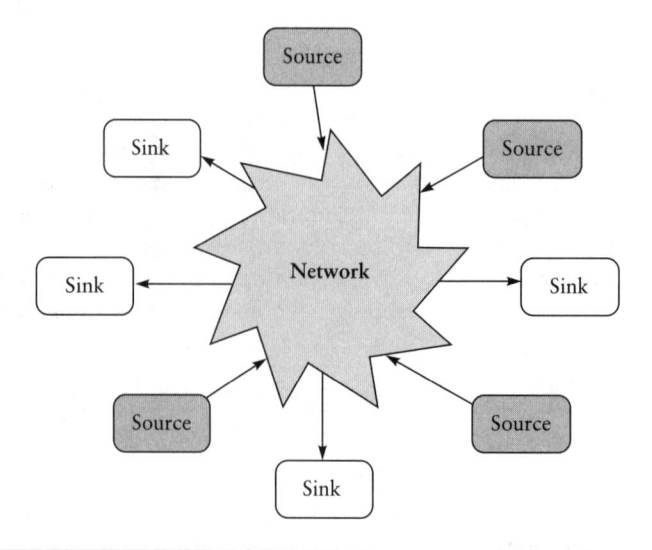

A telecom network with traffic sources and sinks.

real-time constraints, and the network should do its best to keep their loss and delay to a minimum.

An average source exhibits the following probabilities for these different packet types:

- OAM packets: 3%
- RT packets: 20%
- BE packets: 77%

    a. Model the traffic source with S-MoC and select-based processes.

    b. Model the traffic source with S-MoC and consolidation-based processes.

8.2 A *fair merge* process merges two input signals into one output signal. If an input signal is infinite, infinitely many events from this signal eventually appear at the output. If an input is finite, all events from that signal eventually appear at the output.

    a. Instantiate a perfectly synchronous, fair, fully deterministic merge process.

    b. Instantiate a perfectly synchronous, fair, stochastic merge process with a $\tilde{\sigma}^u_{<r_0,[0,1]>}$ process.

    c. Instantiate an untimed, fair, stochastic merge process with a $\tilde{\sigma}^u_{<r_0,[0,1]>}$ process.

chapter nine
Applications

Different models of computation are used in different phases of a design process. We discuss the suitability of MoCs in three design activities.

Performance analysis aims to assess and understand some quantitative properties at an early stage of the product development. As examples we take buffer analysis and delay analysis and find that the U-MoC is suitable for many performance analysis tasks as long as no timing information is required. Buffer analysis is a good case in point. When timing information is required, a synchronous or timed MoC should be used. For many time-related analysis tasks a synchronous MoC is sufficient. A timed MoC is the most general tool and should be used only when a synchronous MoC is not sufficiently accurate.

The functional specification document is used for two, partly contradictory, purposes. On the one hand, a functional model should reflect the first high-level solution to the posed problem. It should help to assess the quality adequacy of this solution. On the other hand, it is the basis for the next design steps and the implementation. As such it should help to assess the feasibility of the solution and its performance and cost properties. Based on an analysis of the objectives of the functional specification, we conclude all three MoCs may be required, but the synchronous MoC is appropriate for most tasks.

In synthesis and design all three MoCs are required. The closer we come to the final implementation, the more accurate the timing information should be. Thus, we see a refinement from an untimed into a synchronous into a timed MoC. Not surprisingly, the general rule is to avoid detailed timing information as long as possible and to use a more accurate timing model whenever necessary.

In the previous chapters we reviewed several important models of computation and concurrency and discussed some of their essential features. On several occasions we have related these features to applications of modeling. For instance, we found that discrete event models are very good for capturing and simulating the accurate timing behavior of existing systems, but they are ill suited to serve as an entry point for synthesis and formal verification tasks. The discussion should have made clear by now that features of computational models are not good or bad, but they are merely more or less useful for a particular purpose.

In this chapter we consider some important applications of modeling, and we shall analyze their requirements on models in a more systematic way. We will find out that even within the same application area some requirements contradict each other. For instance, system specification serves essentially two different purposes: capturing the system functionality and forming the basis for design and implementation. However, requirements derived from these two purposes contradict each other directly to some degree. On the one hand, the system specification model should be as abstract and application oriented as possible to allow an implementation-independent exploration of system features and for a solid basis for contacts with the application-oriented customer. On the other hand, it should be implementation sensitive to allow accurate performance and cost estimation and to have a good starting point for an efficient design and implementation process.

The contradictions between different application areas such as performance modeling and synthesis are insurmountable. Therefore we have a variety of computational models with different features, strengths, and weaknesses.

We have selected some of the most important applications. No larger system design project can avoid any of these. But by no means can we cover all application areas of interest. For instance we do not touch upon the equally important fields of requirements engineering and the modeling and design of analog components. Moreover, we do not attempt an in-depth treatment of the application areas that we do take up. We do not present a thorough account of the theory and techniques of performance modeling or formal verification. Numerous good books are devoted solely to these subjects. We provide a rather superficial introduction to these topics with the objective of extracting and explicitly discussing the requirements and constraints they impose on modeling concepts and techniques. We summarize the requirements from different applications and discuss them in light of the computational models we have available. From this discussion we will obtain a better understanding of what a computational model can accomplish and which model we should select for a given task.

It will also allow us to relate modeling concepts in a different way. In the previous chapter we categorized them with the help of the Rugby metamodel based on the purely technical feature of abstraction levels in different domains. Now we relate them in terms of appropriateness in the context of given tasks and applications. We will notice that there is a certain correlation between the two points of view because specific applications tend to be concerned with only certain abstraction levels. For instance, performance modeling almost exclusively deals with data at the Symbol and Number level. We shall investigate this correlation briefly and also examine how it relates to typical design phases in development projects.

9.1 Performance Analysis

System performance is as important as functionality. The term "performance" is used with two different meanings. In its narrower sense it denotes time behavior, for example, how fast the system can react to events, at which period it can process input signals. In its broader sense it also includes many other nonfunctional properties such as acceptable noise levels of input signals, mean time between failures, reliability, fault tolerance, and other parameters. It is typically limited to properties that can be measured and quantified and does not include features such as "convenience of use." Hence, the terms "performance analysis" and "performance modeling" are restricted to a numerical analysis of some sort.

In general we always prefer closed formulas based on a mathematical model that describes the parameters of interest. For instance, $E[X] = \lambda E[S]$ tells us how the mean queue length $E[X]$ depends on the mean customer arrival rate λ and the mean system processing time $E[S]$ of an arbitrary system. This equation, known as Little's law, is very general and useful because it allows us to dimension system buffers without knowing many of the details such as the scheduling policy of incoming customers, the number and organization of buffers, the number and operation of service units, precise sequence and timing of input events, and so on. Based on a measurement of the mean time of input events and knowing Little's law, we can dimension the input buffers with respect to an expected system timing performance, or we can decide on the buffer sizes and formulate performance requirements for the system design and implementation.

Unfortunately, closed formulas are the exception rather than the rule. Either we do not have suitable mathematical models of the concerned properties that allow us to derive closed formulas, or the formulas are complex differential or difference equations that, in general, cannot be solved accurately. In these cases we have to resort to simulation as the only universally applicable tool of analysis. The main drawback of simulation is that the result depends on the selected inputs. Typically we can only apply a tiny fraction of the total input space in a simulation, which makes the result sensitive to the quality of the inputs.

We restrict the discussion to the buffer size analysis and delay analysis by using the untimed and the synchronous MoC, respectively.

9.1.1 Untimed Analysis

Consider the process network in Figure 9-1 with three processes, A, B, and C. The numbers at the inputs and outputs of the processes denote the number of tokens a process consumes and produces at each invocation. Thus, process B reads one token from its input and emits one token to each of its outputs. Process A can emit one, two, three, or four tokens, depending on its state, which is controlled by the command input. In *state* α the process always emits one token, in *state* β it emits one token in 99.5% of the invocations and two tokens in 0.5%, in *state* γ it emits three tokens in 0.5% of its invocations, and in *state* δ four tokens in 0.5% of invocations. Commands from the command input set the state of the process. If we assume that all processes need equally long for each invocation, a simple analysis reveals that if process A is in *state* α, the system is in a steady state without the need for internal buffers. On the

FIGURE 9-1

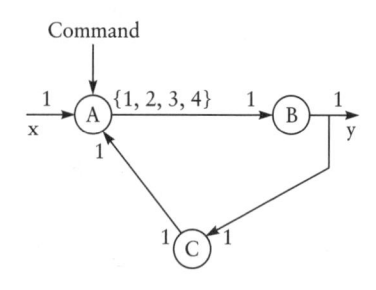

Buffer analysis of a process network.

other hand, when process A is in one of the other states, it produces more tokens than B can consume, and any finite-size buffer between A and B will eventually overflow.

Let us further assume we are dealing with some kind of telecommunication network that receives, transforms, and emits communication packets. *State α* is the normal mode of operation that is active most of the time. However, occasionally the system is put into a monitoring mode, for instance, to measure the performance of the network and identify broken nodes. Monitoring packets are sent out to measure the time until they arrive at a specific destination node, and to count how many are lost along the way. A monitoring phase runs through the following sequence of states: *state β* for 1000 tokens, *state γ* for 1000 tokens, *state δ* for 1000 tokens, *state γ* for 1000 tokens, and finally *state β* for 1000 tokens before *state α* is resumed again. Even if we introduce a buffer between A and B to handle one monitoring phase, the system would still be instable because the system is running under full load even in *state α* and cannot reduce the number of packets in the buffer between the monitoring phases. Consequently, a finite-size buffer would still overflow if we run the monitoring phases often enough.

To make the example more interesting and realistic, we also make some assumptions about the traffic and how it is processed. Some of the packets are in fact *idle packets*, which only keep the nodes synchronized and can be dropped and generated as needed. The packets are represented as integers, and idle packets have the value 0. Processes B and C pass on the packets unchanged, but A processes the packets according to the formula $z = (x + y) \mod 100$, where x and y are input packets and z is the output packet. Consequently, packet values at the output of A are confined to the range $0 \cdots 99$.

To handle the monitoring phases, we introduce a buffer between processes A and B that drops idle packets when they arrive and emits idle packets when the buffer is empty and process B requests packets. What is the required buffer size?

Analysis

Apparently, the answer depends on the characteristics of the input traffic, and in particular how many idle cells there are. Let m be the number of excess packets per normal input packet, introduced by a monitoring process, i is the fraction of idle packets, and N is the total number of input packets in the period we consider. In the first step of the monitoring phase, where process A is in *state β*, we have

$$m = \frac{1}{200} = 0.005$$

$$N = 1000$$

The number of packets we have to buffer is the number of excess packets, introduced by the monitoring activity, minus the number of idle packets:

$$B = \max(0, (m - i) \cdot N) \qquad (9.1)$$

We need the max function because *B* may never become negative; that is, the number of buffered packets cannot be less than 0. For the entire monitoring phase we get

	m	N
state β	0.005	2000
state γ	0.01	2000
state δ	0.015	1000

If we assume a uniform distribution of input packet types, only 1% of the input packets are idle packets, and according to (9.1) we get

$$B = \max(0, 0.005 - 0.01) \cdot 1000 + \max(0, 0.01 - 0.01) \cdot 1000$$
$$+ \max(0, 0.015 - 0.01) \cdot 1000 + \max(0, 0.01 - 0.01) \cdot 1000$$
$$+ \max(0, 0.005 - 0.01) \cdot 1000$$
$$= 5$$

Only *state δ* seems to require buffering, and a buffer of size 5 would suffice to avoid a loss of packets. However, (9.1) assumes that the excess and idle packets are evenly distributed over the considered period of *N* packets. This is not correct; excess packets come in bursts of 1, 2, or 3 at a time, rather than 0.005 at a time. Thus, even in *state β* the buffer has to be prepared to handle the bursts, until consecutive idle packets allow the buffer to empty again. Consequently, we modify (9.1) to take bursts into account, with *b* denoting the longest expected bursts of excess packets:

$$B = \max(0, (m - i) \cdot N) + b \qquad (9.2)$$

For the entire monitoring phase we get

$$B = \max(0, 0.005 - 0.01) \cdot 1000 + 1 + \max(0, 0.01 - 0.01) \cdot 1000 + 2$$
$$+ \max(0, 0.015 - 0.01) \cdot 1000 + 3 + \max(0, 0.01 - 0.01) \cdot 1000 + 2$$
$$+ \max(0, 0.005 - 0.01) \cdot 1000 + 1$$
$$= 14$$

We are quite conservative now in assuming that the effects from bursts accumulate over the entire period. But it is more likely that at the end of a phase, say, *state β*, the buffer is empty although it was not during the entire phase of *state β*. We could try to be more accurate in taking this effect into account by distinguishing the different phases. Let P be all the phases we are looking at and m_p, i_p, N_p, b_p the respective values of phase $p \in P$.

$$B = \sum_p^P (\max(0, (m_p - i_p) \cdot N_p)) + \max(b_p) \tag{9.3}$$

$$\begin{aligned} B = {} & \max(0, 0.005 - 0.01) \cdot 1000 + \max(0, 0.01 - 0.01) \cdot 1000 \\ & + \max(0, 0.015 - 0.01) \cdot 1000 + \max(0, 0.01 - 0.01) \cdot 1000 \\ & + \max(0, 0.005 - 0.01) \cdot 1000 + \max(1, 2, 3) \\ = {} & 8 \end{aligned} \tag{9.4}$$

The assumption seems to be very conservative and unrealistic because it assumes a network load of 99%. However, we have to bear in mind that we investigate the worst-case scenario, and we can conclude that with a buffer size of 8 we will not lose packets up to a network load of 99%; only above this figure will we start to lose packets. This is in fact a reasonable objective for a highly reliable network.

Simulation

In our analysis we have two potential sources of errors. First, there could be a flaw in our mathematical model. We have refined it several times, but perhaps we have not found all the subtleties or maybe we just made simple mistakes in writing down the formulas. Second, we made assumptions about the system and its environment, which must be tested. In particular, we assumed 1% idle packets in the incoming packet stream. Even if this is correct for the system input, it might not be a correct assumption for the input of the buffer we are discussing. The stream is in fact modified in a quite complex way. Process A merges the primary input packets with packets from a feedback loop inside the system. Even though processes B and C do not modify the packets, they delayed the packets on this loop by an amount of time that depends on the state of the buffer, if we ignore the time for processing and communicating the packets for the moment. This is a nonlinear process. Although we might be able to tackle our example with more sophisticated mathematical machinery (see, for instance, Cassandras 1993), in general we will find many nonlinear systems that are not tractable by an analysis aiming at deriving closed formulas. In fact, simulation is our only general-purpose tool, which may help us further in the analysis of systems of arbitrary conceptual complexity. The limits of simulation come from the size of the system and from the size of the input space.

In order to test the assumptions and to validate our model (9.3) and result (9.4), we develop a simulation model based on the untimed model of computation.

We choose untimed process networks because intuitively it fits best to our problem. It is only concerned with sequences of tokens without considering their exact timing. By selecting the relative activation rate of the processes, it should be straightforward to identify the buffer requirements. A discrete event system seems to be overly complicated because it forces us to determine the time instance for each token.

FIGURE 9-2

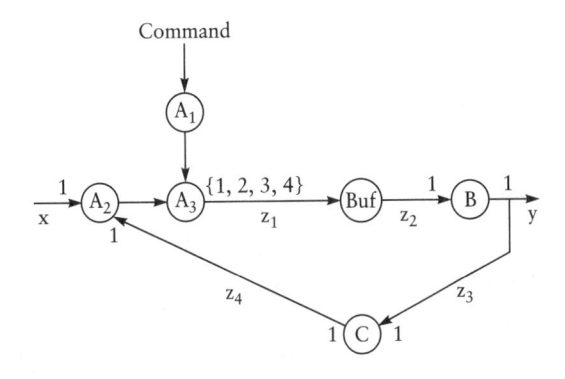

System level of the untimed process network for the example of Figure 9-1.

A synchronous model seems to be inappropriate too because it binds the processes very tightly together, which seems to contradict our intention of investigating what happens in between processes A and B when they operate independently of each other.

We implement the processes of Figure 9-1 as untimed processes and introduce an additional process Buf. We make our model precise by using the generic process constructors (see Figures 9-2 and 9-3).

Process A is composed of three processes (A_1, A_2, and A_3), and the provided functions define how the processes act in each firing cycle. A_1 has a state consisting of two components: a counter counting from 0 through 199, and a state variable remembering the last command. Whenever the counter is 0, a 1, 2, 3, or 4 is emitted depending on the state variable; otherwise a 1 is emitted. This output is used by A_3 and function f_{ctrl}. Depending on this value, A_3, emits 1, 2, 3, or 4 copies of its input to its output. Process A_2 merges the input streams x and z_4 by means of function f_{merge}.

A few remarks are in order. Because an untimed model does not allow us to check for the presence or absence of an input value, we model the cmnd input as a stream of values with a specific value, NoUpdate, if nothing changes. This is a rather unnatural way to model control signals with rare events.

The same property of the untimed model causes some trouble for the buffer process. Process A can emit 1, 2, or any number of tokens during one activation. But Buf has to know in advance how many tokens it wants to read. If it tries to read two tokens, but only one is available, it will block and wait for the second. During this time it cannot serve requests from B. We can rectify this by bundling a number of packets into a single token between A and Buf. Buf always reads exactly one token, which may contain 1, 2, 3, or 4 packets. Consequently, in Figure 9-2 function f_{ctrl} produces a sequence of values at each invocation, and function f_{nbuf} takes a sequence as second argument.

While this solves our problem, it highlights a fundamental aspect of untimed models as we have defined them. Due to the deterministic behavior with a blocking read it is not obvious how to decouple two processes A and B completely from each other. A truly asynchronous buffer is cumbersome to realize. However, what we can do, and what we want to do in this case, is to fix the relative activation rate of A and B

FIGURE 9-3

$$A(cmnd, x, z_4) = A_3(A_1(cmnd), A_2(x, z_4)))$$
$$A_1 = \texttt{mooreU}(1, g_{ns}, f_o, 0)$$

$$f_o(state) = \begin{cases} 1 & \text{if } state = (0, stateA) \\ 2 & \text{if } state = (0, stateB) \\ 3 & \text{if } state = (0, stateC) \\ 4 & \text{if } state = (0, stateD) \\ 1 & \text{otherwise} \end{cases}$$

$$g_{ns}(cmnd, state) = \begin{cases} (x+1 \bmod 200, s) & \text{if } state = (x,s) \wedge cmnd = \text{No Update} \\ (0, StateA) & \text{if } cmnd = CmndA \\ (0, StateB) & \text{if } cmnd = CmndB \\ (0, StateC) & \text{if } cmnd = CmndC \\ (0, StateD) & \text{if } cmnd = CmndD \end{cases}$$

$$A_2 = \texttt{zipWithU}(1, 1, f_{merge})$$
$$f_{merge}(x, y) = (x + y) \bmod 100$$
$$A_3 = \texttt{zipWithU}(1, 1, f_{ctrl})$$

$$f_{ctrl}(c, x) = \begin{cases} \langle x \rangle & \text{if } c = 1 \\ \langle x, 100 \rangle & \text{if } c = 2 \\ \langle x, 100, 101 \rangle & \text{if } c = 3 \\ \langle x, 100, 101, 102 \rangle & \text{if } c = 4 \end{cases}$$

$$B = C = \texttt{mapU}(1, f_{id})$$
$$f_{id}(x) = x$$
$$Buf = \texttt{mealyU}(1, g_{nbuf}, f_{read}, \langle \rangle)$$
$$g_{nbuf}(buf, x) = tail(buf) \oplus x$$
$$f_{read}(buf, x) = head(buf \oplus x)$$

Definitions of processes and functions for the example in Figure 9-2.

and investigate how the buffer in between behaves. In this respect, untimed models resemble synchronous models. In synchronous models we cannot decouple processes A and B either, but we can observe the buffer in between by exchanging varying amounts of data in each activation cycle. If we want to decouple two processes in a nondeterministic or stochastic way, neither the untimed nor the synchronous model can help us.

The simulation of the model yields quite different results depending on the input sequence. Table 9-1 shows the result of 11 different runs. As input we have used 5000 integers in increasing order. For the first row in the table the input started with 0, 1, 2,...; for the second row it started with 1, 2, 3,...; and so on. The last row, marked with a *, the input sequence started with 1, 2, 3,..., but the function f_{merge} was not merging the streams x and x_4 but just passing on x without modification. The objective for this row was to get a very regular input stream without an irregular distribution of 0s. The second column gives the number of 0s dropped by the buffer; the third column gives the number of 0s inserted by the buffer when it was empty and process B requested data.

Table 9-1 shows that the maximum buffer length is very sensibly dependent on the input sequence. If it is a regular stream, like in the last row, the simulation corresponds to our analysis above. But for different inputs we get maximum buffer lengths

TABLE 9-1: *Simulation runs with 5000 samples of consecutive integers.*

Sequence start	Dropped 0s	Inserted 0s	0s %	Buffer length
0	109	69	2.16	14
1	55	11	1.09	13
2	45	8	0.89	13
3	46	4	0.91	9
4	37	2	0.73	16
5	75	31	1.49	13
6	61	18	1.21	12
7	55	12	1.09	11
8	41	3	0.81	13
9	56	21	1.11	15
10	50	11	0.99	16
*	50	6	0.99	8

of anywhere between 8 and 16. Even though the fraction of 0s on the input stream is always 1%, it varies between 0.73% and 2.16% before the buffer. But even when the fraction is similar, like in the last two rows, the maximum buffer length can deviate by a factor of 2.

Besides the great importance of realistic input vectors for any kind of performance simulation, this discussion shows that an untimed model can be used for the performance analysis of parameters unrelated to timing, or rather, to be precise, where we can keep the relative timing fixed in order to understand the relation between other parameters. In this case we examine the impact of control and data input on the maximum buffer size.

The two problems we have encountered when developing the untimed model would be somewhat easier to handle in a synchronous model. The *NoUpdate* events on the control input, which we had to introduce explicitly, would not be necessary, because we would use the more suitable ⊔ events, which represent absent events, indicating naturally that most of the time no control command is received. The situation of a varying data rate between process A_3 and Buf could be handled in two different ways. Either we bundle several packets into one event, as we did in the untimed model, or we reflect the fact that A_1 and Buf have to operate at a rate four times higher than other processes. In *state* δ these two processes have to handle four packets when all other processes have to handle only one. Thus, if we do not assume other buffers, these processes have to operate faster. In the synchronous model we would have four events between A_1 and Buf for each event on other streams. Since most of the time only one of these four events would be valid packets, most of the events would be absent (⊔) events. This had the added benefit that we could observe the impact of different relative timing behavior. In an untimed model we can achieve the same, but we have to model absent events explicitly. However, the differences for problems of this kind are not profound, and we record that untimed and synchronous models are equivalent in this respect, but require slightly different modeling techniques.

The timed model confronts us with the need to determine the detailed timing behavior of all involved processes. While this requires more work and is tedious, it gives

FIGURE 9-4

$$A(cmnd, x, z_4) = A_3(A_1(cmnd), A_2(x, z_4)))$$
$$A_1 = \mathtt{mooreS}(1, g_{ns}, f_o, 0)$$
$$A_2 = \mathtt{zipWithS}(1, 1, f_{merge})$$
$$A_3 = \mathtt{zipWithS}(1, 1, f_{ctrl})$$

$$f_{ctrl}(c, x) = \begin{cases} \langle x \rangle & \text{if } c = 1 \\ \langle x, 100, 100 \rangle & \text{if } c = 2 \\ \langle x, 100, 100, 101, 101 \rangle & \text{if } c = 3 \\ \langle x, 100, 100, 101, 101, 102, 102 \rangle & \text{if } c = 4 \end{cases}$$

$$B = C = \mathtt{mapU}(1, f_{id})$$
$$\mathrm{Buf} = \mathtt{mealyU}(1, g_{nbuf}, f_{read}, \langle \rangle)$$

Redefinitions of processes of Figures 9-2 and 9-3 to obtain a synchronous model.

us more flexibility and generality. We are not restricted to a fixed integer ratio between execution times of processes but can model arbitrary and data-dependent time behavior. Furthermore, we do not need to be concerned with the problem of too tight coupling of the buffer to its adjacent processes. This gained generality, however, comes at the price of less efficient simulation performance and, if a δ-delay-based model is used, a severe obstacle to parallel distributed simulation as discussed in Chapter 5.

9.1.2 Timed Analysis

We have already started to indicate how different models of computation constrain our ability for analyzing timing behavior that is at the core of performance analysis. To substantiate the discussion, we again use the example of the previous section. We will take the example of Figure 9-2, and replace the definitions of Figure 9-3 with the ones in Figure 9-4.

All processes behave identically with the exception of A_3, which introduces a delay for the maintenance packets. We have replaced the untimed model with a synchronous one in order to measure the delay of a packet through the network.

Table 9-2 shows the result. The maximum delay of packets through the network is significant. If there is only a single buffer in the system, as in our case, the delay becomes equal to the maximum buffer size. The necessary buffer size has increased compared to Figure 9-1 because B cannot process packets as fast as before and relies on additional buffering. This kind of analysis is natural in a synchronous model because inserting delays that are integer multiples of the basic event execution cycle is straightforward.

9.2 Functional Specification

In a product development process the specification is typically the first document, where the extensive discussion of many aspects of the problem leads to a first proposal

TABLE 9-2: *Simulation runs with 5000 samples for Figure 9-4.*

Sequence start	Buffer length	Max packet delay
0	50	50
1	56	56
2	58	58
3	54	54
4	48	48
5	48	48
6	47	47
7	51	51
8	59	59
9	55	55
10	52	52

of a system that will solve the given problem. Hence, the purpose of the specification document is twofold:

P-I. It is a means to study if the proposed system will indeed be a solution to the posed problem with all its functional and nonfunctional requirements and constraints, that is, to make sure to make the right system.

P-II. It defines the functionality and the constraints of the system for the following design and implementation phases.

From these two purposes we can derive several general requirements for a specification method:

A. *To support the specification process:* To write a specification is an iterative process. This process should be supported by a technique that allows the engineer to add, modify, and remove the entities of concern without a large impact on the rest of the specification.

B. *High abstraction:* The modeling concepts must be at a high enough abstraction level. The system engineer should not be bothered with modeling details that are not relevant at this stage.

C. *Implementation independent:* The system specification should not bias the design and implementation in undesirable ways. System architects must be given as much freedom as possible to evaluate different architecture and implementation alternatives. Products are frequently developed in several versions with different performance and cost characteristics. Ideally the same functional specification should be used for all versions.

D. *Analyzable:* The specification should be analyzable in various ways, for example, by simulation, formal verification, performance analysis, and so on.

E. *Base for implementation:* The specification should support a systematic technique to derive an efficient implementation. This is in direct conflict with requirements B and C.

In this section we concentrate on the functional part of the specification for the purpose of developing and understanding the system functionality. Item E on the above list, "base for implementation," will be discussed in the following section on design and synthesis. Some aspects of item D, "analyzable," have been investigated in the section on performance analysis, and other aspects belong to the realm of formal verification, which is beyond the scope of this book. Hence, here we elaborate on the first three items.

The specification phase comes after the requirements engineering and before the design phase. During requirements engineering all functional and nonfunctional requirements and constraints are formulated. After that, the functional specification describes a system that meets the given constraints. To cut out the space for the functional specification, we introduce two examples and take both from a requirements definition over a functional specification to a design description. The first example is a number sorter, and the second example is a packet multiplexer. After that we discuss the role of the functional specification and what kind of computational model best supports this role.

9.2.1 Example: Number Sorter

Figure 9-5 shows the requirements definition of a functional number sorter that converts an indexed array A into another indexed array B that is sorted. The requirements only state the conditions that must hold in order that B is considered to be sorted, but it does not indicate how we could possibly compute B.

Figure 9-6 gives us two alternative functional models for sorting an array. The first, sort1, solves the problem recursively by putting sorted subarrays together to form the final sorted array. First comes the sorted array of those elements that are smaller than the first array element (first(A)), and finally the sorted array of those elements that are greater or equal to first(A). The second function, sort2, is also recursively defined and takes all elements that are smallest and puts them in front of the sorted rest of the array. Both functions are equivalent, which can be proved, but they exhibit important differences that affect performance characteristics of possible implementations. One important feature of both function definitions is that potential parallelism is only constrained by data dependences. For example, sort2 does not define if smallest(A) is computed before, in parallel with, or after the rest of the array. In fact, it does not matter for the result

---

FIGURE 9-5

$$\text{sort} :: (\text{IntArray}A) \Rightarrow (\text{IntArray}B)$$

$$\text{Precondition: true}$$

$$\text{Postcondition: } \forall a : a \in A \to a \in B$$

$$\land \forall b : b \in B \to b \in A$$

$$\land \forall i, j \in \mathbb{Z}, b_i, b_j \in B : i < j \to b_i \leq b_j$$

Functional requirements of a number sorter.

9.2 Functional Specification 315

FIGURE 9-6

$$\text{sort} :: (\text{IntArray}A) \Rightarrow (\text{IntArray}B)$$

$$\text{sort1}([\,]) = [\,]$$

$$\text{sort1}(A) = \text{sort1}(\text{select\_less}(\text{first}(A), \text{tail}(A))) + + \text{first}(A)$$
$$+ + \text{sort1}(\text{select\_geq}(\text{first}(A), \text{tail}(A)))$$

$$\text{sort2}([\,]) = [\,]$$

$$\text{sort2}(A) = \text{smallest}(A) + + (\text{sort2}(\text{select\_not}(\text{smallest}(A), A)))$$

Two functional models of a number sorter.

FIGURE 9-7

```
sort2_alg (IntArray& A)
{ for (i = 2; i <= length(A); i++)
    { for (j = length(A); j >= i; i--)
        { if (B [j-1] > B [j])
            {
                int tmp = input [j-1];
                B [j-1] = B [j];
                B [j] = tmp;
            }
        }
    }
}
```

An algorithm for a number sorter.

in which order the computations are performed. However, the computation is divided into two uneven parts. smallest(A) is inherently simpler to compute than deriving and sorting the rest of the array. This is also obvious from the fact that smallest(A) is even a part of the second computation. Thus, the data dependences will enforce a rather sequential implementation. sort1, on the other hand, has a much higher potential for a parallel implementation because the problem is broken up more evenly. Unless A is not almost sorted, the problems of sorting those elements that are less than first(A) will be as big as sorting those elements that are greater or equal to first(A). Which one of the two functional models is preferable will depend on what kind of implementation will be selected. sort1 is better for a parallel architecture such as an FPGA or a processor array. For a single-processor machine that executes purely sequential programs, sort2 should be preferred because sort1 might be hopelessly inefficient. Figure 9-7 shows a sequential algorithm of function sort2. Several design decisions were taken when refining sort2 into sort2_alg. The amount of parallelism has been fixed, basically by removing the potential parallelism. The representation of the data and how to access it has been determined. Finally, the details of how to compute all functions such as smallest( ) have been worked out.

9.2.2 Example: Packet Multiplexer

The packet multiplexer merges four 150 packets/second streams into one 600 packets/second stream as shown in Figure 9-8. The requirements definition would be mostly concerned with the constraints on the interfaces and the performance of the input and output streams. The functional requirements would state that all packets on the inputs should appear on the output and that the ordering on the input streams should be preserved on the output stream.

A functional model fulfilling these requirements is shown in Figure 9-9. We assume that we have a process constructor $zipWith4U$, which is a generalization of the previously defined $zipWithU$ and merges four signals into one.

Since the functionality we are asking for is simple, the functional model is short. But a further refined design model could be more involved, as is shown in Figure 9-10. It is based on three simple two-input multiplexers that merge the input stream in two steps. It is refined, although it implements the same functionality, because we made a decision to realize the required function by means of a network of more primitive elements.

Let's assume that our requirements change and the output does not have a capacity of 600 packets/second but only 450. That means we have to drop one out of four packets

FIGURE 9-8

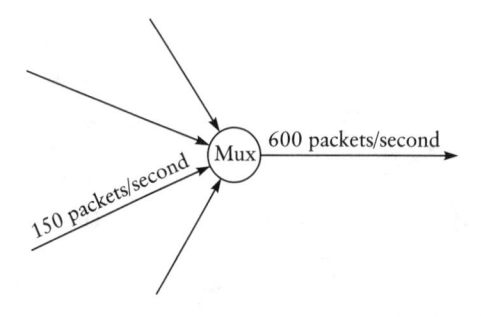

Packet multiplexer.

FIGURE 9-9

$$\text{Mux}(s_1, s_2, s_3, s_4) = zipWith4U(1, 1, 1, 1, f)$$
$$f(x_1, x_2, x_3, x_4) = \langle x_1, x_2, x_3, x_4 \rangle$$

A functional model of the packet multiplexer as an untimed process.

FIGURE 9-10

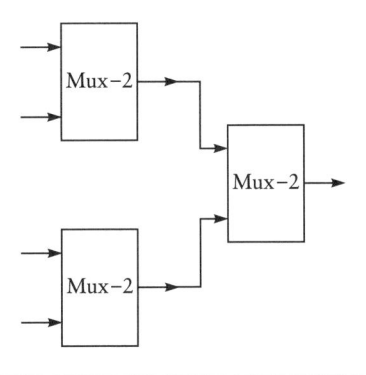

A refined packet multiplexer based on simpler multiplexer units.

FIGURE 9-11

$$\text{Mux2}(s_1, s_2, s_3, s_4) = \mathtt{zipWith4U}(1, 1, 1, 1, f)$$
$$f(x_1, x_2, x_3, x_4) = \langle x_2, x_3, x_4 \rangle \text{ if } x_1 = idle$$
$$\langle x_1, x_3, x_4 \rangle \text{ if } x_2 = idle$$
$$\langle x_1, x_2, x_4 \rangle \text{ if } x_3 = idle$$
$$\langle x_1, x_2, x_3 \rangle \text{ otherwise}$$

A modified functional model of the packet multiplexer.

and, according to the new requirements, we should avoid buffering packets and drop one packet in each multiplex cycle. If there is an idle packet, we should drop it; otherwise it does not matter which one we lose. A new functional description, which meets the new requirements, is shown in Figure 9-11. The function f outputs only three of the four input packets, and it drops the packet from stream s_4 if none of the other packets is idle. When we propagate the modification down to the design model, we have several possibilities, depending on where we insert the new functionality. It could become part of modified Mux-2 blocks, or it could be realized by a separate new block before the output. We could also distribute it and filter out all idle packets at the inputs, and either insert idle packets or filter out a packet from stream s_4 before the output. The solution we have chosen, as shown in Figure 9-12, inserts two new blocks, filter_idle, which communicate with each other to filter out the right packet.

9.2.3 Role of a Functional Specification

The borderlines between requirements definition, functional specification, and design are blurred because in principle it is a gradual refinement process. Design decisions are

FIGURE 9-12

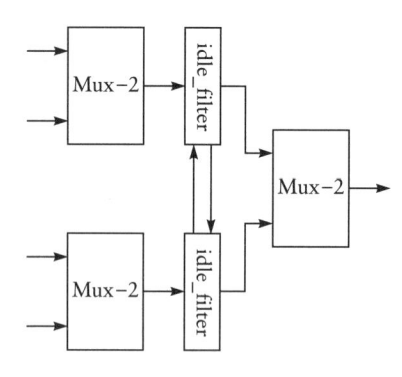

A modified design of the packet multiplexer.

added step by step, and the models are steadily enriched with details, which eventually leads hopefully to a working system that meets the requirements. Hence, to some degree it is arbitrary which model is called the requirements definition, functional specification, or design description, and in which phase which decisions are taken. On the other hand, organizing the design process into distinct phases and using different modeling concepts for different tasks is unavoidable.

In the number sorter example, we have made several design decisions on the way from the requirements definition to a sequential program. We have not made all the decisions at once but rather in two steps. In the first step, which ended with a functional model (Figure 9-6), we broke the problem up into smaller subproblems and defined how to generate the solution from the subsolutions. The way we do this constrains the potential parallelism. In addition, the breaking the problem up into subproblems will also constrain the type of communication necessary between the computations solving the subproblems. This may have severe consequences for performance if a distributed implementation is selected.

While these observations may be self-evident, for the number sorter it is less obvious that we need a functional model at all. It might be preferable to develop a detailed implementation directly from the requirements definition because the best choice on how to break up the problem only becomes apparent when we work on the detailed implementation with a concrete architecture in mind. Neither sort1 nor sort2 is clearly better than the other; it depends on the concrete route to implementation.

Although this is correct and might render a functional model for simple examples useless, the situation is more difficult for the development of larger systems. First, with the introduction of a functional model the design decisions between requirements definition and implementation model are separated into two groups. Since the number of decisions may be very large and overwhelming, to group them in a systematic way will lead to better decisions. This is true in particular if certain issues are better analyzed and explored with one kind of model than with another. One issue that is better investigated with a functional model, as we have exposed with the number sorter example, is the potential parallelism and, as a consequence, the constraints on dataflow and

9.2 Functional Specification

communication between the computations of subproblems. However, this does not rule out taking into account the different architectures for the implementation. On the contrary, in order to analyze sensibly different functional alternatives, we must do this with respect to the characteristics of alternative architectures and implementations.

Another issue where a functional model may support a systematic exploration is revealed by the second example, the packet multiplexer. The functional design phase must explore the function space. Which functions should be included or excluded? Can a given set of functions meet user needs and requirements? These user needs and requirements are sometimes only vaguely specified and may require extensive analysis and simulation. For instance, the convenience of a user interface can often only be assessed when simulated and tried by a user. The impact of adding or deleting a function on other parts of the system is sometimes not obvious. For instance, the change of functionality in the packet multiplexer example, which means that up to 100% of the packets in stream s_4 are lost, may have undesirable consequences for the rest of the telecommunication network. It might be worthwhile to evaluate the impact of this change by an extensive simulation. A functional model can greatly facilitate these explorations provided we can quickly modify, add, and remove functions. In contrast, the consecutive design phase is much more concerned with mapping a functionality onto a network of available or synthesizable components, evaluating alternative topologies of this network, and, in general, finding efficient and feasible implementations for a fixed functionality. If these two distinct objectives are unconsciously mixed up, it may lead to an inefficient and unsystematic exploration of alternatives. In the case of the packet multiplexer it is unnecessary to investigate with great effort the best place to insert the new filter functionality or to modify the existing Mux-2 blocks if a functional system simulation reveals that the new functionality has undesirable consequences and must be modified or withdrawn.

From these considerations we can draw some conclusions on how best to support purpose **P**-I of a functional specification. In order to efficiently write and modify a specification document, we should avoid all irrelevant details. Unfortunately, what is relevant and what is not relevant is not constant and depends on the context.

- *Time:* Frequently timing is not an issue when we explore functionality, and consequently information on timing should be absent. But for almost every complex system, not just real-time systems, there comes a point when timing and performance must be considered as part of the functional exploration. As we have seen in the previous section on performance analyses, most analysis of nonfunctional properties will involve considerations about time. In light of this argument, we may reason that a synchronous model may be best because, on the one hand, it makes default assumptions about time that alleviate the user from dealing with it when not necessary. On the other hand, it allows us to make assumptions about the timing behavior of processes that can be used both as part of the functional simulation and can serve as timing constraints for the consecutive design phases.

- *Concurrency:* As we have seen in the number sorter example, the functional specification phase breaks up the problem into subproblems and explores potential parallelism. The models of computation that we have reviewed are all models with explicit concurrency. But for functional exploration, the boundaries between concurrent processes should not be too distinct, and functionality should be easily moved from one process to another. Thus, either we should avoid explicit concurrency altogether, only using models of implicit concurrency such as a purely functional model, or we should use the same modeling concepts inside processes as between processes. For instance, if we have a hierarchical untimed process network without a distinction between a coordination and a host language, the process boundaries can be easily moved up and down the hierarchy (Skillicorn and Talia 1998a).

- *Primitive functions:* The primitive elements available are as important as the computational model and are often application specific. For instance, various filters and vector transformation functions are required for signal-processing applications, while control-dominated applications need control constructs, finite state machines, exceptions, and similar features. Obviously, the availability of an appropriate library makes the modeling and design process significantly more efficient.

- *Avoiding details:* The kind and the amount of details in a model is determined by three factors: (a) the abstraction levels in the four domains of computation, communication, time, and data; (b) the primitive functions and components available; (c) the modeling style of the designer. Given (a) and (b), the designer still has significant choice in modeling and how to use languages and libraries.

With respect to purpose **P**-II we refer to the following discussion of design and synthesis.

9.3 Design and Synthesis

Manual design and automatic synthesis are at the core of the engineering task of creating an electronic system. After having determined the functionality of the system, we must create a physical device that implements the functionality. Essentially, this is a decision-making process that gradually accumulates a huge amount of design information and details. It can be viewed as going from a high abstraction level, which captures the basic functionality and contains relative few details, to ever lower abstraction levels with ever more details. This is quite a long way, stretching from a system functionality model to a hardware and software implementation. The hardware implementation is a description of the geometry of the masks that are direct inputs to the manufacturing process in a semiconductor foundry. For software it is sequence of instructions that control the execution of a processor. A huge number of different design decisions have to be taken along this way, for instance, the selection of the target technology, the selection of the components and libraries, the architecture, the memory system, the hardware/software distribution, and so on.

In general we can observe that we use a combination of two techniques:

- Top-down: synthesis and mapping
- Bottom-up: use of components and libraries

If our problem is such that we can find a component in our library that solves the problem directly, we skip the top-down phase, select the right component, and we are done. In this case, although we might be very happy, we would hardly call our activity a system design task. On the other hand, we always have a set of primitive components, from which we must select and assemble to get a device solving our problem. These components may be simple and tiny, such as transistors or instructions for a processor, or they may be large and complex, such as processors, boards, databases, or word processors. Curiously, the synthesis and mapping phase tends to be rather shallow. During the last 30 years the complexity of the "primitive" components has tightly followed the complexity of the functionality of electronic systems. While the "system functionality" developed from combinatorial functions to sequential datapaths and complex controllers to a collection of communicating tasks, the "primitive library components" evolved from transistors to complex logic gates to specialized functional blocks and general processor cores. Attempts to cross several abstraction levels by means of automatic synthesis, such as high-level synthesis tools, have in general not been very successful and have typically been superseded by simpler synthesis tools that map the functionality onto more complex components. The latest wave of this kind has established processor cores as "primitive components."

In the following we discuss in general terms a number of system design problems in order to understand what kind of information has to be captured by different models and how one kind of model is transformed into another. This will allow us to assess how adequate the models of computations are to support these activities.

9.3.1 Design Problems and Decisions

We use a simple design flow as illustrated in Figure 9-13 to discuss the major design tasks and their relation. This flow is simplified and idealistic in that it does not show feedback loops, while in reality we always have iterations and many dependences. For instance, some requirements may depend on specific assumptions of the architecture, and the development of the task graph may influence all earlier phases. But this simple flow suffices here to identify all important design and synthesis activities and to put them in the proper context.

Components and Architecture

In addition to the functional specification, the architecture has to be defined. The architecture is not derived from the functional specification but is the second major independent input to the consecutive design and implementation phases. The architecture cannot be derived from a given functionality for two reasons. First, the number of possible architectures (i.e., the design space) is simply too large. There is no way that a tool or a human designer can systematically consider and assess a significant part of all potentially feasible architectures. But this is not really necessary either because we have experience and a good understanding of which kinds of architectures are efficient and cost effective for a given application. Hence, we only need to assess a few variants of these proven architectures. Second, designers typically face constraints about the architectures they can use. These constraints may spring up from a number

FIGURE 9-13

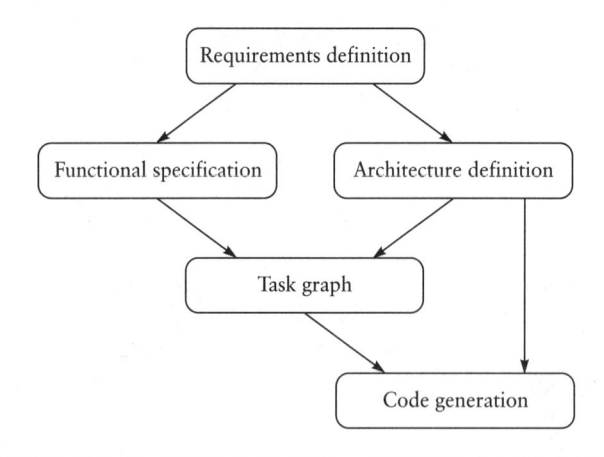

```
                    ┌─────────────────────────┐
                    │ Requirements definition │
                    └─────────────────────────┘
                        ↙             ↘
      ┌─────────────────────────┐  ┌─────────────────────────┐
      │ Functional specification│  │ Architecture definition │
      └─────────────────────────┘  └─────────────────────────┘
                        ↘             ↙
                         ┌────────────┐
                         │ Task graph │
                         └────────────┘
                                ↓
                         ┌─────────────────┐
                         │ Code generation │
                         └─────────────────┘
```

System design flow from requirements engineering to code generation of hardware and software.

of sources. Perhaps we must design only a subpart of a system that has to fit physically and logically to the rest of the system. Or maybe our company has special deals with a particular supplier of a processor, a board, or an operating system, forcing us to use these elements rather than something else. Another source of constraints is that we have to design not just one product but a family of products, and we have to consider all of them at the same time. Consequently, the architecture is always defined almost independently from the definition of the functionality, and it is an important input to the following steps of task graph design and code generation. To define an architecture basically means to select a set of components and a structure to connect them.

The selection of the components and the design of the topology can be a very intricate problem. First, component selection is often hierarchical both for software and hardware components. We have to choose a library from a particular vendor or for a specific technology. From the libraries we select the components for our design. The alternative libraries are usually not equivalent or directly comparable. Certain components may be faster in one library than in another; others may be slower. Some components may not be available in a library, forcing us to find alternative solutions based on a combination of other components. Second, the selection of individual components is not independent from the selection of other components. The choice of one particular microprocessor core determines or constrains the kind of bus we can use, which in turn constrains the selection of other components. The processor also imposes constraints on the memory system and caches, which may have far-reaching consequences for system performance and software code size, which in turn may affect the task partitioning, software component selection, and scheduling policy.

Not only are components organized hierarchically into libraries, but also the assembly of components are organized in a hierarchy of architectures. We may have a system architecture, a board architecture, and a chip architecture. The software may also be organized in a hierarchical architecture comprising several layers.

These observations illustrate that the selection of components and the design of architectures are indeed intricate tasks. Many alternatives must be considered, assessed, and compared. This demands that we are able to model the components and their composition with respect to functional and nonfunctional features.

Tasks and Code Generation

The system functionality must be partitioned into tasks, which in turn must be mapped onto the network of allocated components. Again, this involves many delicate, interdependent decisions. Some of the important activities are the partitioning into tasks and the mapping of tasks onto resources such as processors, micro controllers, DSPs, FPGAs, custom hardware, and so on. For tasks mapped to software, the scheduling policy must be determined. Scheduling can be static and decided during design time, or dynamic during run time. In the latter case a run-time scheduler must be included. Each task must then be implemented, which means the generation of executable code for software and a layout description for hardware. The design and synthesis of interfaces and communication code between tasks and components is one of the most troublesome and error-prone activities.

Constraints and Objectives

All these design activities and decisions have to be done with a permanent awareness of the overall constraints and objectives with respect to nonfunctional properties.

9.3.2 Consequences for Modeling

This brief review of design problems allows us to identify some requirements on modeling. One obvious observation is that the assessment of nonfunctional properties is an integral task of most design activities. Unfortunately, nonfunctional properties, with the notable exception of time, are not an integral part of most modeling techniques and modeling languages. Since our theme is the analysis of computational models, we can only acknowledge this situation and restrict further discussion to issues of functionality and time.

To model components and networks of components realistically, we should employ a timed model because it is the only metamodel that can capture the timing behavior with arbitrary precision. This may not always be necessary; in fact in many cases it is not. When we only deal with the functionality, the delay of individual components is irrelevant. In other cases we are content with a more abstract notion of time corresponding to a synchronous model. For instance, in hardware design the very popular synchronous logic partitions a design into combinatorial blocks separated by latches or registers, as illustrated in Figure 9-14. The edge-triggered registers at the inputs and outputs of the combinatorial block are controlled by the rising edge of a clock signal. Thus, when the clock signal rises, the inputs a, b, c, and d to the combinatorial block change and the gates react by changing their respective outputs as appropriate. When the clock signal rises the next time, the outputs y and z are stored in the output registers and are thus available for the next combinatorial block.

FIGURE 9-14

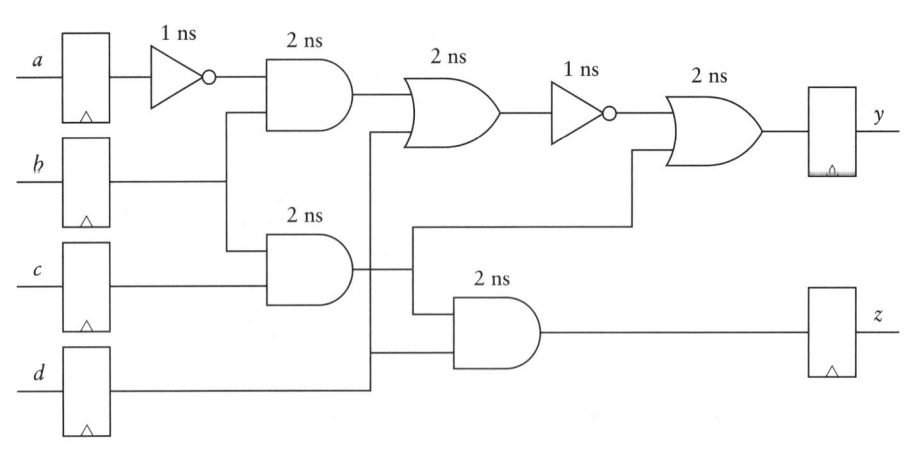

A combinatorial network of gates.

This allows us to separate timing issues from functionality issues; in the first step we assume that all combinatorial blocks are fast enough to deliver a result within a clock period, and in the second step we verify that this assumption is in fact correct. Thus, the first step does not require a timed model but only an untimed or a synchronous model. It is indeed an order of magnitude more efficient to simulate such designs in a synchronous timing model as the cycle-true simulators have proven. Only the second step requires a timed model, but even there we can separate timing from functionality by means of a static timing analysis. For the example in Figure 9-14 the static timing analysis would calculate the longest path, which is from input a to output y, and takes 8 ns. Thus, when any of the inputs change, all outputs would be available after 8 ns and the clock cycle needs to be no longer than 8 ns.

Consequently we could describe the functionality of a component based on an untimed model that just represents the input-output behavior, and store the timing information as a separate annotation. But even though the actual use of accurate timing information is limited, using the timed model for component modeling is still recommended for two reasons. First, almost always comes a situation where it is mandatory, for instance, when the behavior depends on the precise timing or vice versa, and this is relevant. Static timing analysis can only reveal the worst-case delay. However, when this worst case can never happen because the input vector to a combinatorial block is inherently restricted by another block, static timing analysis would be too conservative. For instance, assume that in the example of Figure 9-14 the inputs are constrained such that a is always equal to b. The AND gate that connects NOT a and b would always have an output of 0 and would never change, no matter what the inputs were, if we ignore glitches. The longest path would be 5 ns from d to y. Such situations cannot be detected by static analysis or functional simulation alone. Only when we analyze the functional behavior and the timing of the components together can we detect this. The second reason why we prefer a timed model for modeling components and structure is that timing information can be easily ignored when components are included in a simulation based on another computational model, as has been shown with functional and cycle true simulations.

Another main concern for modeling is the transformation of functionality during the design process. System functions are partitioned into tasks, communication between tasks is refined from high-level to low-level protocols, parallel computations are serialized and vice versa, functions are transformed until they can be mapped onto a given network of primitive components, and so on and so on. If all this is done manually, we do not have much to worry about, at least not for the design process itself. We must still worry about validation. For an automatic synthesis process the situation is more delicate because all the transformations and refinements must be correct. While it is possible—and even popular—to hardwire rules into the synthesis tool, this process can be supported more systematically and in a more general way by providing large classes of equivalences. Whenever we have a case of semantic equivalence a tool can use this to transform a given design into a more optimal implementation. For instance, for a function over integers $f_1(x) = h(x) + g(x)$, the commutative law tells us that this is equivalent to $f_2(x) = g(x) + h(x)$. A synthesis tool could then transform f_1 into f_2 if the hardware structure is such that the latter represents a more efficient implementation. However, if this is not guaranteed by the semantics of the language, such a transformation is not possible and the exploration space of the synthesis tool is significantly restricted. For instance in C this particular equivalence is not guaranteed because g and h may have side effects that may change the result if they are evaluated in a different order.

9.3.3 Review of MoCs

The essential difference of the three main computational models that we introduced in Chapters 3 through 5 is the representation of time. This feature alone weighs heavily with respect to their suitability for design tasks and development phases.

Timed Model

The timed model has the significant drawback that precise delay information cannot be synthesized. To specify a specific delay model for a piece of computation may be useful for simulation and may be appropriate for an existing component, but it hopelessly overspecifies the computation for synthesis. Assume a multiplication is defined to take 5 ns. Shall the synthesis tool try to get as close to this figure as possible? What deviation is acceptable? Or should it be interpreted as "max 5 ns"? Different tools will give different answers to these questions. Synthesis for different target technologies will yield very different results, and none of them will match the simulation of the DE model. The situation becomes even worse when a δ-delay-based model is used. As we discussed in Chapter 5 the δ-delay model elegantly serves the problem of nondeterminism for simulation, but it requires a mechanism for globally ordering the events. Essentially, a synthesis system had to synthesize a similar mechanism together with the target design, which is an unacceptable overhead.

These problems notwithstanding, synthesis systems for both hardware and software have been developed for languages based on timed models. VHDL- and Verilog-based tools are the most popular and successful examples. They have avoided these problems by ignoring the discrete event model and interpreting the specification according to a clocked synchronous model. Specific coding rules and assumptions allow the tool to identify a clock signal and infer latches or registers separating the combinatorial blocks. The drawbacks of this approach are that one has to follow special coding guidelines for synthesis, that specification and implementation may behave differently, and in general that the semantics of the language is complicated by distinguishing between a simulation and a synthesis semantics. The success of this approach illustrates the suitability of the clocked synchronous model for synthesis but underscores that the untimed model is not synthesizable. Apparently, this does not preclude efficient synthesis for languages based on the untimed model, even though it comes at a cost.

Synchronous Model

The synchronous models represent a sensible compromise between untimed and fully timed models. Most of the timing details can be ignored, but we can still use an abstract time unit, the evaluation or clock cycle, to reason about the timing behavior. Therefore it often has a natural place as an intermediate model in the design process. Lower-level synthesis may start from a synchronous model. Logic and RTL synthesis for hardware design and the compilation of synchronous languages for embedded software are prominent examples. The result of certain synthesis steps may also be represented as a synchronous description such as scheduling and behavioral synthesis.

It is debatable if a synchronous model is an appropriate starting point for higher-level synthesis and design activities. It fairly strictly defines that activities occurring in the same evaluation cycle but in independent processes are simultaneous. This imposes a rather strong coupling between unrelated processes and may restrict early design and synthesis activities too much.

Untimed Model

As we saw in Section 3.15, the dataflow process networks and their variants have nice mathematical features that facilitate certain synthesis tasks. The tedious scheduling problem for software implementations is well understood and efficiently solvable for synchronous and boolean dataflow graphs. The same can be said for determining the right buffer sizes between processes, which is a necessary and critical task for hardware, software, and mixed implementations. How well the individual processes can be compiled to hardware or software depends on the language used to describe them. The dataflow process model does not restrict the choice of these languages and is therefore not responsible for their support. For what it is responsible—the communication between processes and their relative timing—it provides excellent support due to a carefully devised mathematical model.

Figure 9-15 summarizes this discussion and indicates in which design phases the different MoCs are most suitable.

9.4 Further Reading

Many good books describe various performance-modeling techniques in detail. Examples are Cassandras (1993) and Severance (2001).

FIGURE 9-15

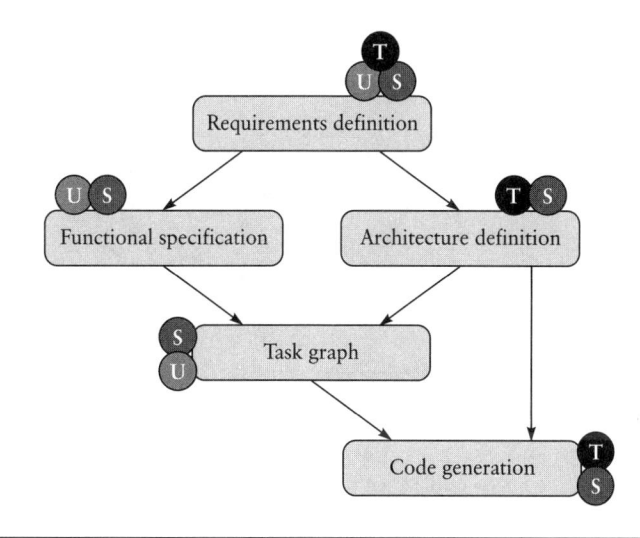

Suitability of MoCs in different design phases.

There are also several texts that discuss the subject of functional system specification. However, the presentation of this subject varies considerably since no standard point of view has been established yet. Gajski and co-workers have taken up an extensive general discussion several times while introducing a concrete language and methodology, SpecChart (Gajski et al. 1994) and SpecC (Gajski et al. 2000). Both books also discuss the design and synthesis aspects of models.

A more formal approach is described by van der Putten and Voeten (1997). Another very formal approach, based on abstract algebra, is given by Ehrig and Mahr (1985).

There are a large number of methodology-oriented books on specification and design. A classic book on structured analysis and specification is DeMarco (1978). Good introductions to object-oriented analysis and specification are Selic et al. (1994), Yourdon (1994), and Coad and Yourdon (1991).

9.5 Exercises

9.1 Given is a switch and a number of n resources, all directly connected to the switch, as shown in Figure 9-16.

Every resource sends packets to another resource with a probability of p_n. The destination resource is selected randomly with a uniform distribution; that is, the probability to select a particular resource is $p_n = \frac{1}{n}$. If destination and source are the same, no packet is transmitted over the network.

The switch operates with a cycle time $c_s = \rho c_r, \rho \geq 1$, where c_r is the cycle time of the resources. Thus, the switch can send and receive ρ packets, while a resource can send and receive only 1 packet.

FIGURE 9-16

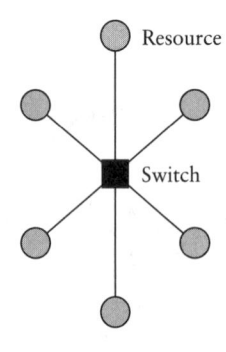

A single switch-based network of resources.

The switch has a central queue with length q. If a new packet arrives when the buffer is full, the new packet is discarded.

What is the probability of losing a packet due to buffer overflow? Make a simulation model with the following parameters:

a. $n = 4, \rho = 2, q = 5$
b. $n = 6, \rho = 2, q = 5$
c. $n = 16, \rho = 2, q = 5$
d. $n = 6, \rho = 2, q = 2$
e. $n = 6, \rho = 2, q = 10$
f. $n = 6, \rho = 2, q = 20$
g. $n = 6, \rho = 3, q = 5$
h. $n = 6, \rho = 4, q = 5$
i. $n = 6, \rho = 6, q = 5$

Simulate with 1000, 10,000, and 100,000 cycles to see which probability is approximated.

9.2 Develop and model the data link protocol layer of the architecture of Exercise 9.1. The switch is connected to each resource node by a 32-bit bus and a few control lines. You may introduce and use a reasonable number of control lines such as "REQUEST," "ACKNOWLEDGE," "DATA VALID," and so on. The switch and the resource run with different clocks at different speeds and with different phases. No assumptions can be made about the relationship of the clocks. Both the switch and the resource can initiate a data transfer.

Develop a protocol to exchange data between the switch and the resource, based on 32-bit words. The package size is fixed with 16 words. The protocol should

FIGURE 9-17

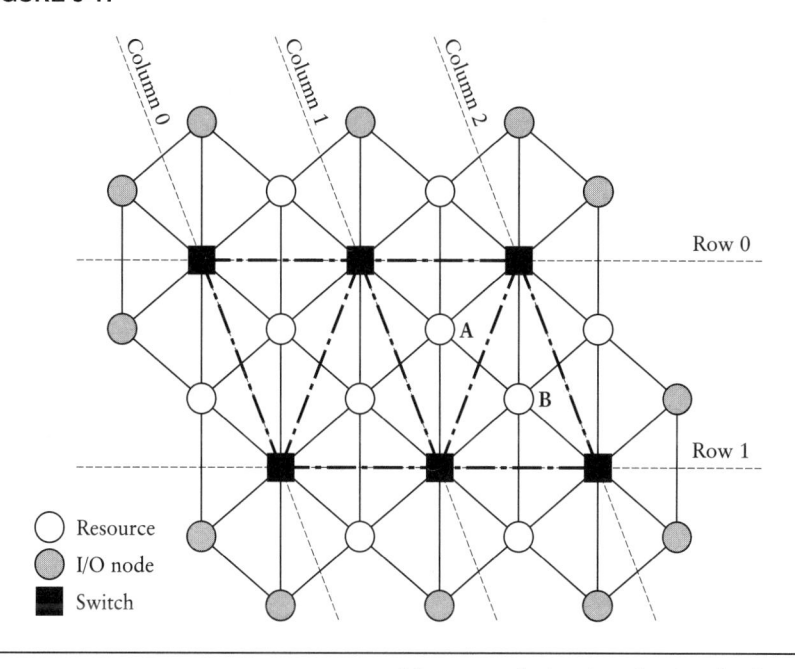

A honeycomb structured network of resources.

be reliable; thus the protocol is not allowed to lose packets. Optimize the latency and the throughput for both very low and very high data rates.

You may use either the synchronous or the untimed MoC.

9.3 Given is a honeycomb network of resources and switches (Figure 9-17) with s switches, r resources, o of which are I/O resources. $r(1) = 6, r(s) = 3s+4, s > 1$; $o(1) = 6, o(s) = s+6, s > 1$.

The switches are organized in rows and columns as shown in Figure 9-17. In this way each switch gets a unique address, which consists of the row and the column number <row#,col#>. A resource gets the coordinates of the switch to which it is connected plus a third number, which determines its position in the hexagon of the switch. This number, called "wind", is between 0 and 5 and is defined as follows: $0 = $ N, $1 = $ NE, $2 = $ SE, $3 = $ S, $4 = $ SW, $5 = $ NW. To avoid multiple addresses for the same resource, we define that only row 0 has wind $= 0, 1,$ and 5; all other rows have only the bottom half of the wind coordinates. Similarly, only column 0 has wind $= 4$ and 5, all others have only the right part of the wind coordinates. In this way all resources have a unique address. For example, resource A in Figure 9-17 has the coordinates <row $= 0$, col $= 1$, wind $= 2$>, while B has <row $= 0$, col $= 2$, wind $= 3$>.

Every resource sends packets to another resource with a probability of p_r. The destination resource is selected randomly with a uniform distribution; that is, the probability to select a particular resource is $p_r = \frac{1}{r}$. If destination and source are the same, no packet is transmitted over the network.

The routing in the network follows some rules. If source and destination are directly connected, they communicate directly; otherwise the source sends the packet to the switch to which it is associated, then the packet is handed from switch to switch until it reaches the switch to which the destination is associated. The routing tables in the switches are set up such that a packet is first transferred to the destination row and then to the destination column.

The switches operate with a cycle time $c_s = \rho c_r, \rho \geq 1$; that is, a switch can send and receive ρ packets, while a resource can send and receive only 1 packet.

The switches have central queues with length q. If a new packet arrives when the buffer is full, the new packet is discarded.

What is the probability of losing a packet due to buffer overflow? Make a simulation model with the following parameters:

a. $s = 30(5\ rows), \rho = 2, q = 5$
b. $s = 30(5\ rows), \rho = 4, q = 5$
c. $s = 30(5\ rows), \rho = 2, q = 10$
d. $s = 30(5\ rows), \rho = 2, q = 15$
e. $s = 100(10\ rows), \rho = 2, q = 5$
f. $s = 100(10\ rows), \rho = 2, q = 10$
g. $s = 100(10\ rows), \rho = 2, q = 15$
h. $s = 100(4\ rows), \rho = 2, q = 15$

Simulate with 1000, 10,000, and 100,000 cycles to see which probability is approximated.

9.4 Repeat Exercise 9.3 with an additional locality assumption. The probability of selecting a particular destination resource is a function of the distance d_r to this resource. The probability that a resource sends a packet to another resource at distance d is

$$p_D(d) = \frac{1}{A(D)2^d}$$

where D is the maximum possible distance and

$$A(D) = \sum_{d=1}^{D} \frac{1}{2^d}$$

Hence, $A(D)$ is a normalization factor that guarantees that the sum of the probabilities is 1. The probability of selecting a particular resource from resources with the same distance from the sending node is uniformly distributed.

9.5 Develop and model the network layer protocol of the NOC architecture of Exercise 9.3 for both the switches and the resources. Use the data link protocol for the communication between individual switches and resources. Provide a reliable service for the next higher layer without acknowledgment; that is, that an application uses the network layer by calling it with a packet and the destination address. Then the application can be assured that the packet will receive the destination, and it does not expect any acknowledgment.

chapter ten
Concluding Remarks

Frameworks for models of computation have two purposes. From the practical point of view, the integration of different computational models provides a modeling environment that allows us to co-simulate models written in different design languages and with distinct timing, communication, and synchronization concepts. This is necessary because today's systems are heterogeneous in nature and consist of parts with radically different characteristics designed by different groups with different tools and design methodologies. Practically oriented frameworks deal as much with the syntactic differences of languages as with the semantics of interaction between MoC domains.

From a theoretical point of view, frameworks can be tools to analyze the essential differences and relations between computational models. A solid understanding is a good basis for building an integrated simulation environment because the interaction semantics between MoC domains will be well understood. It can also lead further to an understanding of the main distinguishing factors between different MoCs and what is common to them. The differences obviously are rooted in distinct perceptions of timing and synchronization. The representation of data and computation, on the other hand, is not a distinguishing factor and can be identical in different MoC domains. If they are different, their transformation and relation should be kept separate from and not confused with the differences of synchronization and timing concepts.

The framework introduced in this book tries hard to separate these issues. The main device for separation is the process constructor, which allows a user to freely define the computation and data representation part of a process, but restricts the interaction of processes with other processes. By selecting process constructors the user defines the model of computation.

The main subject of this book is to represent existing MoCs in our framework. By doing so we also analyzed them, identified the essential parts, and related them to each other. The potential of the framework goes beyond what has been presented here, and it is worth considering further directions for investigation.

In Sections 3.10 and 3.11 we considered the up-rating and down-rating operations. We have only alluded to the possibility of developing them into an algebraic structure that may eventually lead to a powerful formalism to manipulate the synchronization and communication part of a process with a controlled impact on the computation part. It may form the basis of techniques to systematically merge and split processes, formally analyze and verify them, and, perhaps most interesting, to develop new abstractions that isolate interesting properties of time, synchronization, communication, and computation. It will be particularly exciting to see how such techniques can be developed for synchronous and timed MoCs.

It is unsatisfactory that important timed models are not represented well in the framework. The two-level time structure of δ-delay timed models does not match well with the absent-event blocking-read-based framework. In principle it can be modeled, for instance by introducing two time levels and two kind of absent events, one used only in the "δ-time" plane and the other in the "real-time" plane. The number of events in the δ-time plane between every pair of events in the real-time plane would vary according to the activity of processes. The dilemma is that a mechanism is required to monitor all process activities in the system, calculate the number of events in the δ-time plane, and distribute this information to all processes. It would violate the independence property, discussed in Section 5.2.3. In fact, every attempt to embed the δ-delay model in the framework seems to be an ad hoc approach. In Section 5.2.3 we also discussed other variants to represent global time, the local-timer and the time tag approaches. Both have their own strengths and weaknesses, as is the case with the absent-event approach of our framework, and there seems to be no general and natural way to accommodate all desirable features. Moreover, continuous time is not represented in the framework, and there seems to be no obvious and natural way to include it. Thus, it would be most interesting to investigate generalizations of the framework that could encompass all important timed MoCs.

We have only started to investigate the interaction of different MoC domains in Chapter 6. We found that the time structure of an MoC domain is not completely defined by the domain itself because it does not relate to an "absolute time." When two MoC domains interact, their time structure relationship has to be defined. This relationship cannot be directly inferred from the time structures of the involved domains. We also noted that untimed MoC domains import the time structure from timed and synchronous domains, but the domain, interfaces influence the details of this import. Thus, care has to be taken that the resulting combined timing structure is desired. It will be interesting to see where the concept of relative time structures leads and how the interactions of MoC domains can be systematically analyzed, designed, and verified. In the same chapter we also touched upon the issue of migrating functionality from one domain to another. We have only dealt with specific cases, and a general treatment of this problem would be fascinating and useful. However, it may require the involvement of significant application- and context-dependent information.

This short list of research directions is obviously incomplete but exciting because they can only be formulated on the basis of an MoC framework like the one we have developed in this book. Many other topics can be addressed, and many traditional research themes of modeling, simulation, design, synthesis, and verification can perhaps benefit from the concepts elaborated on in this framework. Thus, we hope that this framework will lead to a better understanding of models of computation, that it will be useful for formulating and solving theoretical problems, and that it will assist in addressing the practical improvement of design techniques, tools, and methodologies.

Bibliography

Agerwala, T., and M. Flynn. 1973. Comments on capabilities, limitations and correctness of Petri nets. Hopkins Computer Research Report No. 26, Computer Science Program, Johns Hopkins University, Baltimore, Maryland, July.

Aho, A. V., R. Sethi, and J. D. Ullman. 1988. *Compilers, Principles, Techniques, and Tools*. Addison Wesley Publishing Company, Reading, Massachusetts.

Armstrong, J., R. Virding, and M. Williams. 1993. *Concurrent Programming in Erlang*. Prentice Hall, Englewood Cliffs, New Jersey.

Balarin, F., M. Chiodo, P. Giusto, H. Hsieh, A. Jurecska, L. Lavagno, C. Passerone, A. Sangiovanni-Vincentelli, E. Sentovich, K. Suzuki, and B. Tabbara. 1997. *Hardware-Software Co-Design of Embedded Systems—The POLIS Approach*. Kluwer Academic Publishers, Dordrecht, The Netherlands.

Bening, L., and H. Foster. 2001. *Principles of Verifiable RTL Design*. Second edition. Kluwer Academic Publishers, Dordrecht, The Netherlands.

Benveniste, A., and G. Berry, editors. 1991a. Special Issue on the Synchronous Approach to Reactive and Real-Time Programming. *Proceedings of the IEEE*, 79(9).

Benveniste, A., and G. Berry. 1991b. The synchronous approach to reactive and real-time systems. *Proceedings of the IEEE*, 79(9):1270-1282.

Benveniste, A., P. Caspi, S. A. Edwards, N. Halbwachs, P. le Guernic, and R. de Simone. 2003. The synchronous languages 12 years later. *Proceedings of the IEEE*, 91(1):64-83.

Bergeron, J. 2001. *Writing Testbenches—Functional Verification of HDL Models*. Kluwer Academic Publishers, Dordrecht, The Netherlands.

Berry, G. 1991. A hardware implementation of pure Esterel. *Proceedings of the 1991 International Workshop on Formal Methods in VLSI Design*, January.

Berry, G. 1998. The foundations of Esterel. In G. Plotkin, C. Stirling, and M. Tofte, editors, *Proof, Language and Interaction: Essays in Honour of Robin Milner*. MIT Press, Cambridge, Massachusetts.

Berry, G. 1999. The constructive semantics of pure Esterel—draft version 3. INRIA, 06902 Sophia-Antipolis CDX, France, July 2.

Berry, G., P. Couronne, and G. Gonthier. 1988. Synchronous programming of reactive systems: An introduction to Esterel. In K. Fuchi and M. Nivat, editors,

Programming of Future Generation Computers. Elsevier, Amsterdam, pp. 35-55.

Bhatnagar, H. 2002. *Advanced ASIC Chip Synthesis.* Kluwer Academic Publishers, Dordrecht, The Netherlands.

Bhattacharyya, S. S., P. K. Murthy, and E. A. Lee. 1996. *Software Synthesis from Dataflow Graphs.* Kluwer Academic Publishers, Dordrecht, The Netherlands.

Bilsen, G., M. Engels, R. Lauwereins, and J. A. Peperstraete. 1995. Cyclo-static data flow. *Proceedings of the IEEE International Conference on Acoustics, Speech, and Signal Processing*, pp. 3255-3258.

Bjuréus, P., and A. Jantsch. 2000. MASCOT: A specification and cosimulation method integrating data and control flow. *Proceedings of the Design and Test Europe Conference.*

Bjuréus, P., and A. Jantsch. 2001. Modeling of mixed control and dataflow systems in MASCOT. *IEEE Transactions on Very Large Scale Integration (VLSI) Systems*, 9(5):690-704, October.

Booch, G., and D. Bryan. 1994. *Software Engineering with Ada.* Benjamin/Cummings Publishing Company, Redwood City, California.

Bouali, A. 1997. Xeve: An esterel verification environment, v1_3. Technical report, Institut National de Recherche en Informatique et en Automatique (INRIA), December.

Bouali, A., J.-P. Marmorat, R. de Simone, and H. Toma. 1996. Verifying synchronous reactive systems programmed in ESTEREL. In B. Jonsson and J. Parrow, editors, *International Symposium on Formal Techniques in Real-Time and Fault-Tolerant Systems*, vol. 1135 of *Lecture Notes on Computer Science*, Springer Verlag, Berlin, pp. 463-466.

Brock, J. D. 1983. *A Formal Model for Non-deterministic Dataflow Computation.* PhD thesis, Massachusetts Institute of Technology, Cambridge, Massachusetts.

Brock, J. D., and W. B. Ackerman. 1981. Scenarios: A model of non-determinate computation. In J. Diaz and I. Ramos, editors, *Formalism of Programming Concepts*, vol. 107 of *Lecture Notes in Computer Science*, Springer Verlag, Berlin, pp. 252-259.

Buck, J. T. 1993. *Scheduling Dynamic Dataflow Graphs with Bounded Memory Using the Token Flow Model.* PhD thesis, Department of Electrical Engineering and Computer Science, University of California at Berkeley.

Buck, J., S. Ha, E. A. Lee, and D. G. Messerschmitt. 1992. Ptolemy: A framework for simulating and prototyping heterogenous systems. *International Journal of Computer Simulation*, 25(2):155-182.

Burch, J. R., R. Passerone, and A. L. Sangiovanni-Vincentelli. 2001a. Overcoming heterophobia: Modeling concurrency in heterogeneous systems. In *Proceedings of the second International Conference on Application of Concurrency to System Design*, Newcastle upon Tyne, UK, June 25-29, 2001.

Burch, J. R., R. Passerone, and A. L. Sangiovanni-Vincentelli. 2001b. Using multiple levels of abstractions in embedded software design. In *Proceedings of the first International Workshop on Embedded Software*, Tahoe City, California, October 8-10, 2001.

Caspi, P., A. Girault, and D. Pilaud. 1999. Automatic distribution of reactive systems for asynchronous networks of processors. *IEEE Transactions on Software Engineering*, 25(3): 416-427, May/June.

Cassandras, C. G. 1993. *Discrete Event Systems*. Aksen Associates, Boston, MA.

Chiodo, M., P. Giusto, A. Jurecska, and M. Marelli. 1993. Synthesis of mixed software-hardware implementations from cfsm specifications. *Proceedings of the International Workshop on HW/SW Codesign*, October.

Clarke, E. M., and J. M. Wing. 1996. Formal methods: State of the art and future directions. *ACM Computing Surveys*, 28(4), December.

Clarke, E., O. Grumberg, and D. Long. 1996. Model checking. In M. Broy, editor, *Deductive Program Design*, vol. 152 of *NATO ASI Series, Series F, Computer and System Sciences*, Springer Verlag, Berlin.

Coad, P., and E. Yourdon. 1991. *Object Oriented Design*. Yourdon Press/Prentice Hall, New York and Englewood Cliffs, New Jersey.

Davey, B. A., and H. A. Priestley. 1997. *Introduction to Lattices and Order*. Cambridge University Press, Cambridge.

Davis, J., M. Goel, C. Hylands, B. Kienhuis, Edward A. Lee, J. Liu, X. Liu, L. Muliadi, S. Neuendorffer, J. Reekie, N. Smyth, J. Tsay, and Y. Xiong. 1999. Overview of the Ptolemy project. ERL Technical Report UCB/ERL No. M99/37, University of California, Berkeley, July.

Davis, J., C. Hylands, J. Janneck, E. A. Lee, J. Liu, X. Liu, S. Neuendorffer, S. Sachs, M. Stewart, K. Vissers, P. Whitaker, and Y. Xiong. 2001. Overview of the Ptolemy project. Technical Memorandum UCB/ERL M01/11, Department of Electrical Engineering and Computer Science, University of California, Berkeley, March.

DeMarco, T. 1978. *Structured Analysis and System Specification*. Yourdon, New York.

Dennis, J. B. 1974. First version of a data flow procedure language. In G. Goos and J. Hartmanis, editors, *Programming Symposium*, vol. 19 of *Lecture Notes on Computer Science*, Springer Verlag, Berlin, pp. 362-376.

Dennis, J. B., and G. R. Gao. 1995. Multiprocessor implementation of nondeterminate computation in a functional programming framework. Computation Structures Group Memo 375, Laboratory for Computer Science, Massachusetts Institute of Technology, Cambridge, Massachusetts, January.

De Prycker, M. 1995. *Asynchronous Transfer Mode*. Prentice Hall, Englewood Cliffs, New Jersey.

de Simone, R., and A. Ressouche. 1994. Compositional semantics of ESTEREL and verification by compositional reductions. In D. L. Dill, editor, *Proceedings of the International Conference on Computer Aided Verification*, vol. 818 of *Lecture Notes on Computer Science*, Springer Verlag, Berlin, pp. 441-454.

Dijkstra, E. 1968. Cooperating sequential processes. In F. Genuys, editor, *Programming Languages*, Academic Press, New York, pp. 43-112.

Ecker, W., M. Hofmeister, and S. Maerz-Roessel. 1996. The design cube: A model for VHDL design flow representation and its application. In R. Waxman and J.-M. Berge, editors, *High Level System Modeling: Specification and Design*

Methodologies, vol. 4 of *Current Issues in Electronic Modeling*, Chapter 3, Kluwer Academic Publishers, Dordrecht, The Netherlands.

Ehrig, H., and B. Mahr. 1985. *Fundamentals of Algebraic Specification 1*, vol. 6 of *Monographs on Theoretical Computer Science*, Springer Verlag, Berlin.

Ellsberger, J., D. Hogrefe, and A. Sarma. 1997. *SDL—Formal Object Oriented Language for Communicating Systems*. Prentice Hall, Englewood Cliffs, New Jersey.

Francez, N. 1986. *Fairness*. Texts and Monographs in Computer Science. Springer Verlag, Berlin.

Gajski, D. D. 1997. *Principles of Digital Design*. Prentice Hall, Englewood Cliffs, New Jersey.

Gajski, D., N. Dutt, A. Wu, and Steve Lin. 1993. *High Level Synthesis*. Kluwer Academic Publishers, Dordrecht, The Netherland.

Gajski, D. D., and R. H. Kuhn. 1983. Guest editor's introduction: New VLSI tools. *IEEE Computer*, vol. 16, pp. 11-14, December.

Gajski, D. D., F. Vahid, S. Narayan, and J. Gong. 1994. *Specification and Design of Embedded Systems*. Prentice Hall, Englewood Cliffs, New Jersey.

Gajski, D. D., J. Zhu, R. Dömer, A. Gerstlauer, and S. Zhao. 2000. *SpecC: Specification Language and Methodology*. Kluwer Academic Publishers, Dordrecht, The Netherlands.

Girault, A., B. Lee, and E. A. Lee. 1999. Hierarchical finite state machines with multiple concurrency models. *Integrating Communication Protocol Selection with Hardware/Software Codesign*, 18(6):742-760, June.

Grötker, T., S. Liao, G. Martin, and S. Swan. 2002. *System Design with SystemC*. Kluwer Academic Publishers, Dordrecht, The Netherlands.

Hack, M. Analysis of production schemata by Petri nets. 1972. Master's thesis, Department of Electrical Engineering, Massachusetts Institute of Technology, Cambridge, Massachusetts, February.

Halbwachs, N. 1993. *Synchronous Programming of Reactive Systems*. Kluwer Academic Publishers, Dordrecht, The Netherlands.

Halbwachs, N., P. Caspi, P. Raymond, and D. Pilaud. 1991. The synchronous data flow programming language LUSTRE. *Proceedings of the IEEE*, 79(9):1305-1320, September.

Halbwachs, N., F. Lagnier, and C. Ratel. 1992. Programming and verifying real-time systems by means of the synchronous data-flow language LUSTRE. *IEEE Transactions on Software Engineering*. Special Issue on the Specification and Analysis of Real-Time Systems, September.

Hanselman, D. C., and B. C. Littlefield. 1998. *Mastering MATLAB 5: A Comprehensive Tutorial and Reference*. Prentice Hall, Englewood Cliffs, New Jersey.

Harel, D. 1987. Statecharts: A visual formalism for complex systems. *Science of Computer Programming*, 8:231-274.

Hartmanis, J., and R. E. Stearns. 1966. *Algebraic Structure Theory of Sequential Machines*. Prentice Hall, Englewood Cliffs, New Jersey.

Hoare, C. A. R. 1978. Communicating sequential processes. *Communications of the ACM*, 21(8):666-676, August.

Hopcroft, J. E., and J. D. Ullman. 1979. *Introduction to Automata Theory, Languages, and Computation*. Addison-Wesley, Reading, Massachusetts.

Hu, T.C. 1961. Parallel sequencing and assembly line problems. *Operations Research*, 9:841-848.

Jantsch, A., and P. Bjuréus. 2000. Composite signal flow: A computational model combining events, sampled streams, and vectors. *Proceedings of the Design and Test Europe Conference*.

Jantsch, A., S. Kumar, and A. Hemani. 1999. The Rugby model: A framework for the study of modelling, analysis, and synthesis concepts in electronic systems. *Proceedings of Design Automation and Test in Europe*, Munich, 9-12 March, 1999.

Jantsch, A., S. Kumar, and A. Hemani. 2000. A metamodel for studying concepts in electronic system design. *IEEE Design & Test of Computers*, 17(3):78-85, July-September.

Jantsch, A., I. Sander, and W. Wu. 2001. The usage of stochastic processes in embedded system specifications. *Proceedings of the Ninth International Symposium on Hardware/Software Codesign*, April.

Jensen, K. 1997a. *Coloured Petri Nets: Basic Concepts, Analysis Methods and Practical Use*. Vol. 1, Basic Concepts. Second edition. *Monographs in Theoretical Computer Science*. Springer Verlag, Berlin.

Jensen, K. 1997b. *Coloured Petri Nets: Basic Concepts, Analysis Methods and Practical Use*. Vol. 2, Analysis Methods. Second edition. *Monographs in Theoretical Computer Science*. Springer Verlag, Berlin.

Kahn, G. 1974. The semantics of a simple language for parallel programming. *Proceedings of the IFIP Congress 74*. North-Holland, Amsterdam.

Keller, R. M. 1977. Denotational models for parallel programs with indeterminate operators. In E. J. Neuhold, editor, *Formal Descriptions of Programming Concepts*, North-Holland, Amsterdam, pp. 337-366.

Kern, C., and M. R. Greenstreet.1999. Formal verification in hardware design: A survey. *ACM Transactions on Design Automation of Electronic Systems*, 4(2), April.

Keutzer, K., S. Malik, R. Newton, J. Rabaey, and A. Sangiovanni-Vincentelli. 2000. System-level design: Orthogonalization of concerns and platform-based design. *IEEE Transactions on Computer-Aided Design of Integrated Circuits and Systems*, 19(12):1523-1543, December.

Kloos, C. D., and P. Breuer, editors. 1995. *Formal Semantics for VHDL*. Kluwer Academic Publishers, Dordrecht, The Netherlands.

Kosaraju, S. 1973. Limitations of Dijkstra's semaphore primitives and Petri nets. Technical Report 25, Computer Science Program, Johns Hopkins University, Baltimore, Maryland, May.

Kosinski, P. R. 1978. A straight forward denotational semantics for nondeterminate data flow programs. *Proceedings of the 5th ACM Symposium on Principles of Programming Languages*, pp. 214-219.

Kristensen, L. M., S. Christensen, and K. Jensen. 1998. The practitioner's guide to coloured Petri nets. *International Journal on Software Tools for Technology Transfer*, 2(2):98-132, March.

Lee, B. 2000. *Specification and Design of Reactive Systems*. PhD thesis, Department of Electrical Engineering and Computer Sciences, University of California, Berkeley.

Lee, E. A. 1991. Consistency in dataflow graphs. *IEEE Transactions on Parallel and Distributed Systems*, 2(2), April.

Lee, E. A. 1997. A denotational semantics for dataflow with firing. Technical Report UCB/ERL M97/3, Department of Electrical Engineering and Computer Science, University of California, Berkeley, January.

Lee, E. A. 1998. Modeling concurrent real-time processes using discrete events. Technical Report UCB/ERL Memorandum M98/7, Department of Electrical Engineering and Computer Science, University of California, Berkeley, March.

Lee, E. A., and D. G. Messerschmitt. 1987a. Static scheduling of synchronous data flow programs for digital signal processing. *IEEE Transactions on Computers*, C-36(1):24-35, January.

Lee, E. A., and D. G. Messerschmitt. 1987b. Synchronous data flow. *Proceedings of the IEEE*, 75(9):1235-1245, September.

Lee, E. A., and T. M. Parks. 1995. Dataflow process networks. *Proceedings of the IEEE*, May.

Lee, E. A., and A. Sangiovanni-Vincentelli. 1998. A framework for comparing models of computation. *IEEE Transactions on Computer-Aided Design of Integrated Circuits and Systems*, 17(12):1217-1229, December.

Le Guernic, P., T. Gautier, M. Le Borgne, and C. Le Maire. 1991. Programming real-time applications with SIGNAL. *Proceedings of the IEEE*, 79(9):1321-1336, September.

Lipsett, R., C. F. Schaeffer, and C. Ussery. 1993. *VHDL: Hardware Description and Design*. Kluwer Academic Publishers, Dordrecht, The Netherlands.

Manna, Z. 1974. *Mathematical Theory of Computation*. Computer Science Series. McGraw-Hill, New York.

McFarland, M. C. 1993. Formal verification of sequential hardware: A tutorial. *IEEE Transactions on Computer-Aided Design of Integrated Circuits and Systems*, 12(5):633-654, May.

Merlin, P. 1974. *A Study of the Recoverablity of Computing Systems*. PhD thesis, Department of Information and Computer Science, University of California, Irvine.

Milner, R. 1980. *A Calculus of Communicating Systems*, vol. 92 of *Lecture Notes of Computer Science*. Springer Verlag, Berlin.

Milner, R. 1989. *Communication and Concurrency*. International Series in Computer Science. Prentice Hall, Englewood Cliffs, New Jersey.

Murata, T. 1989. Petri nets: Properties, analysis and applications. *Proceedings of the IEEE*, 77(4):541-580, April.

Olsen, A., O. Færgemand, B. Møeller Pedersen, R. Reed, and J. R. W. Smith. 1995. *Systems Engineering with SDL-92*. North-Holland, Amsterdam.

Palanque, P., and R. Bastide. 1996. Time modelling in Petri nets for the design of interactive systems. *SIGCHI Bulletin*, 28(2):8, April.

Panangaden, P., and V. Shanbhogue. 1992. The expressive power of indeterminate dataflow primitives. *Information and Computation*, 98:99-131.

Park, C., J. Jung, and S. Ha. 2002. Extended synchronous dataflow for efficient dsp system prototyping. *Design Automation for Embedded Systems*, 6(3):295-322, March.

Park, D. 1979. On the semantics of fair parallelism. In G. Goos and J. Hartmanis, editors, *Abstract Software Specification*, vol. 86 of *Lecture Notes on Computer Science*, Springer Verlag, Berlin, pp. 504-526.

Park, D. 1983. The "fairness" problem and nondeterministic computing networks. In J. W. De Baker and J. van Leeuwen, editors, *Foundations of Computer Science IV, Part 2: Semantics and Logic*, vol. 159, Mathematical Centre Tracts, pp. 133-161. Amsterdam, The Netherlands.

Parks, T. M., J. L. Pino, and E. A. Lee. 1995. A comparison of synchronous and cyclo-static dataflow. *Proceedings of the 29th Asilomar Conference on Signals, Systems and Computers*, vol. 1, pp. 204-210.

Parrow, J. 1985. *Fairness Properties in Process Algebras with Applications in Communication Protocol Verification*. PhD thesis, Department of Computer Science, Uppsala University, Uppsala, Sweden.

Patil, S. 1970. *Coordination of Asynchronous Events*. PhD thesis, Department of Electrical Engineering, Massachusetts Institute of Technology, Cambridge, Massachusetts, May.

Perrin, D. 1994. Finite automata. In J. van Leeuwen, editor, *Handbook of Theoretical Computer Science*, vol. B: Formal Models and Semantics, Chapter 1, Elsevier, Amsterdam.

Peterson, J. L. 1981. *Petri Net Theory and the Modelling of Systems*. Prentice Hall, Englewood Cliffs, New Jersey.

Poigné, A., M. Morley, O. Maffeïs, L. Holenderski, and R. Budde. 1998. The synchronous approach to designing reactive systems. *Formal Methods in System Design*, 12(2):163-188, March.

Rabaey, J. M. 1996. *Digital Integrated Circuits*. Electronics and VLSI Series. Prentice Hall, Englewood Cliffs, New Jersey.

Ramchandani, C. 1973. *Analysis of Asynchronous Concurrent Systems by Timed Petri Nets*. PhD thesis, Massachusetts Institute of Technology, Cambridge, Massachusetts.

Rashinkar, P., P. Paterson, and L. Singh. 2001. *System-on-a-Chip Verification*. Kluwer Academic Publishers, Dordrecht, The Netherlands.

Reisig, W. 1985. *Petri Nets*. Springer Verlag, Berlin.

Roth, C. H, Jr. 1998. *Digital Systems Design Using VHDL*. PWS Publishing Company, Boston, MA.

Sander, I., and A. Jantsch. 1999. Formal design based on the synchronous approach, functional models and skeletons. *Proceedings of the 12th International Conference on VLSI Design*.

Sander, I., and A. Jantsch. 2002. Transformation based communication and clock domain refinement for system design. *Proceedings of Design Automation Conference*, June.

Sander, I., A. Jantsch, and Z. Lu. 2003. The development and application of formal design transformations in ForSyDe. *Proceedings of the Design Automation and Test Europe*, March.

Saraco, R., and P. A. J. Tilanus. 1987. CCITT SDL: An overview of the language and its applications. *Computer Networks & ISD Systems*, Special Issue on CCITT SDL, 13(2):65-74.

Selic, B., G. Gullekson, and P. T. Ward. 1994. *Real-Time Object-Oriented Modeling*. John Wiley & Sons, New York.

Severance, F. L. 2001. *System Modeling and Simulation*. John Wiley & Sons, New York.

Sifakis, J. 1977. Use of Petri nets for performance evaluation. In H. Beilner and E. Gelembe, editors, *Measuring, Modelling and Evaluating Computer Systems*, North-Holland, Amsterdam, pp. 75-93.

Skillicorn, D. B., and D. Talia. 1998. Models and languages for parallel computation. *ACM Computing Surveys*, 30(2):123-169, June.

Sriram, S., and S. S. Bhattacharyya. 2000. *Embedded Multiprocessors: Scheduling and Synchronization*. Marcel Dekker, New York.

Staples, J., and V. L. Nguyen. 1985. A fixedpoint semantics for nondeterministic data flow. *Journal of the Association for Computing Machinery*, 32(2):411-444, April.

Stoy, J. E. 1989. *Denotational Semantics*. Fifth edition. MIT Press, Cambridge, Massachusetts.

van der Putten, P. H. A., and J. P. M. Voeten. 1997. *Specification of Reactive Hardware/Software Systems*. Technische Universiteit Eindhoven.

Yakovlev, A., L. Gomes, and L. Lavagno, editors. 2000. *Hardware Design and Petri Nets*. Kluwer Academic Publishers, Dordrecht, The Netherlands.

Yourdon, E. 1994. *Object Oriented Systems Design*. Prentice Hall, Englewood Cliffs, New Jersey.

Zwoliński, M. 2000. *Digital System Design with VHDL*. Prentice Hall, Englewood Cliffs, New Jersey.

Index

Page numbers in **bold** *denote whole sections and subsections that address a topic*

32-bit adder, Rugby metamodel, 22-3

A
absent events, global time, 234-5
abstraction
    design phases, 38
    Rugby metamodel, 21-5, 35
activation cycle, untimed MoC, 116
Ada, tightly coupled process networks, 280
after process, validation, 213
algorithmic level, domains, 29-32
alwaysSince, validation, 213
Amplifier
    process merge, 152-4
    process signatures, 137-8
    process up-rating, 141, 143
    scan-based processes, 143
    signal partitioning, 114-15
    untimed MoC, 110-11, 116-24
analysis, design methods and methodology, 39
Analyzer process, digital equalizer system, 250-3
applications, 303-31
    design and synthesis, **320–6**
    functional specification, **312–20**
    MoC interfaces, **265–7**
    performance analysis, **305–12**
    timed MoC, **240–1**
assembler model software, design project, 44
asynchronous interfaces
    MoC interfaces, **253–7**
    stochastic processes, 257
asynchronous transfer mode (ATM), 41
automata models, FSMs, 67

B
BDF *see* boolean dataflow
bibliography, 335-42

blocking read, tightly coupled process networks, 269-70, **274–6**
blocking synchronization, tightly coupled process networks, 274-6
blocking write, tightly coupled process networks, **274–6**
boolean dataflow (BDF), untimed MoC, 174-5
bottom-up design and synthesis, 320-1
boundedness, Petri nets, 87, 95
buffer analysis, process network, 305-8

C
Calculus of Communicating Systems (CCS), nondeterminism, 289-92
causality level, 34
CCS *see* Calculus of Communicating Systems
channel delay, tightly coupled process networks, 271-2
characteristic functions, **135–6**
clocked synchronous models
    extended characteristic function, 201-6
    synchronous MoC, **199–201**
clocked time level, 34
codesign FSMs (CFSMs), 69
colored Petri nets, 99
Communicating Sequential Processes (CSP), nondeterminism, 289-92
communication
    computation, **7–8**
    domains, 34-6
complete partial order (CPO), process properties, 126
composite signal flow model, MoC interfaces, 265-6
composition operators, **129–33**
    feedback operator, 130-3
    parallel composition, 129-30
    sequential composition, 130

344 Index

computation
    communication, **7–8**
    domains, 25-32
    logic gate level, 25-9
    transistor level, 25-9
concatenation of two languages, FSMs, 60-1, 63
concatenation, signals and processes, 112-13
concluding remarks, 333-4
concurrency and sequence, Petri nets, 80-1, 83
concurrency, functional specification, 319-20
conflict, Petri nets, 81, 82
conservation, Petri nets, 87-8, 95-6
consolidation-based constructors, nondeterminism, 296-7
constant partitionings, SDF, 159
constraining purpose, nondeterminism **292–3**, 299
continuity, process properties, 125-8
continuous state systems, 15-16
continuous time systems, 15-16
continuous value level, data, 33
coverability, Petri nets, 90-1
coverability tree, Petri nets, 91-8
CPN *see* customer premises network
CPO *see* complete partial order
critical region, Petri nets, 82
CSDF *see* cyclo-static dataflow
CSP *see* Communicating Sequential Processes
customer premises network (CPN), 40
cyclo-static dataflow (CSDF), untimed MoC, 174-5

D
data, domains, 32-3
dataflow networks, nondeterminism, 286-9
dataflow process network, MoC interfaces, 265
datapath (FSMD), 67-8
δ-delay-based time structure, timed MoC, **236–9**
δ-delay model, **236–9**
deadlock, Petri nets, 85, 86, 88-90, 97
decoupled sender/receiver, tightly coupled process networks, 270
description of behavior models, **47–8**
descriptive purpose, nondeterminism, **284–92**, 298-9
design and synthesis
    applications, **320–6**
    architecture, 321-3
    bottom-up, 320-1
    code generation, 323
    combinatorial network of gates, 324
    components, 321-3
    consequences for modeling, 323-5
    constraints, 323
    decisions, 321-3
    design methods and methodology, 38-9
    models/modeling, 2

    objectives, 323
    problems, 321-3
    review of MoCs, 325-6
    synchronous MoC, 326
    tasks, 323
    timed MoC, 325-6
    top-down, 320-1
    untimed MoC, 326
design methods and methodology, **36–9**
    analysis, 39
    design and synthesis, 38-9
    design phases, 37-8
design project
    case study, **39–45**
    hardware, 44-5
    requirements definitions, 41-2
    software: assembler model, 44
    software: C model, 43-4
    synthesised netlist, 44
    system model, 42-3
    VHDL model, 44
details, functional specification, 320
determinate models, 284-5
determinism
    deterministic systems, 17-18
    tightly coupled process networks, 270
    see also nondeterminism
digital equalizer system, integrated MoC, 250-1
dining philosophers, Petri nets, 85, 86
discrete event models based on d-delay
    event-driven simulation cycle, 239, 240
    timed MoC, **236–9**
    two-level time structure, 236-9
diversity, heterogeneous models, 5
domain polymorphism, 7
domains, **25–36**
    communication, 34-6
    computation, 25-32
    data, 32-3
    MoC interfaces, **244–6**
    MoCs, 5-7, 108-11
    Rugby metamodel, 21-5
    terminology, 6
    time, 33-4
down-rating, process, **149**
dynamic scheduling, SDF, 160

E
embedded control systems, synchronous MoC, 183
ϵ-moves, FSMs, 56-7
Erlang, tightly coupled process networks, 280
Esterel compiler, feedback loops, 196
evaluation cycle, untimed MoC, 116
event-based process invocation (mealyST), timed MoC, 230-1

event count with time-out (mealyTT), timed MoC, 231
event-driven simulation cycle, discrete event models based on δ-delay, 239, 240
event-driven systems, 18-19
events
    symbols, 111-12
    systems, 18
    types, 111-13
exhaustive simulation, validation, 216
extended characteristic function
    clocked synchronous models, 201-6
    synchronous MoC, **201-6**
extended Petri nets, **98-100**
extensions, FSMs, 67-9

F
feasibility analysis, 40
feedback loops, synchronous MoC, **188-96**
feedback operator
    composition operators, 130-3
    Scott order over event values, 191
    synchronous MoC, 188-96
FIFO-based communication, tightly coupled process networks, 276-80
finite state machines (FSMs), **48-69**
    automata models, 67
    basic definition, 49-52
    codesign (CFSMs), 69
    concatenation of two languages, 60-1, 63
    datapath (FSMD), 67-8
    ϵ-moves, 56-7
    equivalence, 57-8
    extensions, 67-9
    general state machine, 52
    hierarchy, 68-9
    infinite state machine, 52
    Kleene closure, 60-1
    Mealy machine, 65-7
    Moore machine, 63-7
    nondeterministic FSMs, 52-7
    output, 63-7
    Petri nets, 76-80
    positive closure, 61, 63
    push-down automata, 67
    regular sets and expressions, 60-3
    state aggregation, 57-60
    statecharts, 68-9
    state machine, defined, 52
firing cycle, untimed MoC, 116
firing vector, Petri nets, 74
fork and join, Petri nets, 81, 82
formal verification, nondeterminism, 294-5
ForSyDe methodology, MoC interfaces, 266-7
frameworks, MoC interfaces, 265-7
frameworks, MoCs, 5-7, **108-11**, 333

free-choice Petri nets, 100
FSMs *see* finite state machines
functional modeling, 2
functional specification
    applications, **312-20**
    concurrency, 319-20
    details, 320
    number sorter, 314-16
    packet multiplexer, 316-17, 318, 319
    primitive functions, 320
    role, 317-20
    timing, 319
function extractor, 135-6

G
general state machine, 52
global time, 231-6
    absent events, 234-5
    distribution, 231-6
    independence property, 231
    local timer, 231-2
    order-isomorphism, 235
    representation, 231-6
    time tags, 232-4

H
heterogeneous models, **5-7**
    diversity, 5
hierarchical model of computation (HMoC), 249
hierarchy
    FSMs, 68-9
    process family, 137
    Rugby metamodel, 21-5
    untimed MoC, 115-16
HMoC *see* hierarchical model of computation
Hu levels, precedence graph, 172-3

I
implies process, validation, 213
incidence matrix, Petri nets, 74
independence property, global time, 231
infinite state machine, 52
inhibitor arcs, Petri nets, 98
init-based processes, untimed MoC, 121-2
initial buffer conditions, SDF, 164-5
inputs/outputs
    Petri nets, 75-6
    signal partitioning, 114-15
input variables, systems, 9
insert-based interface constructors, MoC interfaces, 248
instruction set level, domains, 29
integrated MoC
    digital equalizer system, 250-1
    MoC interfaces, **249-53**

346   Index

interfaces, MoC *see* MoC interfaces
interprocess communication level, 35

J
join and fork, Petri nets, 81, 82

K
Kahn's process network model, nondeterminism, 284-5
Kleene closure, FSMs, 60-1

L
limitations, Petri nets, 96-8
linear/non-linear systems, 16-17
liveness, Petri nets, 88-90
local timer, global time, 231-2
logic gate level, computation, 25-9
logic value level, data, 33

M
map-based processes
    process merge, 150, 199
    process signatures, 137
    process up-rating, 139-41
    synchronous MoC, 185-6
    untimed MoC, 116-17, 123-4
marked graphs, Petri nets, 99-100
Mascot modeling technique, MoC interfaces, 266
mathematical model, systems, 10
Matlab, MoC interfaces, 266
Mealy-based processes
    process merge, 151-4, 198-9
    process properties, 126-7
    process signatures, 137
    process up-rating, 144-7
    synchronous MoC, 186-7
    timed MoC, 227-8
    untimed MoC, 115-16, 118, 123-4
Mealy machine, FSMs, 65-7
mealyPT *see* timer-based process invocation
mealyST *see* event-based process invocation
mealyT-based processes, timed MoC, 227-8
mealyTT *see* event count with time-out
migration of processes *see* process migration
minimum buffer scheduling, synchronous dataflow (SDF), 165-6
MoC interfaces, 243-68
    applications, **265-7**
    asynchronous interfaces, **253-7**
    composite signal flow model, 265-6
    dataflow process network, 265
    different computational models, **246-8**
    domains of the same MoC, **244-6**
    ForSyDe methodology, 266-7
    frameworks, 265-7
    insert-based interface constructors, 248
    integrated MoC, **249-53**
    Mascot modeling technique, 266
    Matlab, 266
    process migration, **257-65**
    Ptolemy project, 265, 267-8
    strip-based interface constructors, 246-8
MoCs (models of computation)
    domains, 108-11
    frameworks, 5-7, **108-11**, 333
    review, 325-6
    synchronous MoC defined, 196
    timed MoC, defined, 230
    untimed MoC, defining, **133-5**
modeling templates, Petri nets, 80-5
models/modeling
    characteristics, 2-5
    consequences for, 323-5
    constraints, 3-4
    defined, 2-3
    description of behavior, **47-8**
    design and synthesis, 2
    functional modeling, 2
    minimal feature, 3
    performance modeling, 2
    simplification feature, 3
    systems, **8-20**
    temperature controller, 11-13
    terminology, 4
    validation and verification, 2
models of computation *see* MoCs
monitors, validation, 211-15
monotonicity, process properties, 124-5
Moore-based processes
    process merge, 198
    process properties, 126-7
    process signatures, 137
    synchronous MoC, 186-7
    traffic light controller, 207
    untimed MoC, 115-16
Moore machine, FSMs, 63-7
motivation, **2-5**
multiple inputs, process up-rating, 147-8
multiple processes, SDF, 160
multiprocessor schedule, SDF, 166-73
mutual exclusion, Petri nets, 81-3

N
natural extension, feedback loops, 196
network terminal (NT) design project, 39-45
nonblocking read, tightly coupled process networks, 270-4
nondeterminism, 283-301
    CCS, 289-92
    consolidation-based constructors, 296-7
    constraining purpose, **292-3**, 299
    CSP, 289-92
    dataflow networks, 286-9
    descriptive purpose, **284-92**, 298-9

determinate models, 284-5
formal verification, 294-5
Kahn's process network model, 284-5
nondeterminate models, 285-92
process algebras, 289-92
process constructors, **295-7**
select-based constructors, 295-6
σ processes, **293-4**
stochastic processes, 293, 298-9
stochastic skeletons, **297-9**
synthesis, 294-5
tightly coupled process networks, 271-2
weak confluence, 291-2
weak determinacy, 290-1
nondeterministic FSMs, 52-7
defined, 53
ϵ-moves, 56-7
examples, 54-6
theorem, 55, 57
nondeterministic systems, 17-18
non-linear/linear systems, 16-17
notation, xxi-xxii, **36**
Rugby metamodel, 36, 37
NT (network terminal) design project, 39-45
number sorter, functional specification, 314-16
number value level, data, 33

O

onceSince, validation, 213
order-isomorphism, global time, 235
output, FSMs, 63-7
output variables, systems, 9
oversynchronization, tightly coupled process networks, **276-80**

P

packet multiplexer, functional specification, 316-17, 318, 319
PAPS *see* periodic, admissible, parallel schedule
parallel composition
composition operators, 129-30
Petri nets, 78, 80-1, 85, 86
partitioning, signal *see* signal partitioning
PASS *see* periodic, admissible, sequential schedule
perfectly synchronous MoC, **196**
perfect match, process merge, 149-54
perfect synchrony, synchronous MoC, **182-5**
performance analysis
applications, **305-12**
simulation, 308-12
timed analysis, 312
untimed analysis, 305-12
performance modeling, 2
periodic, admissible, parallel schedule (PAPS), SDF, 167, 172-3

periodic, admissible, sequential schedule (PASS), SDF, 161-3, 167-72
persistence, Petri nets, 91
Petri nets, 21, **69-101**
analysis methods, 85-91
arc-timed, 99
boundedness, 87, 95
colored, 99
concurrency and sequence, 80-1, 83
conflict, 81, 82
conservation, 87-8, 95-6
coverability, 90-1
coverability tree, 91-8
critical region, 82
deadlock, 85, 86, 88-90, 97
defined, 70-1
dining philosophers, 85, 86
extended, **98-100**
firing vector, 74
fork and join, 81, 82
free-choice, 100
FSMs, 76-80
incidence matrix, 74
inhibitor arcs, 98
inputs/outputs, 75-6
limitations, 96-8
liveness, 88-90
marked graphs, 99-100
modeling templates, 80-5
mutual exclusion, 81-3
parallel composition, 78, 80-1, 85, 86
persistence, 91
producer/consumer relationship, 83-5
reachability, 73-4, 90-1
restricted, **98-100**
safeness, 95
SDF, 159-60, 161-5
sequence and concurrency, 80-1, 83
sequential composition, 77, 80-1
state machines, 99
temporal arcs, 99
tokens, 70
transition, 70-5, 76, 82-3
untimed MoC, **155-8**
weighting vector, 88
zero test, 98
physical time level, 34
POLIS, tightly coupled process networks, 280-1
positive closure, FSMs, 61, 63
precedence graph
Hu levels, 172-3
SDF, 168-73
primitive functions, functional specification, 320
probability, 283-301
procedure calls, 35
process algebras, nondeterminism, 289-92
process composition, process up-rating, 148-9

process constructors
   nondeterminism, **295–7**
   synchronous MoC, **185–8**
   timed MoC, **227–36**
   untimed MoC, **115–24**
process core, 7-8
process down-rating, **149**
processes and signals, **111–13**
process family hierarchy, process signatures, 137
process merge, **149–55**
   Amplifier, 152-4
   map-based processes, 150, 199
   Mealy-based processes, 151-4, 198-9
   Moore-based processes, 198
   perfect match, 149-54
   rational match, 154-5
   scan-based processes, 150-1, 197
   synchronous MoC, **197–9**
process migration
   MoC interfaces, **257–65**
   timing regimes, 258
   untimed to synchronous, 260-2
   untimed to timed, 262-5
process network, buffer analysis, 305-8
process network model, Kahn's, 284-5
process properties, **124–9**
   continuity, 125-8
   CPO, 126
   Mealy-based processes, 126-7
   monotonicity, 124-5
   Moore-based processes, 126-7
   sequentiality, 125
   sequential processes, 128-9
   sink-based processes, 125
   source-based processes, 125
   zip-based processes, 128-9
process shell, 7-8
process signatures, **136–8**
   Amplifier, 137-8
process up-rating, **138–49**
   Amplifier, 141, 143
   map-based processes, 139-41
   Mealy-based processes, 144-7
   multiple inputs, 147-8
   process composition, 148-9
   scan-based processes, 141-3
   zip-based processes, 147-8
producer/consumer relationship, Petri nets, 83-5
Ptolemy project, MoC interfaces, 265, 267-8
push-down automata, FSMs, 67

R

Rank Test, SDF, 161-3
rational match, process merge, 154-5
reachability, Petri nets, 73-4, 90-1
reactive systems, synchronous MoC, 183

regular sets and expressions, FSMs, 60-3
relations, domains, 30-2
remainder, signal partitioning, 113-14
rendezvous-based communication, tightly coupled process networks, 274-6
request merger, transaction server, 292-3
requirements definitions, design project, 41-2
restricted Petri nets, **98–100**
review of MoCs, design and synthesis, 325-6
roadmap, this book's, xvi-xviii
Rugby coordinates, 36, **155**
   synchronous MoC, **209**
   tightly coupled process networks, **280**
   timed MoC, **239–40**
Rugby metamodel, **20–5**
   abstraction, 21-5, 35
   domains, 21-5, **25–36**
   hierarchy, 21-5
   notation, 36, 37

S

safeness, Petri nets, 95
scan-based processes
   Amplifier, 143
   process merge, 150-1, 197
   process signatures, 137
   process up-rating, 141-3
   synchronous MoC, 186-7
   untimed MoC, 115, 117-18, 186
scand-based processes, untimed MoC, 120-1
Scott order over event values, feedback operator, 191
SDF *see* synchronous dataflow
SDL *see* Specification and Design Language
select-based constructors, nondeterminism, 295-6
SELECT, BDF, 174-5
sequential composition
   composition operators, 130
   Petri nets, 77, 80-1
sequentiality, process properties, 125
sequential processes, process properties, 128-9
signal partitioning, **113–15**
   Amplifier, 114-15
   defined, 113
   inputs/outputs, 114-15
   remainder, 113-14
signals and processes, **111–13**
simulation, performance analysis, 308-12
single-processor schedule, SDF, 161-6
sink-based processes
   process constructors, 121, 188
   process properties, 125
   synchronous MoC, 188
   timed MoC, 229
   untimed MoC, 121

software: assembler model, design project, 44
software: C model, design project, 43-4
source-based processes
    process properties, 125
    synchronous MoC, 188
    timed MoC, 229
    untimed MoC, 121
SPDF *see* synchronous piggy-backed dataflow
Specification and Design Language (SDL), 42-3
    tightly coupled process networks, 280
σ processes, nondeterminism, **293-4**
state aggregation
    algorithm, 58-60
    defined, 57-8
    equivalence, 57-8
    example, 59-60
    FSMs, 57-60
statecharts, FSMs, 68-9
state-less/state-full systems, 13-16
state machines
    defined, 52
    Petri nets, 99
static scheduling, SDF, 160
stochastic processes
    asynchronous interfaces, 257
    nondeterminism, 293, 298-9
stochastic skeletons, nondeterminism, **297-9**
stochastic systems, 17-18
strip-based interface constructors, MoC
        interfaces, 246-8
subway, U-turn section controller, 209-11
SWITCH, BDF, 174-5
symbols
    data, 33
    events, 111-12
synchronous dataflow (SDF), **158-73**
    constant partitionings, 159
    defined, 159
    dynamic scheduling, 160
    initial buffer conditions, 164-5
    minimum buffer scheduling, 165-6
    multiple processes, 160
    multiprocessor schedule, 166-73
    PAPS, 167, 172-3
    PASS, 161-3, 167-72
    Petri nets, 159-60, 161-5
    precedence graph, 168-73
    Rank Test, 161-3
    single-processor schedule, 161-6
    static scheduling, 160
    unroll factor J, 167-8
synchronous events, 111-13
synchronous MoC
    clocked synchronous models, **199-201**
    defined, 196
    design and synthesis, 326

embedded control systems, 183
extended characteristic function, **201-6**
feedback loops, **188-96**
map-based processes, 185-6
Mealy-based processes, 186-7
Moore-based processes, 186, 186-7
perfectly synchronous MoC, **196**
perfect synchrony, **182-5**
process constructors, **185-8**
process merge, **197-9**
reactive systems, 183
Rugby coordinates, **209**
scan-based processes, 186-7
sink-based processes, 188
source-based processes, 188
telecommunication backbone networks, 183
traffic light controller, **206-9**
validation, **209-16**
wireless communication devices, 183
zip-based processes, 187
synchronous piggy-backed dataflow (SPDF),
        untimed MoC, 175-6
synthesis
    nondeterminism, 294-5
    see also design and synthesis
synthesised netlist, design project, 44
system functions, domains, 30-2
system model, design project, 42-3
systems, **8-20**
    classification summary, 19-20
    continuous state, 15-16
    continuous time, 15-16
    defined, 8
    deterministic, 17-18
    event-driven, 18-19
    events, 18
    linear/non-linear, 16-17
    mathematical model, 10
    nondeterministic, 17-18
    properties, 13-19
    state-less/state-full, 13-16
    stochastic, 17-18
    system properties, 13-19
    terminology, 8
    time-driven, 18-19
    time-varying/time-invariant, 14

T
telecommunication backbone networks,
        synchronous MoC, 183
temperature controller, models/modeling, 11-13
temporal arcs, Petri nets, 99
terminology, models/modeling, 4
tightly coupled process networks, 269-82
    Ada, 280
    blocking read, 269-70, **274-6**

tightly coupled process networks (*continued*)
  blocking synchronization, 274-6
  blocking write, **274–6**
  channel delay, 271-2
  decoupled sender/receiver, 270
  determinism, 270
  Erlang, 280
  FIFO-based communication, 276-80
  nonblocking read, 270-4
  nondeterminism, 271-2
  oversynchronization, **276–80**
  POLIS, 280-1
  rendezvous-based communication, 274-6
  Rugby coordinates, **280**
  SDL, 280
timed analysis, performance analysis, 312
timed events, 111-13
timed MoC, 223-42
  applications, **240–1**
  δ-delay-based time structure, **236–9**
  defined, 230
  design and synthesis, 325-6
  discrete event models based on d-delay, **236–9**
  event-based process invocation (mealyST), 230-1
  event count with time-out (mealyTT), 231
  mealyT-based processes, 227-8
  process constructors, **227–36**
  Rugby coordinates, **239–40**
  simulation tool, 240-1
  sink-based processes, 229
  source-based processes, 229
  timer-based process invocation (mealyPT), 230
  unzipT-based processes, 229
  variants, 230-1
  zip-based processes, 228-9
time, domains, 33-4
timed Petri nets, 99
time-driven systems, 18-19
timer-based process invocation (mealyPT), process constructors, 230
timer-based process invocation (mealyPT), timed MoC, 230
time tags, global time, 232-4
time-varying/time-invariant systems, 14
timing, functional specification, 319
tokens, Petri nets, 70
top-down design and synthesis, 320-1
traffic light controller, 207-16
  Moore-based processes, 207
  synchronous MoC, **206–9**
  validation, **209–16**
transaction server, request merger, 292-3
transistor level, computation, 25-9

transition, Petri nets, 70-5, 76, 82-3
two-level time structure, discrete event models based on δ-delay, **236–9**

U
unroll factor J, SDF, 167-8
untimed analysis, performance analysis, 305-12
untimed events, 111-13
untimed MoC, 134-5
  activation cycle, 116
  Amplifier, 116-24
  BDF, 174-5
  characteristic functions, **135–6**
  composition operators, **129–33**
  CSDF, 174-5
  defining, **133–5**
  design and synthesis, 326
  evaluation cycle, 116
  firing cycle, 116
  hierarchy, 115-16
  init-based processes, 121-2
  map-based processes, 116-17, 123-4
  Mealy-based processes, 115-16, 118, 123-4
  MoC framework, **108–11**
  Moore-based processes, 115-16
  Petri nets, **155–8**
  process constructors, **115–24**
  process down-rating, **149**
  processes and signals, **111–13**
  process merge, **149–55**
  process properties, **124–9**
  process signatures, **136–8**
  process up-rating, **138–49**
  Rugby coordinates, **155**
  scan-based processes, 115, 117-18
  scand-based processes, 120-1
  SDF, **158–73**
  signal partitioning, **113–15**
  sink-based processes, 121
  source-based processes, 121
  SPDF, 175-6
  unzip-based processes, 121
  variants, **174–6**
  zip-based processes, 118-22
untimed process network, performance analysis, 309
unzip-based processes, untimed MoC, 121
unzipT-based processes, timed MoC, 229
up-rating *see* process up-rating
U-turn section controller, validation, 209-11

V
validation
  after process, 213

alwaysSince, 213
exhaustive simulation, 216
implies process, 213
monitors, 211-15
onceSince, 213
strategies, 215-16
synchronous MoC, **209–16**
traffic light controller, **209–16**
U-turn section controller, 209-11
and verification, 2
variants
    timed MoC, 230-1
    time-varying/time-invariant systems, 14
    untimed MoC, **174–6**
VHDL, 239

W

weak confluence, nondeterminism, 291-2
weak determinacy, nondeterminism, 290-1
weighting vector, Petri nets, 88
wireless communication devices, synchronous MoC, 183

Z

Zeno behavior, 238
zero test, Petri nets, 98
zip-based processes
    process properties, 128-9
    process up-rating, 147-8
    synchronous MoC, 186-8
    timed MoC, 228-9
    untimed MoC, 118-22